IET CONTROL, ROBOTICS AND SENSORS SERIES 100

Organic Sensors

Other volumes in this series:

Volume 8 **A History of Control Engineering, 1800–1930** S. Bennett
Volume 18 **Applied Control Theory, 2nd Edition** J.R. Leigh
Volume 20 **Design of Modern Control Systems** D.J. Bell, P.A. Cook and N. Munro (Editors)
Volume 28 **Robots and Automated Manufacture** J. Billingsley (Editor)
Volume 33 **Temperature Measurement and Control** J.R. Leigh
Volume 34 **Singular Perturbation Methodology in Control Systems** D.S. Naidu
Volume 35 **Implementation of Self-Tuning Controllers** K. Warwick (Editor)
Volume 37 **Industrial Digital Control Systems, 2nd Edition** K. Warwick and D. Rees (Editors)
Volume 39 **Continuous Time Controller Design** R. Balasubramanian
Volume 40 **Deterministic Control of Uncertain Systems** A.S.I. Zinober (Editor)
Volume 41 **Computer Control of Real-Time Processes** S. Bennett and G.S. Virk (Editors)
Volume 42 **Digital Signal Processing: Principles, devices and applications** N.B. Jones and J.D.McK. Watson (Editors)
Volume 44 **Knowledge-Based Systems for Industrial Control** J. McGhee, M.J. Grimble and A. Mowforth (Editors)
Volume 47 **A History of Control Engineering, 1930–1956** S. Bennett
Volume 49 **Polynomial Methods in Optimal Control and Filtering** K.J. Hunt (Editor)
Volume 50 **Programming Industrial Control Systems Using IEC 1131-3** R.W. Lewis
Volume 51 **Advanced Robotics and Intelligent Machines** J.O. Gray and D.G. Caldwell (Editors)
Volume 52 **Adaptive Prediction and Predictive Control** P.P. Kanjilal
Volume 53 **Neural Network Applications in Control** G.W. Irwin, K. Warwick and K.J. Hunt (Editors)
Volume 54 **Control Engineering Solutions: A practical approach** P. Albertos, R. Strietzel and N. Mort (Editors)
Volume 55 **Genetic Algorithms in Engineering Systems** A.M.S. Zalzala and P.J. Fleming (Editors)
Volume 56 **Symbolic Methods in Control System Analysis and Design** N. Munro (Editor)
Volume 57 **Flight Control Systems** R.W. Pratt (Editor)
Volume 58 **Power-Plant Control and Instrumentation: The control of boilers and HRSG systems** D. Lindsley
Volume 59 **Modelling Control Systems Using IEC 61499** R. Lewis
Volume 60 **People in Control: Human factors in control room design** J. Noyes and M. Bransby (Editors)
Volume 61 **Nonlinear Predictive Control: Theory and practice** B. Kouvaritakis and M. Cannon (Editors)
Volume 62 **Active Sound and Vibration Control** M.O. Tokhi and S.M. Veres
Volume 63 **Stepping Motors, 4th Edition** P.P. Acarnley
Volume 64 **Control Theory, 2nd Edition** J.R. Leigh
Volume 65 **Modelling and Parameter Estimation of Dynamic Systems** J.R. Raol, G. Girija and J. Singh
Volume 66 **Variable Structure Systems: From principles to implementation** A. Sabanovic, L. Fridman and S. Spurgeon (Editors)
Volume 67 **Motion Vision: Design of compact motion sensing solution for autonomous systems** J. Kolodko and L. Vlacic
Volume 68 **Flexible Robot Manipulators: Modelling, simulation and control** M.O. Tokhi and A.K.M. Azad (Editors)
Volume 69 **Advances in Unmanned Marine Vehicles** G. Roberts and R. Sutton (Editors)
Volume 70 **Intelligent Control Systems Using Computational Intelligence Techniques** A. Ruano (Editor)
Volume 71 **Advances in Cognitive Systems** S. Nefti and J. Gray (Editors)
Volume 72 **Control Theory: A guided tour, 3rd Edition** J.R. Leigh
Volume 73 **Adaptive Sampling with Mobile WSN** K. Sreenath, M.F. Mysorewala, D.O. Popa and F.L. Lewis
Volume 74 **Eigenstructure Control Algorithms: Applications to aircraft/rotorcraft handling qualities design** S. Srinathkumar
Volume 75 **Advanced Control for Constrained Processes and Systems** F. Garelli, R.J. Mantz and H. De Battista
Volume 76 **Developments in Control Theory towards Glocal Control** L. Qiu, J. Chen, T. Iwasaki and H. Fujioka (Editors)
Volume 77 **Further Advances in Unmanned Marine Vehicles** G.N. Roberts and R. Sutton (Editors)
Volume 78 **Frequency-Domain Control Design for High-Performance Systems** J. O'Brien
Volume 80 **Control-Oriented Modelling and Identification: Theory and practice** M. Lovera (Editor)
Volume 81 **Optimal Adaptive Control and Differential Games by Reinforcement Learning Principles** D. Vrabie, K. Vamvoudakis and F. Lewis
Volume 83 **Robust and Adaptive Model Predictive Control of Nonlinear Systems** M. Guay, V. Adetola and D. DeHaan
Volume 84 **Nonlinear and Adaptive Control Systems** Z. Ding
Volume 88 **Distributed Control and Filtering for Industrial Systems** M. Mahmoud
Volume 89 **Control-Based Operating System Design** A. Leva *et al.*
Volume 90 **Application of Dimensional Analysis in Systems Modelling and Control Design** P. Balaguer
Volume 91 **An Introduction to Fractional Control** D. Valério and J. Costa
Volume 92 **Handbook of Vehicle Suspension Control Systems** H. Liu, H. Gao and P. Li
Volume 93 **Design and Development of Multi-Lane Smart Electromechanical Actuators** F.Y. Annaz
Volume 94 **Analysis and Design of Reset Control Systems** Y.Guo, L. Xie and Y. Wang
Volume 95 **Modelling Control Systems Using IEC 61499, 2nd Edition** R. Lewis & A. Zoitl
Volume 96 **Cyber-Physical System Design with Sensor Networking Technologies** S. Zeadally and N. Jabeur (Editors)
Volume 99 **Practical Robotics and Mechatronics: Marine, Space and Medical Applications** I. Yamamoto
Volume 102 **Recent Trends in Sliding Mode Control** L. Fridman JP. Barbot and F. Plestan (Editors)
Volume 105 **Mechatronic Hands: Prosthetic and Robotic Design** P.H. Chappell
Volume 107 **Solved Problems in Dynamical Systems and Control** D. Valério, J.T Machado, A.M. Lopes and A.M. Galhano

Organic Sensors

Materials and Applications

Edited by Eduardo Garcia-Breijo,
Berta Gómez-Lor Pérez
and Piero Cosseddu

The Institution of Engineering and Technology

Published by The Institution of Engineering and Technology, London, United Kingdom

The Institution of Engineering and Technology is registered as a Charity in England & Wales (no. 211014) and Scotland (no. SC038698).

First published 2016

The Institution of Engineering and Technology
Michael Faraday House
Six Hills Way, Stevenage
Herts, SG1 2AY, United Kingdom

www.theiet.org

British Library Cataloguing in Publication Data
A catalogue record for this product is available from the British Library

ISBN 978-1-84919-985-8 (hardback)
ISBN 978-1-84919-986-5 (PDF)

Typeset in India by MPS Limited

Contents

Chapter 1

Conducting polymers in sensor applications

Vibha Saxena[1] and D.K. Aswal[2]

Abstract

Conducting (or conjugated) polymers (CPs) are widely used for fabricating chemical and biosensors. These materials enable fabrication of devices in short lead time and at relatively low costs owing to their unique advantages of light weight, easy processability, compatibility with biological systems, tunable electronic and optic properties, large area fabrication and potential for flexible or wearable sensors. This chapter describes the basic concepts underlying the interaction of analyte with CPs as well as the other components of sensors. The operating principles of transducers used in the sensor are discussed in detail. Sensors based on conducting polymers are classified in accordance with the changes in the properties of conducting polymers. These changes involve changes in (i) doping levels, (ii) optical properties and (iii) changes due to weak interactions. Furthermore, the sensors have been reviewed for various conducting polymers such as polyaniline, polythiophene, polypyrrole, polydaicetylene for analysis of wide variety of analysts including toxic gases, radiation, pH, temperature, external stimuli; and recent advances in sensors based on conducting polymers have been discussed. Future directions and thrust in emerging field have been projected in the area of molecular engineering, material synthesis, film preparation and transducers, circuits and algorithms for developing futuristic sensors especially wearable sensors.

1.1 Introduction

Detection or identification of specific composition, external stimuli (pressure, temperature, touch), chemical activity, concentration, presence of specific element or ion is one of the most sought after requirements in a variety of areas such as medicine, environment, Industry, offices and homes. The most common and conventional methods developed for detection of specific analytes are based on high-performance liquid chromatography and gas chromatography. However, these

[1]Technical Physics Division, Bhabha Atomic Research Centre, Trombay, Mumbai 400 085, India
[2]CSIR-National Physical Laboratory, Dr K.S. Krishnan Marg, Pusa, New Delhi 110 012, India

techniques are time consuming and expensive. Therefore, a simple method is required for enabling rapid detection of analytes efficiently and effectively at low cost. A sensor is a device that transforms information such as chemical composition, chemical activity, concentration and stress into detectable physical signals (such as current, mass, acoustic, optic) that could be analyzed using the available methods. As a result of interaction of the analyte with sensing material, information may be collected owing to changes in the physical properties of the sensing material. An ideal sensor should be portable, affordable, easy to fabricate and should respond to the particular analyte quickly and selectively, independent of ambient conditions. Such ideal sensors are, however, yet to be realized although several successful sensors are commercially available. Therefore, there is a constant demand for developing sensors, by using new sensing materials, which may result in an improvement in selectivity and sensitivity of the device. Limiting features for a typical sensor are as follows: selectivity, sensitivity, detection limit, response and recovery time and packaging size. In order to realize a high-performance sensor having good sensitivity and selectivity, there is a need to consider several issues in material synthesis.

Conducting (or conjugated) polymers (CPs) are promising materials for application in various sensors including chemical and biosensor owing to their light weight, easy processing, tunable electronic and optic properties, large area fabrication and especially potential for fabricating flexible or wearable sensors [1]. Intrinsic/doped as well as functionalized conducting polymers have been used as an active sensing element for fabricating sensors. Sensor fabrication for specific analyte is mainly dependent on the application requirement and existing sensor technology. The following section discusses the main components required for the fabrication of sensor.

1.1.1 Key components of a sensor

As shown in Figure 1.1, a typical sensor fabrication requires two main components (i) sensing material (i.e. conjugated polymer) and (ii) transducer. In general, the fabrication of a sensor requires a good quality thick or thin film that is uniform and free of pin-holes. Several techniques such as dip coating, drop casting, spin coating, inkjet printing, gravure printing, slot die coating and screen printing were used to prepare high-quality films for use in sensors. In addition, ultra-thin films of conducting polymers were also prepared using the Langmuir–Blodgett (LB) or self-assembly technique. The CP interacts with the analyte and the transducer detects

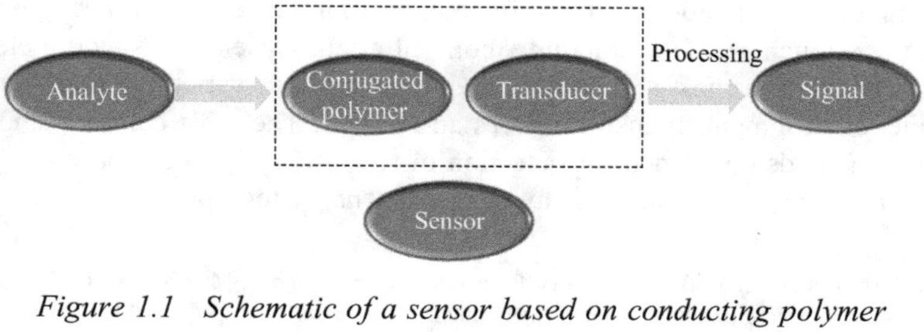

Figure 1.1 Schematic of a sensor based on conducting polymer

the resultant changes in the physical properties of the CP and converts to a measurable parameter which is then processed using a suitable electronic system to generate the final signal. The interaction of CP with the analyte and transducers is discussed in detail in the following section.

1.1.2 Interaction of CP with analyte CP films

The interaction of the analyte with CP depends on many parameters such as film morphology, structure, work function and ionization potential (IP, electron affinity). Therefore, many factors have to be optimized for the synthesis of CPs to be used as sensing material. These include high surface-to-volume ratio for enhanced response, chemical functionality to improve selectivity, crystallinity or structural ordering to have efficient charge transfer and fast adsorption/desorption kinetics for quick response/recovery and are depicted in Figure 1.2.

1.1.2.1 Langmuir model

The Langmuir model describes the relationship between the available sites in CP which can interact with a number of adsorbed molecules and is useful to describe the adsorption of gas on monolayers. In this model, it is assumed that (i) molecules are held at specific sites on the surface, and the total coverage is limited to a single layer and (ii) there is no interaction between adsorbed molecules/atoms on different sites and the enthalpy of desorption is independent of the number of sites and is the same for all the sites.

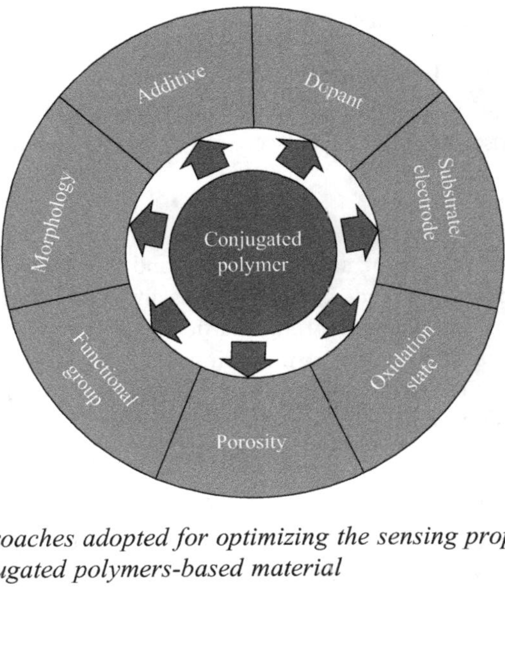

Figure 1.2 Approaches adopted for optimizing the sensing properties of conjugated polymers-based material

Based on these assumptions, the fundamental kinetic equation for the Langmuir isotherm is given by the following equation:

$$\frac{p}{\theta} = \frac{1}{\chi} + p \tag{1.1}$$

where $\chi = (\tau_a/n_o)I/p$, τ_a, n_o, I and p being the surface lifetime of gas molecules after absorption, total number of sites available for adsorption, impingement rate, the pressure of the adsorbate, respectively, and θ is the number of occupied sites as a fraction of total sites. This model was used to describe many CP-based sensors and was also used in conjunction with other models [2,3].

1.1.2.2 Brunauer, Emmett and Teller (BET) model

The Langmuir model does not hold good if there is any interaction between adsorbed species. In this case, adsorbed species are not confined to single monolayer as a result of which, the heat of adsorption of an adsorbate molecule adsorbing on CP will differ significantly and accordingly, can be described by the following equation:

$$V = \frac{YV_{max}P/P_V}{\left(1 - \frac{P}{P_V}\right)\left(1 + \frac{(Y-1)P}{P_V}\right)} \tag{1.2}$$

where V is the adsorbed gas volume, V_{max} denotes the volume needed to form a monolayer on the whole surface, P_V denotes the vapor pressure of the adsorbed gas, Y is a constant number and $Y \approx \exp[((\Delta H_{des} - \Delta H_{vap})/(RT))]$, where ΔH_{des} is the heat of desorption, ΔH_{vap} is the heat of vaporization, R is the ideal gas constant and T is the temperature. This model has also been used widely to describe the mechanism in CPs [4].

1.1.2.3 The Freundlich model

This adsorption model is an empirical function that expresses the relationship between the magnitude of adsorption and the pressure at a constant temperature as follows:

$$\frac{\chi}{m} = Kp^{1/n} \tag{1.3}$$

where χ and m are masses of adsorbed vapor and adsorbent at an equilibrium pressure p, respectively. n and K are the constants that are dependent on the adsorbent and vapor intrinsic nature at a particular temperature. This model was used to explain the sensitivity of ultra-thin films of CP for NO_2 [5,6]. Other than the above models, modified models were also proposed such as diffusion limited [7] and reaction–diffusion model [8].

1.1.3 Operating principles of transducers

Besides the interaction of the analyte with CP, the transducer is another important part of a sensor. A variety of sensors have been developed by using conducting polymers in different transduction modes, as depicted in Figure 1.3.

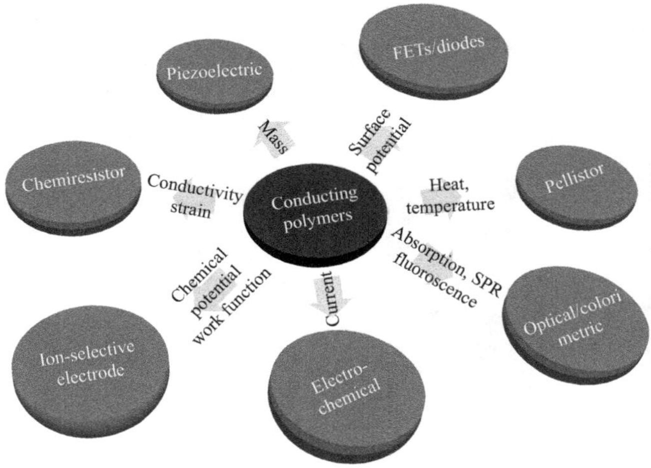

Figure 1.3 Most commonly used transducers used in sensors and the arrow indicates the change in physical parameter upon interaction with conducting polymer

The operating principle of the transducer utilized in the sensor could be any of the combination of the parameters that are presented in Figure 1.3 and are discussed as follows.

1.1.3.1 Chemiresistors

In a chemiresistor, which consists of one or several pairs of electrodes, as depicted in Figure 1.4(i), electrical resistance change is measured upon interaction with the analyte and a simple ohmmeter may be used to collect the data. In general, the conductivity of CPs ranges from semiconductive to insulating regions in the pristine state, which can be altered by controlled addition of chemical species in small non-stoichiometric quantity termed as *dopants*. Doping generally refers to the oxidation/reduction of the π-electronic system of CPs. Upon exposure of analytes to CP, the change in conductance/resistance could arise due to either (i) change in doping levels or (ii) conformational changes. This change can happen in solid form or in solution and can be bought about chemically or electrochemically. Since this redox process is reversible, sensitive transducers can be developed using this parameter. Other than redox processes, physiological parameters, such as pressure, temperature and moisture, also lead to the change in the resistance of CPs. The efforts are being made to develop pressure sensors based on CPs with sensitivity on a par with human skin and are discussed in detail in Chapter 4. Besides pressure, the piezoresistive property of CPs was also used to transduce mechanical strain into an impedance/conductivity change. The main parameter of stress is gauge factor (GF) defined as follows:

$$GF = \varepsilon\left(\frac{\Delta R}{R_0}\right) \tag{1.4}$$

where ΔR is the resistance change with strain, R_0 is the initial resistance and ε is the applied strain. A highly sensitive strain sensor requires a GF of ~3. Although metallic

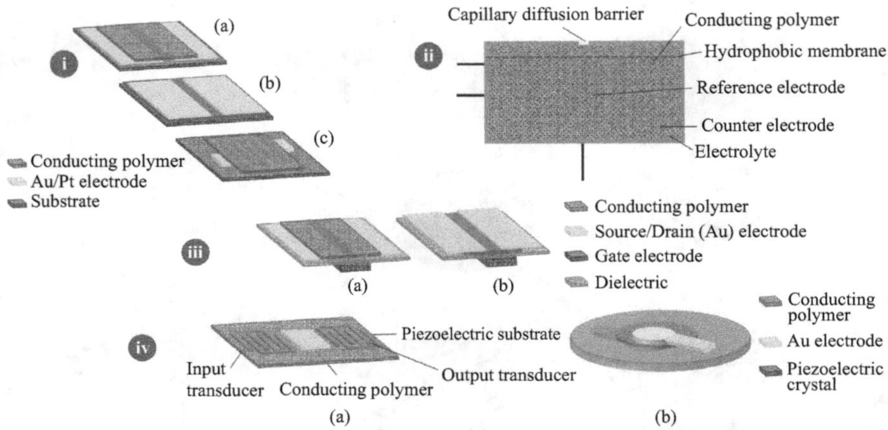

Figure 1.4 Common transducers used for the fabrication of sensors based on conducting polymer: (i) chemiresistor – planar (a, b) or interdigitated (c); (ii) electrochemical sensor; (iii) field-effect transistor (a) bottom-contact (b) top contact and (iv) piezoelectric sensor – (a) SAW and (b) QCM

conductors have a high GF value, stretchability alongwith high GF is a requirement in a touch sensor. The common approach to obtain this is to fill the elastomeric matrix with conduction path. Conjugated polymer composite has a potential to develop such kind of touch (pressure or strain) sensors on flexible conducting or non-conducting substrates and will be discussed in this chapter later on.

1.1.3.2 Ion-selective electrodes (ISEs)

The chemical potential of CP may alter in the presence of an analyte, leading to a change in the open-circuit potential or work function. This change is proportional to the logarithmic of concentration of the analyte and is given by the Nernst equation as follows:

$$E = E_0 + 2.303\left(\frac{RT}{zF}\right)\log a_{\mathrm{H+}} \tag{1.5}$$

where E and E_0 are the half-cell and standard half-cell reduction potential, respectively, with R, T, F, z and $a_{\mathrm{H+}}$ being the universal gas constant, temperature, Faraday constant number of electrons transferred in the cell reaction, and proton activity, respectively. The sensitivity of ion-selective electrodes (ISEs) is therefore limited to 59 mV for a tenfold sample activity change in case of a monovalent ion. Based on the change in the chemical potential, ion sensors having extremely high selectivity and detection limits down to nM levels [9,10] were developed for in situ sensing of ions and biological substances in appropriate aqueous media and physiological backgrounds. The ISE-based sensor will not be discussed here and are reviewed in another chapter of this book.

1.1.3.3 Electrochemical sensors

In these sensors, changes, either in current or potential, are observed upon interaction with the analyte. The interaction of CP with the analyte can give rise to a current, which is governed by Faraday's law. Accordingly, amperometric, potentiometric or conductometric sensors can be fabricated as depicted in Figure 1.4(ii). The transducers based on electrochemical methods can easily be fabricated using microelectrodes. Because of this reason, this method offers several advantages for detection of ions or biological molecules as compared to macroscopic electrodes because (i) they provide higher spatial resolution because of small geometric area (i.e. selectivity) and (ii) have faster electron transfer and mass transport rate and so have quick response. The excellent redox activity of conjugated polymers has been utilized in fabricating many bio- and chemical sensors to detect ions as well as gases [11–14]. A detailed discussion on biosensor and electrochemical sensor can be found in Chapters 2 and 8, respectively, and we shall restrict ourselves to the usage of conjugated polymer for other sensors in this chapter.

1.1.3.4 Optical sensors

Optical sensors use analyte-sensitive conducting polymer to detect changes in absorbance, luminescence or fluorescence, reflectance, etc. upon interaction with chemical species. A typical optical sensor consists of an optical fiber with a conducting polymer immobilized on the fiber tip. By immobilization of different polymer layers onto the fiber tip, the sensor can be tuned to distinguish between a particular analyte. Other than this, optical sensors were also developed based on measuring the RGB%, absorbance, refractive index, reflectance, etc. Different methods of detecting optical signals using an optical fiber as well as other transduction methods are discussed hereunder.

1. ***Optical absorption:*** Although the absorption characteristics of a conjugated polymer are mainly determined by the local electronic structure, it can also be tuned by different conformations of the conducting polymer such as coil-like or expanded form. Therefore, in addition to changes related to π–π^* transition, changes in the conformation of CP with interaction of the analyte can alter the band gap and therefore the absorption characteristics.
2. ***Fluorescence:*** Among optical techniques, fluorescence is a very powerful transduction mechanism that allows for detection of even trace element based on a different outcome of fluorescence such as lifetime, wavelength, anisotropy or intensity of the emission of fluorescent probes and energy transfer. The absorption of photons in the fluorescent CPs creates excitons, giving rise to strong fluorescence. The interaction of analytes (quencher) with these fluorescent polymers (fluorophore) can disrupt the extent of delocalization and result in fluorescence quenching. Fluorescence quenching is measured quantitatively with the Stern–Volmer equation:

$$\frac{F_0}{F_{[Q]}} = 1 + K_{SV}\,[Q] \tag{1.6}$$

where $[Q]$ is the quencher concentration, F_0 and $F_{[Q]}$ are the measured fluorescence intensities without and with quencher present, respectively, and K_{SV} is the Stern–Volmer constant. The Stern–Volmer equation does not contain variable for the fluorophore concentration and, therefore, suggests that fluorescence quenching is independent of the fluorophore concentration. Equation (1.6) applies in case of a quencher inhibiting either a photochemical reaction or a photophysical process by a single reaction and has been applied to CPs very frequently.

3. ***Colorimetric:*** In contrast to fluorescence quenching, colorimetric techniques have advantages of low cost, requirement of inexpensive equipment and simplicity. The colorimetric sensors have been developed for both qualitative analyte identification and quantitative analysis. The colorimetric response (CR) has been used quite often to define the response of CPs to the analyte and is defined as follows:

$$\mathrm{CR} = \frac{\mathrm{PB}_0 - \mathrm{PB}_1}{\mathrm{PB}_0} \times 100 \tag{1.7}$$

where $\mathrm{PB} = A_{\mathrm{Blue}}/(A_{\mathrm{Blue}} + A_{\mathrm{Red}})$, in which A_{Blue} and A_{Red} is the absorbance at the blue and the red component, respectively, in the UV–Vis spectrum and PB_0 and PB_1 are the values without and with analyte, respectively. Other than CR%, RGB% has also been used to determine the CR of the sensor.

1.1.3.5 Transistor and diode sensors

Despite several advantages of electrochemical sensors, the determination of analyte remains by using a potentiostat hinders the miniaturization of these devices. Thin-film transistors or electrochemical transistor offers alternatives for overcoming such drawbacks by providing the possibility of developing a low-cost miniaturized device. Upon interaction with the analyte, the local electrostatic charge (electrochemical) or electric field (field effect) of the sensitive gate/channel is modulated. This effect is reversible and can also be tuned for selectivity by proper functionalization. Figure 1.4(iii) shows a typical configuration of bottom-contact and top-contact field-effect transistors (FETs) using CPs as sensors.

1.1.3.6 Gravimetric sensors

Analyte–CP interaction may also induce weight changes in the conducting polymer owing to absorption/desorption of ions. This type of transduction is based on the change in mass of the piezoelectric sensor material owing to analyte absorption, which then results in a change in the resonant frequency. Two types of devices can be used (Figure 1.4(iv)): surface acoustic wave (SAW) device and quartz crystal microbalance (QCM). A SAW device is composed of a piezoelectric substrate with an input (transmitting) and output (receiving) inter-digital transducer deposited on the top of a substrate, such as ZnO, lithium niobate or quartz. A sensitive membrane is placed between the transducers and an acoustic 2D wave is allowed to propagate along the surface of the crystal at a depth of one wavelength by applying an ac signal across the input transducer. The mass of the sensitive material changes upon interaction with the analyte, leading to a change in the wave frequency. QCM, on the other hand, is a piezoelectric resonator, made of quartz with one or both surfaces covered with a sensitive material. The device when connected to an amplifier acts as an oscillator having 3D wave travelling through the crystal at a frequency of 10–30 MHz. In the presence of an analyte, there is a process of absorption/desorption of chemical species that changes the properties of sensitive material and hence the resonant frequency of the QCM device. Acoustic sensors have several advantages over other types of transducers such as high sensitivity, quick response, low power consumption and size. However, acoustic sensors are affected by temperature and humidity, due to this reason QCM has been extensively used to develop good quality sensors specifically sensors based on redox reaction. In such cases, the Sauerbrey equation is used to correlate the frequency change (Δf) and mass change per unit area ($\Delta m/A$) as follows:

$$\Delta f = -\frac{2f_0^2 \Delta m}{AC_f} \tag{1.8}$$

where f_0 is the fundamental frequency of oscillator, C_f (= $\sqrt{\mu\rho}$) is the sensitivity factor for the crystal which depends on the density ($\rho = 2.648$ cm^{-3}) and shear modulus ($\mu = 2.947 \times 10^{11}$ g/cm s^2) of the crystal. The equation is valid under the assumption that the density and thickness of the sensitive material is homogeneous throughout the film and is equal to that of quartz, and that the viscoelasticity of the materials remains unchanged during the measurement. Since these conditions have been found to generally hold for thin CP films, QCM is widely accepted as a sensitive and accurate technique for determination of biological as well as chemical species.

As discussed earlier, a variety of transducers and film deposition techniques are available for developing good quality sensors based on CPs. Upon exposure to an analyte, the CP may undergo either physisorption or chemisorption. In the chemisorption process, chemical bonds such as ionic covalent or hydrogen bonds are formed as a result of shared electrons between the CP and the analyte which are influenced by the analyte, the surfaces species valency and geometric requirements. The interaction of the conducting polymer with many analytes results in a change of doping level as a result of chemical reaction. Physisorption, on the other hand, involves weak Van der Waals forces such as dipole–dipole interactions, dispersion forces and π–π interactions. Since Van der Waals forces exist between all atomic/molecular species, physisorption always occurs and therefore is not limited to a single molecular layer as in case of chemisorption. The ability of a molecule (or analyte) to form an ion, either by accepting or donating an electron, is indicated by the Mulliken electronegativity coefficient that is defined as the average of its molecular affinity (EA) and its IP [15]:

$$\mathrm{E}\eta = 0.5(\mathrm{EA} + \mathrm{IP}) \tag{1.9}$$

Molecules with a large IP and a very negative EA will have a large positive value and so their electronegativity is high. On the other hand, molecules with a small IP and a small EA have a small positive value and therefore they have a low electronegativity. Table 1.1 summarizes various physical parameters of gas molecules as well as work function of many CPs. Upon absorption of gas molecules on CP, the direction of charge transfer depends on the values of Eη and Fermi level (or work function) of the CP, e.g. for Eη lying above E_F, the electron will transfer from CP to gas molecules and vice versa [23]. The nature of these charge-transfer complexes formed between analyte and CPs are not fully understood. Nevertheless, several sensors based on CPs have been developed. Based on the changes in the properties of conducting polymers, sensors can be categorized as follows.

Table 1.1 Physical parameters of some oxidizing and reducing gases

Gas	Electron affinity (eV) [16]	Ionization potential (eV)	Mulliken electronegativity (Eη)	HOMO (eV)	LUMO (eV)	Polymer
NO_2	2.1	10	6	−5.27	−2.51	Polyaniline [17]
O_2	0.45	12.07	6.02	−4.6	−3.6	Polypyrrole [18]
SO_2	1.1	12.3	6.7	−4.76	−2.74	P3HT [19]
PH_3	−1.9	10	4.1	−5.5	−3.6	PCDTBT [19]
CO_2	14.12	–	7.18	−5.15	−3.37	PTB7 [19]
NH_3	−5.6	10.7	2.6	−5.2	−2.4	PEDOT:PSS [20]
H_2	15.86	–	7.33	−4.9	−2.8	MEH-PPV [21]
H_2O	−6.4	12.6	3.1	−2.5	−2.34	PDA [22]

1. **Sensors based on changes involving doping levels**
   The physical properties of conducting polymers are dependent on the extent of doping as well as dopant ions. In general, conducting polymers can be doped/dedoped (called *Primary doping*) by using either redox reaction (Figure 1.5(a)) or acid base reaction, leading to the transfer of electrons from/to the analyte (Figure 1.5(b)) and can be achieved chemically or electrochemically. This electron transfer changes the resistance and work function of the material, which is the basis of the sensing principal for sensors based on doping levels. Additionally, conducting polymers-based p–n junction, Schottky junction as well as metal-oxide-semiconductors (MOSs) were also developed for being used as a sensor. The influence of an analyte on junction properties could be utilized to fabricate efficient sensors. Figure 1.5(c) illustrates the p–n junction formed between a conjugated polymer and n-type material (e.g. metal oxide). The barrier height for holes is decreased in the presence of oxidizing gas and results in decreased resistance. On the other hand, the capacitance (or impedance) of the MOS device is influenced (Figure 1.5(d)), resulting in a response that is dependent on analyte concentration.
2. **Sensors based on changes owing to weak interactions**
   The physical properties of doped conducting polymers may also change due to the weak interactions between neutral analyte molecules and CP. These weak interactions do not change the doping levels of the conjugated polymer; however, they influence the electronic, optical or magnetic properties of CPs owing to the change in the π-system. The interaction is influenced by the molecular volume of analyte, affinity of the analyte to CP, surface voids in CPs and the physical state of CP [24,25]. In specific cases, the increase in the conductivity in CPs owing to this weak interaction, referred to as "*secondary doping*", is the result of a change in the molecular conformation of the CP from compact to expanded coil. This secondary dopant differs from the primary dopant in the sense that a partial charge transfer between the dopant and the CP takes place. This charge transfer might result in dipole formation at CP's surface, resulting

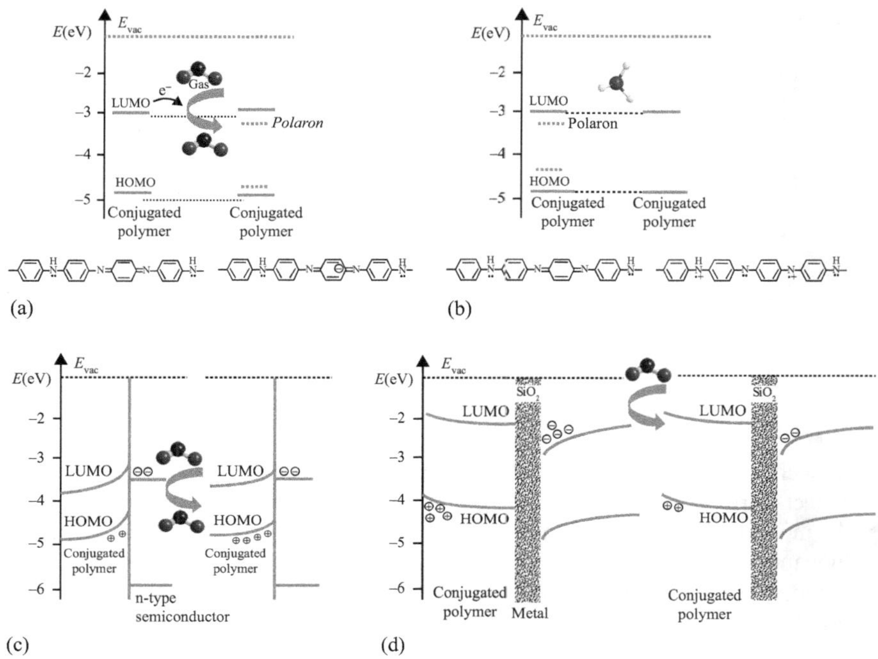

Figure 1.5 Several sensing mechanisms for sensors based on CP, the interaction of the analyte with CP can result in (a) redox reaction: electron (or hole) transfer from LUMO (or HOMO) of polyaniline to NO_2 gas results in appearance (or disappearance) of the polaron level, (b) acid–base reaction: proton transfer from CP (e.g. emeraldine salt of polyaniline) to NH_3 gas results in removal of the polaron level (e.g. emeraldine base of polyaniline), (c) p–n (or Schottky) junction between CP and metal oxide (or metal) and (d) MOS junction: analyte exposure (e.g. NH_3) results in a decrease in the number of accumulated holes in CP due to the transfer of holes from HOMO of CP to the analyte and therefore the barrier height, (e) weak interaction: polar analyte may result in partial charge transfer from the polypyrrole backbone to the absorbed analyte, thereby creating a dipole at the CP surface and therefore, change in work function, (f) fluorescence quenching: presence of the acceptor (A) type of analyte (oxidizing) will result in delocalized π^* excited in CP which is donor (D) type. This D–A electron transfer mechanism facilitates excitation migration process and therefore results in quenching

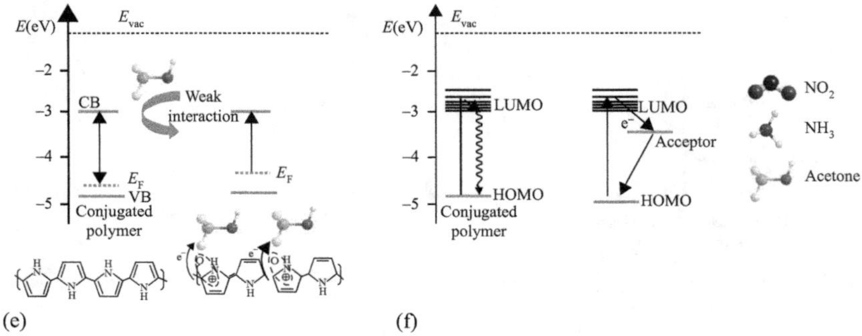

Figure 1.5 (Continued)

in a change in conductivity as depicted in Figure 1.5(e). In case of a pronounced effect of polar surface formation, the change in work function occurs which can be monitored by the Kelvin probe (KP). The affinity of the CP for the secondary dopant may be enhanced by introducing specific binding sites, by synthesizing template-synthesized nanostructures or by using post-treatments.

3. **Sensors based on changes involving optical properties**
When conjugated polymers are doped, electronic defects such as soliton, polaron or bipolaron are created and consequently, new energy levels are formed within the highest occupied molecular orbital (HOMO)–lowest unoccupied molecular orbital (LUMO) level of the CPs. Besides the π–π^* transition, the transitions between energy levels involving new intra-gap energy levels are also possible. The emission as well as absorption properties of CP upon interaction with the analyte may change owing to a change in these intra-gap energy levels as shown in Figure 1.5(a) and (b). In case of some fluorescent-conjugated polymers, the interaction with analytes can produce intra-molecular charge transfer resulting in significant changes in fluorescence. As shown in Figure 1.5(e), in the presence of electron acceptor (analyte), electron transfer occurs from photoinduced excited state of the polymer to the analyte, which leads to fluorescent quenching.

Based on these sensing principles, several conjugated polymers have been utilized as sensors for toxic gases, chemical warfare agents, toxic elements, ions, bending motion [26]. Different approaches adopted for developing a sensor technology are discussed in the following sections.

1.2 Sensor based on doping levels

Conjugated polymers are intrinsically conducting due to conjugated double bonds and their conductivity can be varied by doping upon exposure to the analyte. The doping will proceed if:

$$\Delta E = I_D - A_p - E_{coul} < 0 \tag{1.10}$$

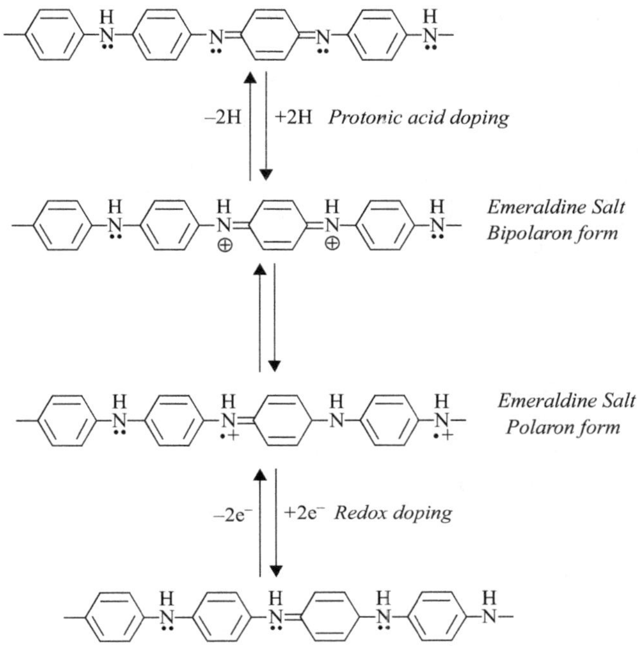

Figure 1.6 Redox and protonic doping of leucoemeraldine base and emeraldine salt of polyaniline, respectively

where I_D is the *ionization potential* of the analyte, *Ap* is the *effective* electron affinity of the polymer, which includes the relaxation energy involved when the analyte is accommodated in the polymer chain and E_{coul} is all electrostatic interactions between charges in the doped state. In addition to this, the kinetic of the analyte is also dependent upon the morphology and structure of the conjugated polymer, e.g. ion-diffusion constant in crystalline part may be very small (10^{-9}–10^{-12} cm^2/s) as compared to that in amorphous part; in case of a solvent, the ion-diffusion constant may be higher for polar solvent as compared to non-polar solvent.

As mentioned in introduction part, this doping may lead to changes in resistance, optical properties and work function and is the basis for the development of sensors based on CP. The changes in doping levels may further be categorized according to the reaction involved: (i) redox reaction and (ii) proton doping. A redox reaction involves the transfer of electrons from conducting polymer to the analyte or vice versa, which results in a change in optical and electronic properties. Proton doping, on the other hand, does not involve a change in the number of electrons associated with the polymer backbone. However, this process results in the rearrangement of energy levels, leading to changes in conductivity and/or optical properties. Both types of doping are illustrated for polyaniline (PANI) in Figure 1.6 and sensors based on these reactions are discussed as follows.

1.2.1 Doping/undoping by redox reactions

As mentioned earlier, the doped (or undoped) conjugated polymer may be reduced (or oxidized) by the redox active analyte. In general, CPs are p-type, and therefore the doping level is enhanced or reduced depending upon the nature of analyte, e.g. the presence of electron acceptor, such as NO_2, I_2, will result in removal of electrons from the aromatic ring, leading to an enhancement in the doping level. Based on the doping, several sensors have been fabricated using polymers such as PANI, polypyrrole (PPy) and polycarbazole as summarized in Table 1.2 and discussed in following sections.

1.2.1.1 Polyaniline

Ultra-thin as well as nanostructured PANI films were prepared to use as NO_2 sensors. The PANI nanofibers film synthesized by an interfacial polymerization method has been used as an efficient NO_2 sensor in the 10–100 ppm range [27]. The change in resistance upon exposure to NO_2 was attributed to oxidation of emeraldine base. Highly ordered as well as ultra-thin structures were prepared using the LB technique. It was suggested that high surface-to-volume ratio resulted from such structures would result in enhanced interaction between sensing materials and analytes, leading to enhanced sensitivity as well as fast adsorption/desorption kinetics for analytes. It was shown that a composite PANI film prepared by the LB method exhibited a faster recovery property for NO_2 [28]. In addition, the response time to NO_2 and the relative change of resistance films increased with the increase in the number of film layers. Ram *et al.* studied various composite LB films prepared from SnO_2 and polyaniline, PPy and polythiophene (PTh). Among all the polymers studied, regioregular PTh, its copolymer and their metal-oxide nanocomposite films were found to be highly sensitive to the NO_2 gas. A new design of conductometric chemical sensors was also proposed which comprised six electrodes, of which two electrodes were used to control the redox state of the polymer (PTh or PANI) and remaining four electrodes were used to measure conductance [29]. A reversible sensor was thus demonstrated for the NO_2 gas in the range of 12–100 ppm. Additionally, site-specific synthesis of nanometer-thick PANI film was carried out for the fabrication of FET/chemiresistor sensors [30]. The devices were observed to respond towards NH_3 and NO owing to electrostatic gating being the dominating mode of sensing.

PANI films were also used to develop sensors for H_2. Both thick and thin films deposited on 36° lithium tantalate SAW transducers were used to detect the hydrogen (H_2) gas at room temperature [34]. The response time was observed to be faster for thin film owing to fast diffusion of the target gas to the PANI surface. H_2 concentrations below 0.5% could thus be detected. Further, both doped and undoped PANI nanofibers deposited onto conductometric sapphire transducers were used for detecting H_2 gas [36]. The un-doped nanofiber sensors were found to exhibit better repeatability than the doped sensors. However, high sensitivity and fast response times of 28 and 32 s, respectively, were observed for both types of polymers. The interaction mechanism of PANI with H_2 is not understood well.

Table 1.2 Conducting polymer sensors to several gas analytes based on redox reactions

Analyte	Deposition technique	Sensing material	Range	Response/ recovery time	Sensitivity	Selectivity against	Transducer	Ref.
NO_2	LB	PBTh	1–100 ppm	–	24% ppm^{-1}	NH_3, methanol, ethanol, CO and CH_4	Chemiresistor	[5]
NO_2	Interfacial polymerization	PANI	10–200 ppm	2–15 min	–	–	Chemiresistor	[27]
NO_2	LB	PANI	20–100 ppm	10 s/4 min (20 ppm)	3.9% ppm^{-1}		Chemiresistor	[28]
NO_2	Electrochemical	PANI	12–100 ppm	2–15 min	–	–	Chemiresistor/ FET	[29]
NO_2	Electrochemical	PANI single wire	50–250 ppb	6.5 min (50 ppb)	3,760% ppm^{-1}	–	Chemiresistor	[30]
NO_2	Spin coating	Polypyrrole	10–100 ppm	218 s/374 s (100 ppm)	3,760% ppm^{-1}	NH_3, H_2S, CH_3OH, C_2H_5OH, Cl_2	Chemiresistor	[31]
NO_2	LB + EPZ	PEDOT	500 ppb–20 ppm	170–180 s/<70 s (20 ppm)	–	HCl, H_2S, SO_2	Chemiresistor	[32]
NH_3	LB	Polycarbazole	50–5,000 ppm	–	–	–	Chemiresistor	[33]
H_2	Electrochemical	PANI	0.06%–1%	–	–	–	SAW	[34]
H_2	Chemical	PANI	10 ppm	–	–	–	Chemiresistor	[35]
H_2	Chemical	PANI	1%	38 s	–	–	Chemiresistor	[36]
H_2	Spin coating	PANI/chitosan	0.4%–3%	–	–	$CHCl_3$, CH_3OH, C_6H_6, NH_3, acetone, acetic acid, hexane	Chemiresistor	[37]
AsH3	Chemical	PANI	150–500 ppb	–	–	–	Chemiresistor	[38]
H_2S, NO_2	LB	PANI	10–50 ppm	–	–	–	Chemiresistor	[39]
p-NP	Electrochemical	PPy	0.1 nM to 100 µM	–	–	–	Electrochemical	[40]
NM1	Electrochemical	PPy	0.03–0.38 mM		–	–	Chemiresistor	[41]
Pressure	Chemical	PPy	1,000–7,000 N/m^2	–	0.0007 mS/N	–	Chemiresistor	[42]

(Continues)

Table 1.2 (Continued)

Analyte	Deposition technique	Sensing material	Range	Response/ recovery time	Sensitivity	Selectivity against	Transducer	Ref.
Pressure	Free-standing	PPy	0–30 Pa		56–133 kPa^{-1}		Chemiresistor	[43]
Strain	In situ chemical	PPy	50%		600 per strain unit		Chemiresistor	[44]
Strain	Dip coating	PEDOT:PSS	80%	–	0.18 S/cm per strain unit	–	Chemiresistor	[45]
Strain	Chemical	PEDOT:PSS	260%	–	–	–	Chemiresistor	[46]
Strain	Chemical	PANI	2%–10%	–	–	–	Chemiresistor	[47]
Strain	Chemical	PANI	0–1500%	–	–	–	Chemiresistor	[48]
Strain	Electrospinning	PEDOT:PSS/PVA	0.2%–1.2%	–	GF 396	–	Chemiresistor	[49]
Strain	Spin coat	PEDOT:PSS	50%–80%	–	GF 42	–	Chemiresistor	[50]
Strain	Spin coat	PEDOT:PSS	0%–80%	–	85% (60% strain)	–	Chemiresistor	[51]
Temperature	Dip coating	PEDOT:PSS	15–45 °C	–		–	Chemiresistor	[52]
Strain			35%	–	–	–		
Humidity			25%–95%	–	–	–		
N_2 flow	Spin coat	P3HT/5CB	0.5–3 sccm	15 s (3 sccm)	–		FET	[53]
DMMP		Hydroxylated PEDOT:PSS	10 ppt–50 ppb	–	–	Trimethyl phosphate, methyl dichlorophosphate, trichlorophosphate	Chemiresistor	[54]
NIR	Spin coat	PANI	2.43 W/cm^2	20 s	–	–	Chemiresistor	[55]
γ-ray	PTh	Polythiophene	5–25 mrad	–	–	–	Chemiresistor	[56]
X-ray	Spin coat	PPV	23–69 kGy	–	–	–	FET	[57]
X-ray	Spin coat	MEH–PPV/ poly(9,9-dioctylfluorene)	4–18.5 mGy/s	150 ms	500 nC mGy^{-1} cm^{-3}	–	Schottky diode	[58]

It was suggested by Unsworth *et al.* that a change in resistance is caused by the formation of water. It was also proposed that hydrogen interacts with doped emeraldine salt (ES) of PANI at the charged amine nitrogen sites [59]. New N–H bonds are formed to the amine nitrogen of the PANI chain as a result of H_2 bond dissociation. A charge transfer between adjacent amine nitrogen brings the PANI back to its ES state, which is accompanied by the release of hydrogen, and thus makes this reaction fully reversible. In contrast to this, the mechanism for sensors fabricated using platinum electrodes was found to be different. Better response upon the hydrogen interaction of CP was observed owing to the reaction of H_2 with platinum at the PANI –platinum interface. Platinum was found to convert to platinum hydride when exposed to H_2 [35]. This resulted in the formation of a Schottky barrier between platinum and PANI and therefore, a significant increase in resistance (+65%) upon exposure to H_2. The sensor could be used to detect H_2 at a concentration of 10 ppm and was reversible as platinum hydride converts back to platinum in the presence of oxygen. On the other hand, only a small resistance decrease (−3%) at low concentrations of hydrogen (1%) was obtained under the same conditions by using gold electrodes. Recently, sensing properties of the oxygen plasma-treated thin PANI film were reported [60]. It was observed that the plasma treatment changed the composition, thickness and morphology of the PANI films, which was studied using X-ray photoelectron spectroscopy (XPS), Fourier transform infrared spectroscopy (FTIR) and atomic force microscopy. The sensor treated by oxygen plasma for 60 s showed a reproducible sensitivity to 10 ppm hydrogen pulses via a ~30% increase in resistance. The sensing behavior was attributed to the change in the oxidation state of oxidized carbon species (carbonyl, hydroxyl and hydroxyl peroxide) at the surface of the plasma-treated PANI film as well as to the formation of water.

In order to improve the selectivity and sensitivity of PANI for H_2, composites were also synthesized. It was suggested that graphene nanosheets can provide sufficient active nucleation sites for PANI as well as excellent electron transfer pathways and can be used in many electronic devices. The composite was investigated for H_2 gas sensing performance and it was observed to have a sensitivity of 16.57% toward 1% of H_2 gas, which was much higher than the sensitivity of graphene sheets or PANI nanofibers. The enhanced sensitivity of the graphene/PANI nanocomposite toward H_2 gas was attributed to the higher porosity of the nanocomposite. A highly selective sensor for hydrogen gas was also developed with a composite of PANI nanofibers embedded in the chitosan matrix [37]. It was suggested that the swelling of chitosan controls the hopping conduction between the nanofibers and the highly enhanced selectivity for hydrogen was due to the filtration action of chitosan for large size and polar molecules.

PANI films were also utilized to fabricate sensors for other analytes, such as smoke, arsine gas (AsH_3) and near IR radiation [38,55,61]. AsH_3 sensors based on PANI nanofiber composites with copper salts were investigated in detail [38]. The reaction between AsH_3 and Cu salt additive generates an acid, and reduces the composite metal ion. The acid produced can then dope the PANI resulting in large changes in resistance. A Ni-coated PANI nanowire (NW) was used as potential cigarette smoke detector devices [61]. It was found that composite shows nearly a four-order decrease in ac impedance in the presence of cigarette smoke. It was suggested that the derivatives of polyaromatic hydrocarbon and alkaloid like nicotine present in the cigarette smoke form a charge-transfer complex with Ni as the binding site on the polymer nanowire. This results in enhancement of the mobile charge carrier in the composite and decreases ac impedance.

As mentioned previously, the doping may result in a change in work function or surface potential of the CP also and can be controlled precisely [62,63]. Subsequently, sensing devices were fabricated to measure these parameters using KP or FET measurements. Janata *et al.* have addressed the problem of analyzing H_2 and NH_3 gas mixtures by depositing two layers, palladium and PANI, and measuring the response using KP measurement [64]. A CHEMFET-based sensor array was fabricated to determine single gases (H_2 and NH_3) diluted in dry and humid air. These sensors operate at 90 °C and display good sensitivity, perfect selectivity among the gases tested, very short response time and excellent stability for NH_3.

1.2.1.2 Polypyrrole

Reduced PPy films have been observed to show sensitivity to electron-acceptor gases having electron affinities larger than those used in chemical doping [65]. Accordingly, a decrease in resistance was observed for PCl_3, SO_2, NO_2 and iodine vapor. However, iodine-doped PPy films show different behavior for NO_2, i.e. an increase in resistance upon exposure to NO_2 gas. The phenomenon was explained in terms of the stabilization of gaseous species on adsorption, which was caused by a partial substitution of the doped iodine with exposed NO_2. In addition, by selecting proper dopant ions such as inorganic acid (carboxylic and polymeric acid), polymer can be converted to semiconducting or metallic and thus be suited for detecting *p*-nitrophenol (*p*-NP) [40]. Electrochemically prepared composite film of PPy and CuPc with the cationic surfactant cetyltrimethylammonium bromide (CTAB) was utilized to detect chemical warfare agent nitrogen mustard-1 [41].

1.2.1.3 Polycarbazole

Polycarbazole (PCz) has excellent electrochemical activity and stability yet there are hardly any reports about the usage of this polymer for fabricating sensors. A Cu (II) selective sensor was fabricated using an electrochemical transistor and ClO_4^--doped PCZ film. A sensor for volatile organic gases was fabricated using PCz deposited on sensor pixel and changes in the impedance were measured [64]. It was observed that the conductivity of PCz decreases on exposure to acetone but increases on exposure to toluene. It was suggested that the electric dipole of acetone interacts with the delocalized electrons of the polymer backbone and leads to an increase in resistance of PCz. On the other hand, toluene, being a non-polar molecule, caused swelling of PCz and increased the effective volume for conduction. An ammonia sensor was fabricated by using the polycarbazole film that had been prepared by using the LB technique. The polycarbazole LB film was observed to behave differently towards ammonia gas as compared to the

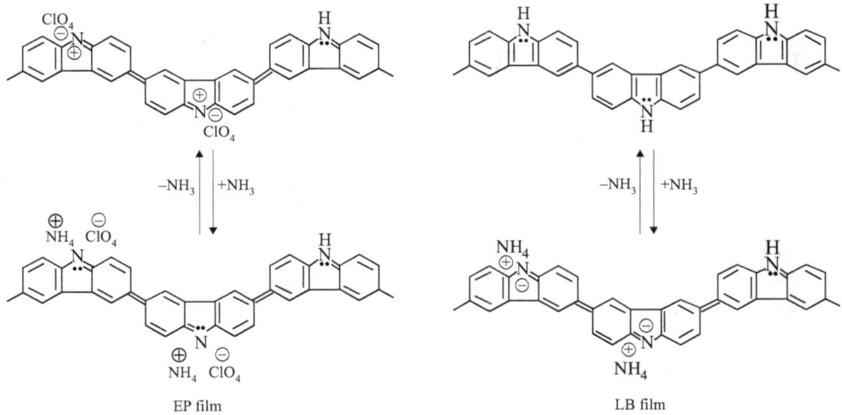

Figure 1.7 Schematic depicting the sensing mechanism of NH_3 for EP and LB film

electropolymerized polycarbazole (EP) film [33]. The polycarbazole LB films showed an increase in the conductance upon exposure to ammonia, while opposite behavior was observed for electropolymerized (EP) PCz films. This behavior was explained on the basis of n-type doping by ammonia molecules in undoped polycarbazole LB films as illustrated in Figure 1.7. For doped EP films, the ammonia interaction results in the reduction of polycarbazole, and therefore a decrease in conductivity.

1.2.1.4 Polythiophene, poly(*p*-phenylene vinylene) (PPV) and polyacetylenes

PTh films were not sensitive enough for NO_2 gases, and therefore various derivatives of PTh were synthesized to develop the sensors for NO_2. LB films of poly [3-(butylthio)thiophene] (PBTh) were found to have better response towards NO_2 gas [5]. An enhanced selectivity of the sensor was observed at a working temperature of 80 °C against ammonia, carbon monoxide, ethanol, methanol and methane. Porous films of poly(3,4-ethylenedioxythiophene):polystyrene sulfonate (PEDOT:PSS) nanocomposite prepared on the reduced graphene oxide (RGO) film were also used as efficient NO_2 chemiresistors. The films exhibited enhancement in sensitivity as well as response and recovery performance by two orders of magnitude, in contrast to bare RGO films. Enhanced selectivity to NO_2 in contrast to other oxidant analyte gases, e.g. HCl, H_2S and SO_2, was attributed to highly uniform distribution of PEDOT porous network and excellent synergetic effect between RGO and porous PEDOT [32]. An excellent response towards NO_2 was obtained for composite of poly(3-hexylthiophene) (P3HT) with ZnO wherein the enhanced selectivity towards NO_2 was attributed to the formation of p–n junction [66]. The P3HT gets reduced by the embedded ZnO network, thereby making a reduced form of P3HT more sensitive toward NO_2 as compared with P3HT films (Figure 1.8). Recently, the microporous polymer network based on a

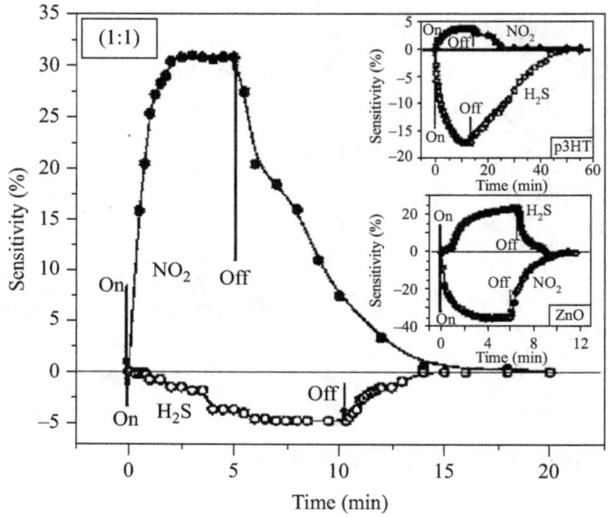

Figure 1.8 Response curves of P3HT:ZnO-nanaowire (1:1) composition hybrid films for NO_2 and H_2S. Insets show response curves of pure P3HT and ZnO-nanowire films. The gas concentration in all the cases was 4 ppm

dithienothiophene derivative (DTTPh2-Th4), possessing four thiophene units at the peripherals, was synthesized [67]. The iodine doped form of polyDTT was shown to exhibit excellent response towards NH_3 which was visible to the naked eye as well.

CPs, such as polyacetylene and PTh, have also been used for radiation detection since their discovery [68]. The change in electrical conductivity and an increase in activation energy upon irradiation were attributed to the irradiation-induced imperfections, such as chain breaks and cross-linking. Further, the existence of dopants in polyacetylene was found to be the cause for cross-linking as non-doped polyacetylene did not exhibit a prominent electrical conductivity decrease upon irradiation. With the discovery of new conjugated polymers, these polymers were further employed for radiation detection by exploiting their luminescent and absorptive properties [69]. A significant increase in conductivity was observed in PTh when irradiated by gamma rays in the presence of SF6 [56]. This was attributed to the radiation-induced doping. FETs of poly(*p*-phenylene vinylene) (PPV) were also fabricated to study the effect of X-ray exposure on conductivity and charge carrier mobility [57]. The enhancement was attributed to the iodine doping after scission of the C–I bond.

1.2.2 Doping/undoping by acid–base reactions

As mentioned earlier, doping can also be achieved by protonic (or acid) doping. Protonic acids are Brönsted acids, which can easily give the H^+ ion; while conjugated polymers with heteroatoms in the polymer chain may be treated as weak Brönsted base, which can take up proton. As a consequence, the number of electrons associated with the polymer backbone remains unaltered; however, the energy levels are rearranged so as to produce a polaron in the CP backbone, leading to a highly conducting form of the polymer [70]. Sensors based on this mechanism are presented in Table 1.3.

1.2.2.1 Polyaniline

Among several conjugated polymers available, PANI possesses a high pK_a value (or basicity) because of which this polymer can be protonated very easily relative to other conjugated polymers. Additionally, reversible doping/dedoping, high conductivity and stability of PANI thin films have been found useful as sensing material for several gases such as NH_3, HCl, N_2H_4, [108,114], humidity sensor [115], toxic chemicals [116] and are discussed as follows.

PANI films deposited by various deposition techniques, such as LB, DC magnetron sputtering, ice as well as DNA template method, electrospinning, were extensively used as ammonia sensors [8,117,118]. PANI LB films have been prepared on a variety of substrates and found to be sensitive for NH_3, NO_2, SO_2 [39] and HCl gases [117]. Nanoscale materials were also synthesized in an effort to improve the performance of the sensors in terms of sensitivity and rapid detection. PANI nanofiber films deposited on commercially available ceramic substrates were fabricated to develop highly sensitive NH_3 and HCl gas sensors [72]. Liu *et al.* prepared various nanostructured PANI films by electrochemical deposition on an interdigitated electrode by controlling the polymerization conditions, and further investigated for NH_3 sensing behavior [119]. Among all nanostructures, a loose 2D nanowire network structure obtained by a polymerization time of 10 min showed the best performance upon exposure to 4 ppm ammonia gas. Further, they also studied the influence of polymerization conditions (polymerization current and time) as well as surface characteristics (hydrophilicity/hydrophobicity and roughness) on the morphology of the deposited PANI [87]. It was suggested that for the application of CP in sensors, the ideal nanostructure should be a horizontally oriented, monolayer, 2D nanostructures. Such structures were grown on 3-Aminopropyltriethoxysilane-treated inter-digitated electrodes. These films when exposed to ammonia gas showed sensitivity to even further lower concentration ~1 ppm of NH_3 gas. Several morphologies of PANI, such as microflakes stacked by 1D nanofibers, porous microwebs, hemispheres and nests piled by nanoparticles, were prepared through an ice-templating method, using different doping acids assisted by secondary solvent [73]. Among all morphologies, the PANI microflakes stacked by nanofibers showed the best performance for 100 ppm NH_3 gas owing to their small nanofiber diameter, high surface area and porous nature. PANI nanostructures were also produced by other chemical methods and compared in terms of their gas sensing behavior [120]. It was observed that among selected methods, the rapid mixing one leads to the most sensitive nanostructures (92 ppm NH_3 gas) as compared to the interfacial polymerization. Moreover, the acidic doping allows nanostructures that are more sensitive than the oxidative doping. On the contrary, the interfacial polymerization method was employed for fabricating highly sensitive NH_3 and HCl gas sensors by using PANI deposited on commercially available ceramic

Table 1.3 Conducting polymer sensors to several gas analytes based on acid–base reaction

Analyte	Deposition technique	Sensing material	Range	Response/ recovery time	Sensitivity	Selectivity against	Transducer	Ref.
NH_3	Electrospin	PANI	20–700 ppm	45 s/64 s(500 ppm)	–	–	Chemiresistor	[8]
NO_2			1–50 ppm	50 s/82 s(50 ppm)	–	–	–	
NH_3	Electrochemical	PANI	10 ppm	–	–	–	Chemiresistor	[71]
NH_3	Chemical	PANI	100 ppm	–	–	–	Chemiresistor	[72]
NH_3	Drop cast	PANI	100 ppm	–	–	–	Chemiresistor	[73]
NH_3	Chemical/Ag template	PANI	5–100 ppm	–	–	Ethanol, acetone, *n*-propanol	Chemiresistor	[74]
NH_3	Chemical	PANI	0–20 ppm 2.5–20 ppm	–	–	–	EIS amperometry	[75]
NH_3	STES	PANI	0.5–50 ppm	10 s/2–4 min	–	–	Chemiresistor	[76]
NH_3	Electrochemical	PANI	200–1,000 ppm		120% ppm^{-1}	–	Chemiresistor	[77]
NH_3	Electrochemical	PANI	25–500 ppm	36 s/170 s	>0.7% ppm^{-1}	–	Chemiresistor	[78]
NH_3	Electrochemical	PANI	100–1,000 ppm	–	2% ppm^{-1}	Ethanol, acetone	Chemiresistor	[79]
NH_3	Electrochemical	PANI	0.1–100 ppb	–	–	–	Chemiresistor	[80]
NH_3	Electrochemical	PPy	8–500 ppm	2 min/2–4 min	116.72 nS/s	–	Chemiresistor	[81]
NH_3	Chemical	PPy	40–300 ppm	–	–	–	Chemiresistor	[82]
NH_3	Electrochemical	PPy	10–100 ppm	–	–	–	Chemiresistor	[83]
NH_3	Dual-nozzle electrospray + VDP	PPy	0.1–100 ppm	1 s/30 s(0.1 ppm)	–	VOCs	Chemiresistor	[84]
NH_3	Interfacial polymerization	PANI	1-200 ppm	–	–	–	Chemiresistor	[85]
NH_3	Spin coat	P3HT	100 ppm	–	–	–	FET	[86]
NH_3	Electrochemical	PANI	1 ppm	–	75% ppm^{-1}	–	Chemiresistor	[87]

NH_3	Paint-brush	PANI	25–100 ppm	–	–	Acetone, ethanol, methanol	Chemiresistor	[88]
H_2	Photolithography + *in situ* polymerization	PANI	100 ppm	–	–	–	Chemiresistor	[89]
NH_3	As-grown	$PANI/SnO_2$	100–300 ppm	12–15 s/80 s (300 ppm)	–	–	Chemiresistor	[90]
NH_3	Self-assembly	$PANI/InO_2$	1–141 ppm	2–3 s/10 s (1 ppm)	–	CO and H_2	Chemiresistor	[91]
NH_3	Drop cast	$PANI/TiO_2$	50–200 ppt	–	–	–	Chemiresistor	[92]
NH_3	LBL	PPy	0.01–1%	2 min (1%)	–	CO, CH_4, O_2, H_2	Chemiresistor	[93]
NH_3	Electrochemical	PPy	0.02–20 ppm	20–40 s	–	–	Chemiresistor	[94]
NH_3	Spin cast	P3HT	10–100 ppm				FET	[95]
NH_3/H_2	–	PANI/Pd	0–10,000 ppm		–154 to –172 mV/decade	–	FET	[64]
NH_3	Micromanipulation system + electrochemical	PTh	00.03	–	–	$CHCl_3$, CH_4, i-C_4H_{10}, C_2H_5OH	Chemiresistor	[96]
$CHCl_3$		Poly(3-*n*-dodecylthiophene)	00.03	–	–	NH_3, CH_4, i-C_4H_{10}, C_2H_5OH	–	
Light	Electrochemical + magnetic alignment	PPy	550 nm	–	–	–	Chemiresistor	[97]
NH_3			75 ppm	–	–	–		
CO	Chemical	PPy	100–300 ppm	<1 s response	–	–	Chemiresistor	[98]
CO	Pellets		5–1,000 ppm	–	–	–	Chemiresistor	[99]

(Continues)

Table 1.3 (*Continued*)

Analyte	Deposition technique	Sensing material	Range	Response/ recovery time	Sensitivity	Selectivity against	Transducer	Ref.
		PANI/polyimide blend						
CO in H_2	Electrochemical	PANI	1–100 ppm	–	–	–	Chemiresistor	[100]
CO_2	Drop cast	PANI	50 ppm to 5%	–	–	–	Chemiresistor	[101]
CO_2	Drop cast	PANI	380–2,400 ppm	–	–	–	Chemiresistor	[102]
CO_2	Drop cast	Poly(anthranilic acid)	3×10^2–1.5×10^5 ppm	–	–	–	Impedance	[103]
CO_2	PPy	PPy	300 ppm	–	–	–	Chemiresistor	[104]
CO_2	Screen print	PPy	25–800 ppm	210 s/1,770 s (300 ppm)	139×10^6 Ω/ppm	–	Chemiresistor	[105]
H_2O	Screen print	PANI	0%–100%	–	−1.26 kΩ/% RH	–	Chemiresistor	[106]
H_2O	Electrospin	PANI	11%–98%	8 s/6 s	–	–	QCM	[107]
H_2O		PEDOT:ClO_4	130–260 ppm	–	–	–	Chemiresistor	[108]
pH	Self-assembly	PANI	2–12	–	58 mV/pH	–	Electrochemical	[109]
pH	Electrochemical	PANI	1–12	–	70 mV/pH	–	Electrochemical	[110]
Hydrazine	Electrochemical	PANI/Gr	0.01 μM–0.1 mM	1 s	32.54×10^{-5} A cm^{-2} mM^{-1}	–	Chemiresistor	[111]
Hydrazine	Drop cast	PANI	3 ppm	–	–	–	Chemiresistor	[112]
Hydrazine, mono-methyl-hydrazine	Spin cast	PANI	10–100 ppb	–	–	–	Chemiresistor	[113]

substrates [85]. The excellent sensing properties of PANI nanofibers to a very low concentration of ammonia ~1 ppm were observed and were attributed to the small diameter of nanofiber which provides a high surface-to-volume ratio and a porous structure. A wafer-sized patterned aligned PANI nanowire arrays with a 5-μm resolution was prepared on flexible polyethylene terephthalate (PET) substrates by combining photolithography and in situ polymerization techniques [89]. This array showed good sensitivity for ammonia (1%) and hydrogen gas (100 ppm) detection. A novel route was adopted for the synthesis of PANI nanofibers using ZnO nanoparticles as template [88]. Excellent selectivity toward ammonia gas was achieved using an optimum 30 wt% ZnO powder. PANI nanowire arrays on Si substrates were prepared by using fully stretched DNA as growing templates while applying gentle horseradish peroxidase enzymatic polymerization [121]. The use of stretched and immobilized DNA as template avoids the agglomeration of DNA in polymer/DNA solution and results in a nanowire array which is very sensitive to the proton doping–undoping process. When exposed to HCl vapor, protonation of imine nitrogen occurs resulting in increase in the conductance. Upon exposure of HCl exposed films to NH_3 gas, a decrease of the conductance occurs which is attributed to the de-protonation of PANI by the NH_3 gas. Further, Bai *et al.* reported the transparent conducting films of hierarchically nanostructured PANI networks on PET substrates [74]. Films were prepared by in situ chemical oxidative polymerization of aniline on an Ag NWs template. Owing to quasi-1D confined nanostructure in the network structure film and the large specific surface area, the films showed enhanced sensitivity and selectivity for NH_3 gas than the pure PANI film. NH_3 gas at concentrations as low as 0.5 ppm was detected using oriented PANI nanowires, with diameters of the order of 100 nm [76]. Recently, electrospun PANI fibers doped with different levels of (+)-camphor-10-sulfonic acid (HCSA) are developed as chemiresistive sensors for NH_3 and NO_2 [8]. The prepared films showed a remarkable change in resistance, up to 60-fold for ammonia sensing, and more than five orders of magnitude for nitrogen dioxide sensing, with response times ~1 min. A time-dependent reaction–diffusion model was used to explain the sensing mechanism.

In addition to nanostructured films, individual nanowires were also deposited utilizing a scanned-tip electrospinning (STES) deposition. The response times of sensors with various diameters correspond to radius-dependent differences in the diffusion time of ammonia gas into the wires. Composites of PANI with metal-oxides have also been used to overcome the shortcomings of metal-oxide-based sensors. The composites may consist p–n heterojunction and based on either change in the resistance or work function could be used for better sensing ammonia [90,91]. Accordingly, a significant improvement in sensitivity was achieved by encasing PANI nanograins into an electrospun TiO_2 fiber surface [92]. The resultant p–n junction owing to the p- and n-type nature of PANI and TiO_2 showed enhanced sensitivity to NH_3. The sensitivity was 1,000 times higher than the best PANI sensor reported in the literature, 50 ppt of NH_3 gas in air could be detected using this sensor.

PANI films were also used as humidity and CO_2 sensor [106,107]. CO_2 is not a redox gas; therefore, the mechanism for sensing it is different. It was proposed that sensing takes place in two steps: (i) CO_2 reacts with water to form carbonic acid which dissociates into H^+ and HCO_3^- ions and (ii) H^+ then takes part in the protonic acid doping of the insulated EB to the conducting region [103]. Highly doped PPy was observed to show good sensitivity and fast response towards CO_2 as measured by a voltage drop method at room temperature. A detailed analysis by various techniques, such as X-ray diffraction (XRD), scanning electron microscopy (SEM), FTIR and thermo gravimetry, suggested that CO_2 molecules formed weak bonds with π-electrons of PPy and resulted in an increase in resistance of PPy [105]. Three conducting polymers with a different doping percentage at pH were analyzed thoroughly by Chen *et al.* [122]. Based on the molecular dynamics model, the relationship between doping percentage and pH was derived. Further, it was suggested that the polymer having a high doping level at the same pH is more sensitive to CO_2 sensors. A selective sensor based on PANI boronic acid was reported to measure CO_2 levels in the range of 380–2,400 ppm and evaluated at different temperature (10–55 °C) as well as relative humidity (20%–70%) [101, 102]. The sensor was found to respond to changes in CO_2 concentration at these humidity and temperature levels.

In addition to CO_2, a response towards CO was observed for PANI-based sensor. The mechanism of this sensing is not yet clear. It was proposed that the surface potential, barrier height and depletion width are reduced as a result of oxidation occurring at the polymer grain surfaces in the presence of CO. As a result, a decrease in the resistance of the PANI film was observed. In another report, it was proposed that the interaction site is at the site of NH or amine nitrogen where CO withdraws an electron and results in a change in electrical conductivity [99]. A gas sensor based on a horizontally oriented, monolayered PANI nanofiber network was developed to detect CO in hydrogen [100]. The results demonstrated that the sensor had an excellent response to even a low concentration of CO (1 ppm) in hydrogen. The response mechanism was attributed to the different binding sites of CO and H_2 with PANI: H_2 interacted with the protonated amine nitrogen atoms, whereas CO with the unprotonated amine nitrogen atoms. Recently, density functional theory (DFT) studies were carried out to evaluate the ability of PANI ES as sensor for NH_3, CO_2 and CO [123]. PANI ES oligomers with 2−8 phenyl rings were used to calculate parameters, such as interaction energy, natural bond orbital and Mulliken charge analysis for different analytes. Interaction energies were calculated and corrected for BSSE. Interaction energy analyses suggested that the inertness of CO towards PANI ES was due to its stable resonance structure $^-C{\equiv}O^+$ (unavailability of lone pair of electrons of oxygen to PANI ES) as compared to PANI ES and NH_3 complex which shows large forces of attraction. The selectivity of PANI ES toward NH_3 as compared to CO_2 and CO was consistent with the experimental observations.

PANI was also used extensively as pH, H_2S and hydrazine sensor due to its unique proton doping ability as well as excellent electrochemical properties. The layer-by-layer (LbL) films of PANI and poly(vinyl sulfonic acid) films on gold electrodes were used as sensing membranes for pH sensor. The sensor displayed sensitivity close to the expected Nernstian behavior, around 58 mV/pH with a small voltage drift [109]. Self-doped PANI showed response as high as 70 mV/pH [110].

Further, PANI-modified glassy carbon and screen-printed electrode showed a reversible response of 90 mV/pH in the pH range of 3–9 [124]. A sensitive, selective and fast H_2S sensor based on PANI nanowires–gold nanoparticles in a wide dynamic range (0.1–100 ppb) has also been reported [80]. The formation of AuS as well as the increase in the doping level of PANI resulted in to enhanced selectivity and sensitivity. PANI doped with sodium dioctyl sulfosuccinate was also used to detect hydrazine detection with a detection limit of 0.24 ppm [125]. UV–Vis and spectroscopic analyses revealed that the quinoid/benzenoid peak changes upon exposure to hydrazine owing to the changes in the chemical structure of PANI. Devices reusable for up to 10 cycles, with response and recovery time of 0.12 and 0.08 min, respectively, were thus demonstrated. In this case, hydrazine acts as a strong reducing agent and converts the emeraldine form of PANI to its leucoemeraldine oxidation state resulting in decreased conductivity. However, an increase in the conductivity upon exposure to hydrazine was observed when PANI was processed using sodium dioctyl sulfosuccinate [112]. The enhanced sensing ability of PANI was achieved for low concentration of hydrazine detection (3 ppm). This additive reacts with hydrazine to produce a strong acid, hydrofluoric acid, which, in turn, protonates the PANI-EB leading to large increases in conductivity (25 times). Recently, PANI/Gr composite modified electrode was fabricated towards the detection of hydrazine [111]. A very high sensitivity (~0.3 mA cm^{-2} mM^{-1}) and short response time (10 s) observed for the composite film were attributed to electrostatic interaction between graphene and hydrazine through positively charged nitrogen atoms of hydrazine.

Recently, a wearable two-ply yarn supercapacitor made from the composite of a metal filament-reinforced carbon nanotube yarn with in situ polymerized PANI nanowires was reported [46]. The wearable two-ply yarn exhibited a very high capacitance (91.67 mF/cm^2) and energy density (12.68 $\mu W\ h/cm^2$) as well as excellent long-term cycling stability for charging–discharging and flexing deformation. The supercapacitors connected in series and in parallel were shown to power miniature electronic devices in smart textiles and will be discussed in another chapter of this book.

1.2.2.2 Polypyrrole

Sensors based on protonation/de-protonation changes were also developed using PPy. The pK_a of the acid ionization of the NH group in pyrrole monomer is 17.5 [59, 126]. Therefore, it was proposed that proton transfer was possible in PPy upon exposure to strong acids. However, the protonation has only a small effect on the electronic structure of the chain, therefore the conductivity of PPy. In basic solution, the de-protonation of PPy chain causes transformation of conducting PPy into a non-conducting polymer of quinoid structure accompanied by de-intercalation of the original counter anions.

NH_3 sensors were developed using the ability of PPy to give up proton [59, 126]. Reversible protonation/de-protonation reaction and subsequent formation of $NH_4^+C10_4^-$ complexes was proposed for explaining the interaction of ammonia with pyrrole. It was observed that if the formed ammonium ion binds close to the counter anion, the proton transfer process was reversible. Contrary results were obtained in the presence of water where irreversible changes occurred due to the formation of $NH_4^+C10_4^-$ complexes as well as OH^- species [127]. Similar results were proposed by others, where anion exchange was found to occur in reversible reaction of oxidized PPy with the hydroxide anion depending on the anion basicity [128]. In both cases, the conjugation length is decreased and therefore the conductivity [2]. Recently, DFT and time-dependent DFT calculation have been performed to investigate the interaction of NH_3 and PPy [129]. The interaction of NH_3 with oligopyrrole causes certain geometric features to change which results in a decrease in resistance for the movement of electron over the oligomer backbone. Further, it was observed that band gap decreased due to donation of electron to the LUMO (electron acceptor) of nPy oligomers and increased its electronic clouds density and therefore decreased the EA.

PPy, chemically oxidized on a poly(etheretherketone) surface, was used to develop NH_3 sensor for detection at concentrations lower than 10 ppm [64]. Electrochemical polymerized PPy thin films were used to fabricate microgas sensors for NH_3. Reversible and reproducible device sensing in the range of 8–1,000 ppm was achieved at room temperature [81]. The PPy LB film was prepared to fabricate ammonia sensor with a lower detectable limit of 100 ppm [93]. Further, the effect of various doping agents was also studied on the NH_3 sensing behavior of PPy films [94]. The effect of the initial conductivity and dopant on the gas sensing behavior of free-standing doped PPy films was also investigated [71]. It was observed that low conductivity caused an increase in sensitivity of PPy films to NH_3. Therefore, PPy films doped with ClO_4^- counter ions display high sensitivity, in the range of 88%–155% at 10 ppm NH_3, whereas the PPy films doped with TsO^- have a low sensitivity of around 20%. Electrochemically co-polymerized PPy and sulfonated PANI were also used as an ammonia sensor for concentration higher than 20 ppm [3].

Single PPy nanowire was also used to examine gas sensing properties and it was found to exhibit good sensitivity towards ammonia at concentration as low as approximately 40 ppm [82]. Further, a single doped PPy nanowire with metal contacts was prepared by selective maskless metal electrodeposition [83]. Different dopants were also investigated to study the effect of polymer reduction during cathodic deposition of metal. The single PPy wire doped with dodecyl sulfate was found to have improved sensitivity toward ammonia gas in the 10–100 ppm range. A novel approach to assemble conducting polymer nanostructures on prefabricated electrodes by the template-directed synthesis of multisegmented nanowires was adopted to fabricate the sensor [97]. The directionality of nanowires was controlled by an external magnetic field while ferromagnetic electrodes were used for controlled placement of nanowires. An excellent sensitivity towards gas molecules such as NH_3, H_2S and CO_2 was obtained using the PPy wire. A decrease in resistance of PPy was observed upon exposure to the vapor of weak acid such as H_2S and CH_3COOH which was suggested to be attributable to the proton transfer from acidic gas to PPy and an increase in the doping level of the polymer [71]. A similar mechanism was proposed for a composite of PPy and Fe_2O_3 for their gas sensing behavior towards toxic gases [130]. Recently, a new technique combining dual-nozzle electrospray and vapor deposition polymerization (VDP) was employed for fabricating urchin-like PPy nanoparticles with PPy-coated metal-oxide nanoneedles on surfaces [84]. As illustrated in Figure 1.9(a), dual-nozzle electrospray provided nucleated sites for the growth of metal-oxide needles on the surface and the VDP method was used to coat the needles with a PPy layer, resulting in the urchin-like (U-PPy) configuration. The responses of the sensors based on U-PPy were highly sensitive and reversible to NH_3, with a lowest detectable level of 0.01 ppm (Figure 1.9(b)). The enhanced surface area of the U-PPy allowed faster diffusion of the analyte gas molecules for fast detection (30 s for 0.1 ppm of NH_3) and rapid response times, as well as enhanced sensitivity.

Other than NH_3 sensor, H_2S and CO sensors were also developed using PPy. Although H_2S is a reducing gas like NH_3, the conductance of PPy was found to increase instead of the expected increase of resistance [131]. Similar behavior was observed for LB films of PANI for H_2S and was explained on the basis of protonation processes [39]. A systematic study carried out using UV–Vis and FTIR spectroscopy on dodecyl sulfate-doped PPy showed that H_2S dissociates at the PPy surface and subsequently, PPy chain is protonated and at the same time the number of the charge carriers increase due to HS^- anions formation. The conductance increases due to the combined effect of the two opposite phenomena. PPy incorporated with ferrocene during polymerization was synthesized to develop a CO gas sensor. These sensors could detect CO gas at low concentrations (300 ppm) under room temperature conditions [104]. Further, electrochemically functionalized PPy with ferrocenylmethyltrimethylammonium iodide (FTMA-I) was used for the detection of CO with high sensitivity to ppm levels [98]. Fast response of less than a second was obtained for 300 ppm carbon monoxide gas and the device could detect CO gas at concentration as low as 100 ppm. The favorable interaction of CO at the iron atom of the ferrocenyl moiety resulted in the transfer of electron cloud near the carbon atom of CO to the ferrocenyl iron atom, thereby decreasing the number of hole charge carriers in the PPy backbone causing a change in conductivity.

In addition to sensors for gas molecules, strain sensors based on PPy were also developed. A conductive PPy was deposited on a fabric, 83% tactel and 17% lycra, by vapor-phase polymerization and was investigated for strain sensors [44]. The fabric was stretched and relaxed by 50% for up to 10 cycles and found to improve in terms of sensitivity upon heat treatment. A resistive sensors employing a PPy-coated polyurethane foam exhibited a sensitivity of ~0.0007 mS/N [42]. The sensor was further demonstrated for a breath monitor, whereby the foam sample is incorporated into a harness to wrap around the ribcage area. Further, a strain sensor, composed of PPy-coated polyurethane foam, was also reported [132].

The foam was prepared by surface diffusion and *in situ* polymerization of PPy inside and on porous polyurethane (PU) substrates. The resultant net-like micro-crack structures made possible the reversible changes in the electrical resistance of PPy/PU elastomers, under stretching and releasing cycles. The sensor was further used as waistband for breath-in to breath-out for an adult. A sensitivity of resistance

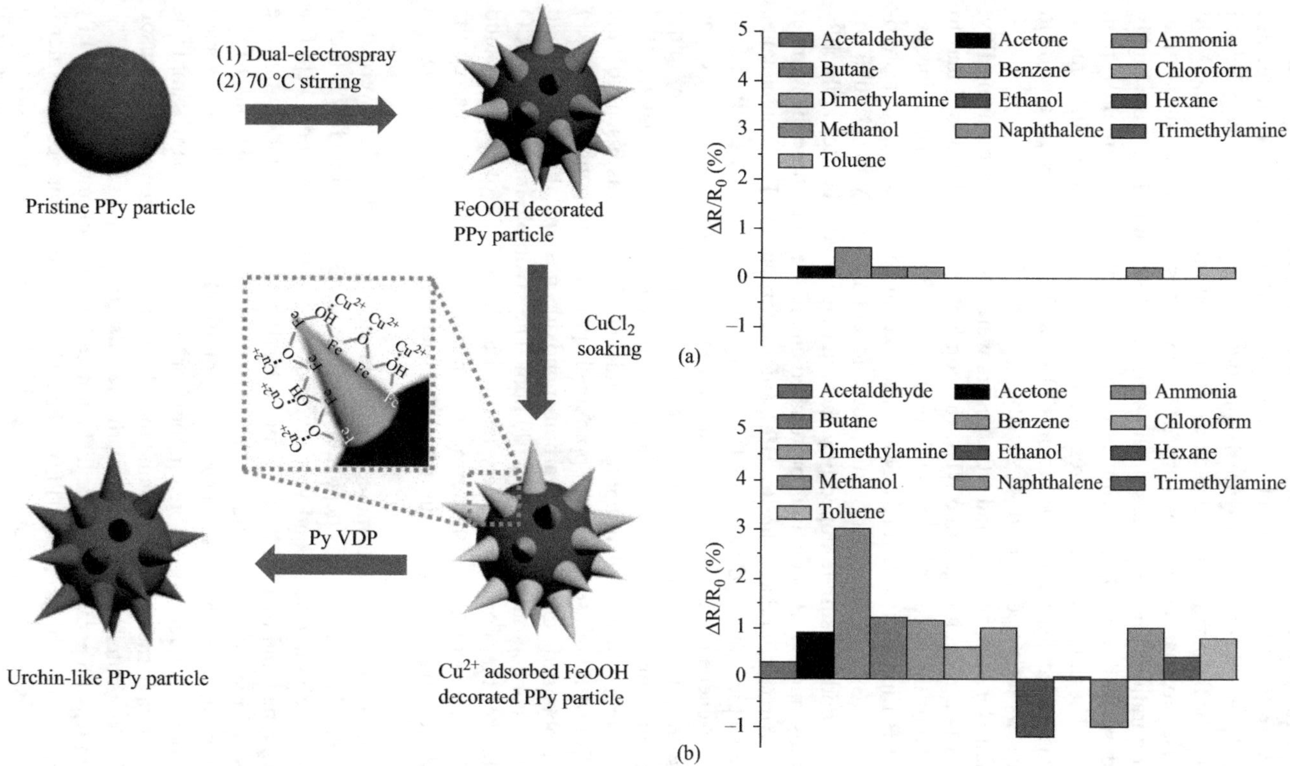

Figure 1.9 Illustrative diagram of the sequential fabrication steps for urchin-like PPy nanoparticles and Sensing performance histogram of 30 nm diameter (a) pristine PPy particles and (b) U_PPy particles to 13 analytes: each analyte concentration was fixed at around 10 ppm. © 2013 IEEE. Reprinted, with permission, from [84]

change up to 35% was observed during ten breathing cycles, demonstrating its potential as a strain sensor for human healthcare applications. Recently, an ultra-sensitive pressure sensor was reported by using a spherical shell structure of PPy that were prepared through a multiphase synthesis technique [43]. Highest sensitivity (~B 56–133.1 kPa^{-1}) of the flexible pressure sensor was reported in the low-pressure regime (<30 Pa). Due to tunable effective elastic modulus, the sensor was able to detect the placement or removal of even tiny weight such as a flower petal corresponding to an extremely low pressure of 0.8 Pa.

1.2.2.3 Polythiophene

Besides PTh and its derivatives, several sensors based on protonation/de-protonation were developed for NH_3, hydrazine and physical/tactile stimulations (pressure, touch) [19,44,45] and are discussed as follows.

The thin films of PTh film were found to exhibit response towards NH_3. The response of film to ammonia was explained as a result of the formation of either irreversible ionic pairs between positively charged electric barrier of NH_4^+ ions and doped anions [96]. Further, regioregular P3HT films were also employed as a sensing active layer for FET-based NH_3 sensor [95]. The reduction of source–drain (I_{SD}) current was attributed to charge carrier trapping, since polarity of NH_3 deteriorates the charge transport due to increased energetic disorder through charge–dipole interactions. In other cases, dielectric was also chosen in order to increase the sensitivity of the sensor, e.g. in ring-opening metathesis polymerized (ROMP)-dielectrics, the mobile NH_4^+ could move within the ROMP dielectrics in the direction of the drain electrode and localize near the drain region at the PVA/semiconductor interface or even in the regioregular P3HT (rr-P3HT) layer [86]. rr-P3HT as an active layer was also used as pH-sensitive and ROMP materials as gate dielectrics and polyvinyl alcohol (PVA) as an intermediate layer between CP and gate dielectrics.

Thin films of P3HT were also found to be ultra-sensitive to hydrazine and monomethyl hydrazine vapor [113]. For optimized film, concentrations of hydrazine in the range of 0.1–100 ppb could be detected as against ammonia, dimethylamine and ethylene diamine. The decrease in conductivity upon exposure to hydrazine vapors was suggested to have resulted from the decrease in delocalized charge carriers on the polymer backbone due to donation of electrons from hydrazine as well as complexation of the dopant counterions.

Tactile sensors based on piezoelectric, optical and resistive and capacitive changes were also developed using PEDOT:PSS. A stretchable e-textile was fabricated from PEDOT–PSS which could be switched between their different colored states, even upon stretching of the Spandex [45]. It was demonstrated that the electrochromic material is still fully functional when the substrate was stretched. The stretching affects the morphology of the spray-coated electrochromic polymer films as shown in Figure 1.10, stretching resulted in gaps in the electrochromic polymer film, thus exposing the PEDOT–PSS soaked fibers underneath. The fibers, on the other hand, no color change showed upon stretching, again confirming the result of a percolated network of PEDOT–PSS particles. The intensity of the red

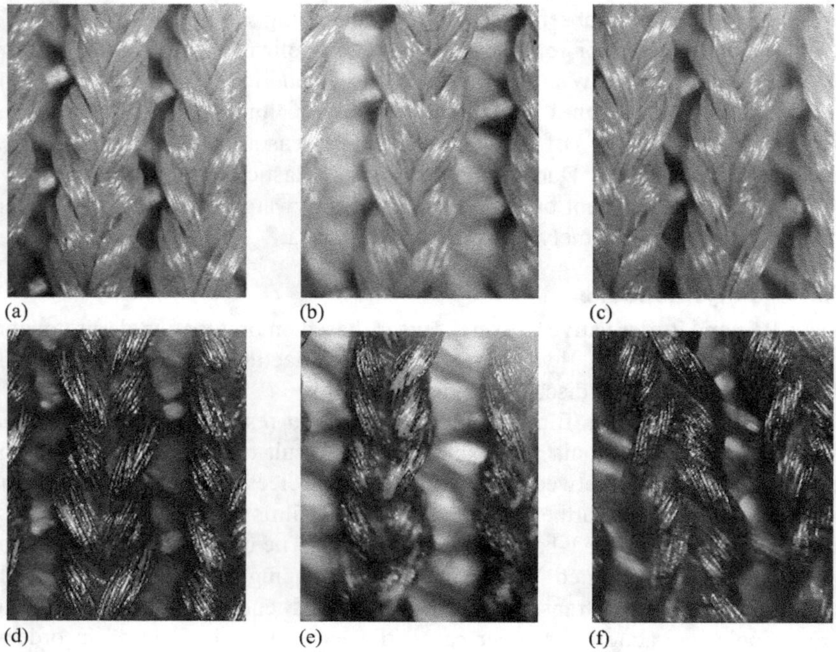

Figure 1.10 Optical microscope images (20×) of (a–c) conducting fabric and (d–f) electrochromic polymer coated fabric. (a, d) Before stretching, (b, e) under ca 20% stretching, and (c, f) after relaxation to its original state. © 2013 IEEE. Reprinted, with permission, from [45]

color diminished because of the exposure of the blue color underneath, and returned to its original state when released. Hwang *et al.* have functionalized multi-walled CNTs (MWCNT) by conductive thiophene molecules and obtained a well-dispersed and non-entangled CNT fillers in a poly(dimethylsiloxane) (PDMS) matrix [133]. The overall resistance of the composite increases upon increasing the external pressure and the piezoresistance response (~pressure range 0–0.12 MPa) was found to be strongly dependent on the concentration of the thiophene molecules. A novel strain sensor based on PEDOT:PSS–PVA nanofibers has also been fabricated on the Kapton substrate by electrospinning [49]. The electrical conductivity of the as-spun PEDOT:PSS–PVA nanofiber network was tuned over a wide range from 4.8×10^{-8} to 1.7×10^{-5} S/cm via controlling the concentration of the additive dimethylsulfoxide (DMSO) (Figure 1.11). The sensor exhibited excellent stability, fast response and a high GF of up to about 396. A highly conductive and stretchable composite of graphene–PEDOT:PSS with PDMS was also reported in [51]. The resistance of the composite remained stable to 50% even after stretching 100 times and bending 5,000 times, thereby potential for next-generation stretchable electronics. Further, Seo *et al.* fabricated using P3HT OFETs, on the top of which liquid crystal (LC) molecules 4-cyano-40-pentylbiphenyl (5CB) was

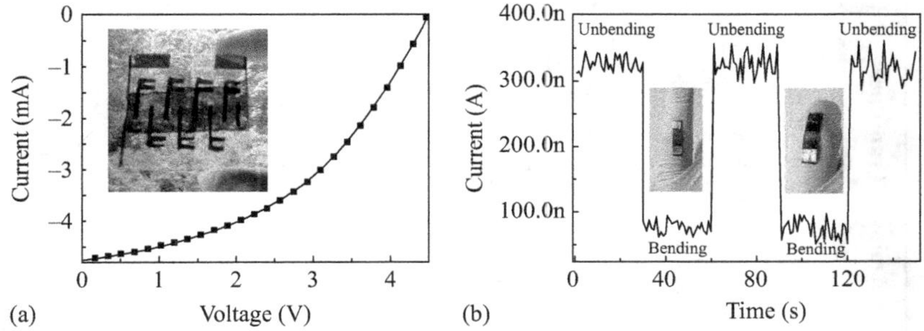

Figure 1.11 *(a) I–V curves of the series connected solar cells under 1 sun, AM 1.5G simulated solar illumination. The inset is a photograph of eight series connected organic bulk-heterojunction solar cells. (b) The current response of the strain sensor (which was driven by the solar cell) with cyclical bending and unbending actions of the finger. © 2013 IEEE. Reprinted, with permission, from [49]*

mounted [53]. The I_{SD} of the LC-on-OFETs increased significantly when nitrogen flowed to the LC layer. The device could sense the gas flow intensity, which could not be detected even by the human skin. Readers are recommended Chapters 4 and 5 for a detailed discussion on these kinds of devices.

1.3 Sensor based on weak interactions

Conducting polymers were widely investigated in sensors based on the Lewis acids and bases detection method; however, their interaction with neutral molecules, such as alcohols, or other volatile organic compounds (VOCs), such as benzene and toluene, is not significant [134]. As discussed earlier, a partial charge transfer between the analyte and conducting polymer may result in the formation of a charge-transfer complex and therefore, leading to the change in the electrical or optical signal [62]. Dipole–dipole interaction or hydrogen bonding may occur between the analyte and the nitrogen atoms. The fabrication of sensors based on these mechanisms is summarized in Table 1.4 and discussed in detail in the following sections.

1.3.1 Weak interactions involving absorbing or swelling

CPs can absorb VOCs or solvents, thereby increasing its volume significantly causing swelling. The swelling may arise due to weak interactions such as polar (dipole/dipole or dipole/induced dipole) and dispersion (induced dipole/induced dipole) or polarizability involved between CP and the analyte. These weak interactions do not change the doping levels of the conjugated polymer; however, they can weakly influence the electronic properties of CPs owing to the change in the organic vapors cause large increases in the conductivity of PANI, PPy and PTh films and, therefore, can be utilized as potential sensors as discussed herein [135].

Table 1.4 Sensors based involving week interactions of conducting polymers to several VOCs and other analytes

Analyte	Deposition technique	Sensing material	Range	Response/ recovery time	Sensitivity	Selectivity against	Transducer	Ref.
Ethanol	Electrochemical	PPy	2,665–25,058 ppm	–	5 μV/ppm	–	Chemiresistor	[3]
Ethanol	Solution cast	PANI	1.3%–100%	–	–	–	Chemiresistor	[135]
Acetone	Pellet	PPy	0.2%–40%	–	0.12 S per %	–	Chemiresistor	[136]
Acetone, ethanol, dimethyl formamide	Pressurized fluid	PANI		5 s	–	Toluene, hexane, ethyl acetate	Chemiresistor	[138]
Phytochemical	Spin coat/drop cast	PTh/carbon, PANI/carbon	10%	–	–	–	Chemiresistor	[139]
1-Hexanol	Drop cast	PTh	100–2,000 ppm	<5 s	–	–	QCM	[140]
$CHCl_3$	Chemical	PANI	30–3,000 ppm	30 s	4×10^{-4} ppm^{-1}	Toluene, CH_2Cl_2, DMSO, acetone, methanol	Chemiresistor	[141]
$CHCl_3$, CH_2Cl_2	Drop cast	PANI	–	–	–	–	Chemiresistor	[142]
VOCs	Pellet	PANI/chromium (III) trioxalate	–	–	–	CCl_4, $CHCl_3$, MeOH, CH_3CN	Chemiresistor	[143]
VOCs	Inkjet printing	PTh	10–3,500 ppm	–	–	PCA analysis	FET	[144]
VOCs	OCVD	Poly(EDOT-co-TAA)	0.1%–12%	100 s/200 s	–	*n*-Pentane, acetone, toluene	Chemiresistor	[145]
$CHCl_3$ NH_3	Soak	PANI	0.1%–12% –	100 s/200 s –	0.017 ppm^{-1} 52 ppm^{-1}	Acetone, ethyl acetate	Chemiresistor	[146]
BTEX	Electrochemical	PANI	1%–4 %	–	–	–	Chemiresistor	[147]
BTEX	Chemical	PPy	20–80 ppm	–	8×10^{-5}–8×10^{-4} Ωppm^{-1}	–	Chemiresistor	[148]
Ethanol	Dip coating	PPy	87–6,000 ppm	175 s/220 s for 87 ppm	–	$CHCl_3$, acetone, benzene	Chemiresistor	[149]

Aroma	Electrochemical	PANI	–	–	–	PCA	Chemiresistor	[150]
Odor	Electrochemical	PANI	300 ppm	–	–	PCA	Chemiresistor	[151]
Diabetes mellitus	Drop cast	PPy	20%–63%	–	–	–	Chemiresistor	[152]
1-Hexanol/ butanol	Drop cast	Poly-DT		–	–	Ethanol, acetone	QCM/FET	[153]
1-Hexanol/ butanol		Poly-DPOT	700–1,500 ppm	–	–	*n*-Heptane, acetone		
Acetone	RF sputtering	PANI/bentonite	3–300 ppm	–	–	Benzene, ethanol, toluene	Chemiresistor	[154]
Butylamine	Electrophoretic	poly(aniline-boronic acid) (PABA)	10–80 ppb	–	–	–	Chemiresistor	[155]
Methanol, H_2O	Electrophoretic	Polyaniline	25–400 ppm	3 s	–	–	Chemiresistor	[156]
VSC (breath analyzer) CH_3CH, H_2S	Electrochemical	PANI	1.5 ppm	–	–	–	Impedance	[157]
Aliphatic amine vapors	In situ PMz	PANI	5–250 ppm	–	–	–	QCM	[158]
3-Methyl-1-butanol	Drop cast	P3HT	25–500 ppm	–	1.45 Hz/ppm	1-Hexanol	QCM	[159]
Alkylating agents	Spin cast	Copolymer	LDL 10^{10}–10^{11} molecules	–	–	Ethanol, Iso-propanol, Water	FET	[160]
Acetone	RF sputtering	PANI/bentonite	3–300 ppm	–	–	Benzene, ethanol, toluene	Chemiresistor	[154]
Acetone	Drop cast	PANI	0.3–40 ppm	1–3 min/3–10 min	5.908×10^{-4} ppm^{-1}	–	Chemiresistor	[161]
DBP vapor	Solution cast	PANI	0.02–0.4 ppm	30 s/5 min	288 Hz/ppm	Acetone, ethanol, acetaldehyde	QCM	[162]
Sevoflurane	Dip coat	PPy	0.1%–1.2%	60 s/20 s	–	–	Chemiresistor	[163]

1.3.1.1 Polyaniline

CSA-doped PANI was observed to respond to alcohol vapors [135]. Based on wide-angle XRD, the changes were attributed to crystallization of conducting polymers. Further, experimental results demonstrated that ethanol and acetone both result in a change in the crystallinity of conducting polymers [136]. It was also observed that short-chain alcohols such as methanol and ethanol interact more efficiently with CPs than longer chain length alcohols and therefore show better sensitivity to methanol and ethanol. Subsequently, PANI and its derivatives such as poly (*o*-toluidine), poly(*o*-anisidine), poly(*N*-methyl aniline), poly(*N*-ethyl aniline) and poly(diphenyl amine) were shown to have sensitivity to a variety of vapors such as methanol, ethanol, propanol, butanol and heptanol vapors [137]. It was found that resistance decreases in the presence of small-chain alcohols but an opposite trend is observed in the case of long-chain alcohols. It was also observed that short-chain alcohols such as methanol and ethanol interact more efficiently with CPs than longer chain length alcohols and therefore show better sensitivity to methanol and ethanol. Thin conducting PANI films were coated on the surfaces of flexible PU to investigate crystallization of conducting polymers upon exposure to ethanol and acetone and found to the change in the crystallinity of conducting polymers [136]. The increase in resistance due to adsorption of chloroform, acetone, ethanol, acetonitrile, toluene and hexane on PANI was attributed to swelling which leads to a higher distance between the PANI chains [164]. The interaction may differ, e.g. for PANI base, a methanol molecule forms two H bonds as a bridge between the PANI chains, twisting the chains and thereby localizing the polaron and cause a decrease in conductivity [165]. The high conductivity of the ES allows the transfer of charge to the polymer, as a result of the hydrogen bonding to methanol. This results in charge distribution along the conjugated chain in the form of polarons and bipolarons and therefore an increase in conductivity. However, for ES protonation of the quinoid moieties' nitrogen hinders the number of hydrogen bonding sites, thereby preventing methanol to form a bridge between two polymer chains.

A DFT study on a series of complexes designed to model the interaction between individual oligoanilines and different alcohol molecules was carried out to explain the PANI response mechanism for alcohols [166]. Cationic oligoaniline containing eight repeating units was used to model single PANI fiber to study its response mechanism and it was found that upon complexing with methanol, ethanol, propanol and isopropanol, the ion–dipole interaction forms. This results in a decrease of the torsion of two adjacent phenyl rings in oligoaniline. Consequently, it was suggested that a high degree in coplanarity of two adjacent phenyl rings in the polymer caused by alcohols exposure is the cause of resistance decrease owing to good electron transfer facilitated along the polymer chain.

Further, polystyrene–PANI core–shell particles were investigated for their responses to different dry gas flow, ethanol vapor, hydrogen chloride and ammonia [167]. It was observed that the sensor performance was greatly influenced by the PANI shells packing in the films. The influence of the specific area and porosity associated with these structures was examined in two different cases. In the first one (referred to as P films), the nanostructures, resulting after removal of PS cores by immersing the core–shell particles dried on a substrate into tetrahydrofuran (THF), were composed of irregular fragments of PANI shells of tens of nanometers. The fast and large response for ethanol was observed for the porous PANI film (obtained by removal of PS cores) than the PS-PANI composite films and was attributed to easy diffusion of the ethanol molecule through the porous film. Gas sensors were also prepared by in situ polymerization of aniline on non-woven porous fabrics to improve the performance of the gas sensor [146]. Sensor fabrication parameters such as acidity of reaction media, precursor and reagent concentrations and the number of modification cycle were optimized. The resulting gas sensor was found to be highly sensitive to a range of VOCs with fast response time (~10 s). The order of sensitivity to VOCs was found to be ethanol > chloroform > toluene > acetone > ethyl acetate.

In addition to alcohols, PANI films were also used to fabricate sensors for other organic vapors, such as benzene, toluene, ethylbenzene, xylene, BTEX compounds, amine vapors and chloroform. The effect of BTEX compounds and chlorobenzene on the change in resistance of nanofibrous PANI films was investigated in detail [168]. Based upon XPS analysis, the decrease in resistance of conducting PANI was attributed to attractive dipole interaction rather than to a change in the number of charge carriers. The dipole interaction results in a decrease in the hopping distance for the carriers causing a decrease in resistance. Thin conducting PANI films were coated on the surfaces of flexible PU fibers through in situ chemical polymerization and were studied for sensing organic vapor [141]. The 55-µm fibers coated with 550-nm PANI layer thickness showed a high sensitivity (4×10^{-4} ppm^{-1}), fast response time (<30 s) and low detection limit (30 ppm) for chloroform gas. A comparative analysis of responses of thin PANI films to the vapors of polar as well as low- and non-polar organic solvents was also reported.

Dopant-induced specificity of PANI-based sensor toward chlorinated hydrocarbon vapors was investigated in detail [143]. It was proposed that the polar solvents, such as CH_3CN, methanol, DMSO and *n*-methyl pyrolidine (NMP), affect the conductivity of the materials via dipolar interaction with the radical cation and the protonated imine N center and by the subsequent de-protonation process. On the other hand, less polar solvent such as CH_2Cl_2, $CHCl_3$ and CCl_4 disrupt the π-electron delocalization of the polymer and therefore, charge carrier transport owing to interaction between radical cations and their negative dipole ends. Further, a detailed investigation was carried out on the sensing behavior of doped PANI with mixture vapors of CH_2Cl_2 and $CHCl_3$ [142]. It was observed that conductivity change observed upon exposure to vapor mixture is significantly higher than the sum of conductance change in PANI upon exposure to individual solvent vapors. Based on spectroscopy (IR, UV–Vis and electron paramagnetic resonance) and structural (XRD) studies, it was suggested that CH_2Cl_2 has a weaker interaction than $CHCl_3$ with PANI materials, but it forms molecular pair with $CHCl_3$ through electronic effects which result in a significant change in the conductivity. This interaction is different for individual vapors where a decrease in conductivity is observed owing to neutralization of polarons.

The on-line detection of primary aliphatic amines vapors was also reported using the PANI/EB film and QCM [158]. The frequency shifts (Δf) of the QCM was found to be linearly correlated with the concentrations of amines vapors ranging from 2 to 25 mg/L. The selectivity of the sensor was attributed to the differences in the diffusion coefficient (D) and the adsorption kinetics of the amine vapors as well as film thickness. The pseudo-first-order rate constant (k) of amines vapor was estimated in the order: methylamine $<$ dimethylamine $<$ trimethylamine $<$ triethylamine. As a result, enhanced selectivity towards methylamine vapors was obtained.

1.3.1.2 Polypyrrole

In addition to PANI, PPy was also investigated for the detection of alcohols, BTEX compounds and chlorinated compounds in their vapor phases. The PPy doped with p-dodecylbenzenesulfonic acid and ammonium persulfate was synthesized and subsequently pretreated to improve sensor performance for VOCs [149]. The sensitivity of the sensor after soaking and thermal pretreatment was found to be better for alcohol gases than other VOCs gases such as acetone and chloroform. An array of sensors based on PPy was used to identify and quantify BTEX compounds [147]. A number of dopants were investigated in terms of their ability to detect the analyte gases with sufficient sensitivity and reproducibility. A pure vapor could be identified and its concentration measured using a group of eight polymers. Further, an adsorption model based on the Langmuir isotherm was found in agreement with experimental results for interaction of BTEX compounds and PPy and was used to estimate the adsorption equilibrium constants, K_m [148]. Further, inverse gas chromatography was used to calculate a specific component of free energy of adsorption, $-\Delta$Gas. The sensing ability of the conducting polymer film was found to be dependent on both the secondary doping level and the affinity of the analyte to CP. A good correlation between K_m, and $-\Delta$Gas, and sensitivity of the gas sensor was obtained.

PPy films were also used extensively to detect a mixture of compounds in vapor phases for possibility as e-nose and e-tongue. Using a variety of dopants, free-grown thin PPy films were used to fabricate aroma sensor for vapors of volatile organic solvents [150]. The percentage decrease in sensitivity was larger for polar compounds as compared to non-polar compounds. The sensitivity of the sensors towards simple organic solvents was analyzed and their reproducibility and aging effect upon the specific responses were studied. A portable gas analyzing system was developed using a PPy sensor array to analyze diabetic patient's breath [152]. The discrimination between patients and normal persons was achieved, and interpreted by the principal component analysis (PCA) with low computational efforts. The swelling mechanism was also the basis of developing an electronic nose based on an array of polymer composite sensors of PPy as a conductive filler in insulating organic polymers matrix [169]. It is observed that the analyte can interact with both the insulating matrix and the conductive polymer and therefore, the composites will swell to varying degrees depending on the polymer–odorant interactions and therefore, a change in the conductivity of the composite film. A detailed discussion on the e-nose and e-tongue is given in another chapter of this book.

1.3.1.3 Polythiophene

PTh and its derivatives were also used to fabricate sensors for VOCs. rr-P3HT-based QCM sensors were also developed to detect alcoholic VOCs associated with spoiled and *Salmonella typhimurium*-contaminated packaged beef headspace [145,159]. The PCA was performed on the sensors sensitivities to discriminate sensed alcohols, viz. 3-methyl-1-butanol from 1-hexanol, could detect 3-methyl-1-butanol from 1-hexanol at low concentrations. It was shown that chemiresistors using regioregular PThs when exposed to VOC vapors exhibit either positive or negative response [151]. The sign and amplitude of the sensing response depends on the analytes, polymer film morphology, chemical structures such as side chains, end groups and secondary chains in a copolymer, suggesting involvement of multiple sensing mechanisms in detection of VOCs [140,144]. Further, it was observed that the chemical sensing responses of organic FETs based on nanostructured regioregular PTh are strongly dependent upon the gate biasing field [170]. With different applied gate voltages, the I_{SD} current response was found to be different both in sign and in magnitude for the same analyte. At zero gate voltage, only acetone and acetonitrile showed negative response and other VOCs such as chloroform, butanol and toluene showed positive response. The sensing mechanisms were attributed to the polymer thin films morphology, where an intra-grain effect yields a positive response, and the grain boundary effect a negative response. Using printable sensor inks, a functional, integrated gas sensor array chip was fabricated from P3HT for a range of volatile organic compounds down to concentrations of 10 ppm [171].

Alkyl-poly-(3,3″-didodecyl-2,2′:5′,2″-terthiophene) (Poly-DT) and alkoxy-substituted (Poly-DPOT) PTh thin films were found to be responsive for selective VOCs sensor, namely, alcohols, alkanes and ketones [153]. The response of Poly-DT is related to the analyte alkyl-chain length, while Poly-DPOT is observed to be dependent on the analyte dipole moment. Charge distribution simulation revealed that Poly-DT polymer has a dominant non-polar alkyl chains interaction, while Poly-DPOT has a much stronger dipole moment. Therefore, Poly-DPOT polar interactions are significant and result in a sensor responsive to ethanol and hexanol. On the other hand, dispersion-type interactions are dominant for Poly-DT which is why it does not respond to ethanol.

Gannot *et al.* synthesized a novel PTh copolymer functionalized with pendant pyridine groups and investigated the response of its OFETs to vaporous alkylating agents [160]. Upon exposure to alkylators, such as methyl iodide, chloromethoxy ethane, 1-chloro-2-ethylsulfanyl ethane and benzyl bromide, the devices showed considerably higher I_{SD} and a resistor-like response without the gate-induced amplifying effect. The responses were attributed to the covalent alkylation/quaternization of the pyridine units of the PTh copolymer, resulting in the formation of electron-deficient pyridinium groups, which attracted charge density from the PTh skeleton, leading to a high-level p-type doping.

The chemical sensing properties of individual PEDOT nanowires were investigated in terms of the effect of dopant on charge transport [108]. It was observed that nanowires doped with ClO_4^- ions show much greater sensitivity to vapor-phase analytes than the one doped with PSS ions. The sensing mechanism was attributed to different charge transport mechanisms operable in PEDOT. Temperature-dependent conductivity measurements show that charge transport in PEDOT:PSS nanowires is in the insulating regime of the metal–insulator transition and dominated by hopping, while PEDOT:perchlorate (ClO_4) nanowires are on the metallic side of the critical regime. Subsequently, the PEDOT:ClO_4 nanowires show enhanced sensitivity for VOCs.

1.3.2 Hydrogen bonding

The interaction between analyte, e.g. acetone, and CPs, such as PANI and PPy, may result in hydrogen bond formation between the NH group of CP and the C=O group of the analyte. This kind of bonding was shown in pyrrole units on the basis of impedance spectroscopy [136]. A similar hydrogen bonding was observed to occur when PANI was exposed to methanol [161]. PPy- and PANI-conducting polymers synthesized on inter-digitated and four-electrode Au/Al_2O_3 electrodes by chemical oxidation-casting were employed to monitor the level of acetone in the gas phase. The sensitivity of acetone gas sensor based on PANI (5.908×10^{-4} ppm^{-1}) was found to be higher than that of PPy. Acetone of concentration ~29 ppm could be detected using the sensor along with response and recovery times of 1–3 and 3–10 min, respectively. QCM gas sensor based on PANI was also fabricated to detect dibutyl phthalate (DBP) vapor. The sensor was extremely sensitive to DBP vapors and could detect concentrations as low as 20 ppb, and could be easily recovered by N_2 purging. The interaction of PANI with DBP results from the strong tendency of DBP to form hydrogen bonding, which occurs between the PANI imine groups and the carboxyl groups of DBP. The process results in an increase of the mass of the film, and therefore, the frequency of QCM changes. Hydrogen bonding is destroyed when the film is purged with nitrogen gas. As a result, DBP is desorbed, making the process reversible.

In addition, the weak intermolecular force was also used to detect enantiomer of chiral gas by PPy with a chiral side group [172]. A new type of gas sensor for measuring resistance was developed to detect sevoflurane concentrations at room temperature using photopolymerized pyrrole [163]. A hydrogen bond, N–H–F, was suggested to form between PPy and sevoflurane based upon molecular dynamic calculations. The sensor showed response in a wide range of 0.1%–10%, with fast response and recovery times of 60 and 20 s, respectively [154]. The composite of Ag with PANI was also synthesized to enhance the sensor performance [2]. It was proposed that the incorporation of Ag generated more sensitive responses to triethylamine than other metals due to the affinity of amines for Ag. Consequently, the response of the sensor was analyzed and found to be fit to chemisorption according to the Langmuir isotherm as well as diffusion models.

In another approach, gold nanoparticles were covalently attached to the functional groups of the copolymer films, poly(3,4-ethylenedioxythiophene-co-thiophene-3-aceticacid) Poly(EDOT-co-TAA), using the oxidative chemical vapor deposition (oCVD) technique [145]. The sensor was found to display selectivity to methanol over *n*-pentane, acetone and toluene. Different sensing mechanisms were also proposed to support the experimental data as opposed to the conventional mechanism where the analyte influences the work function of the metal nanoparticles and thus changing the electric resistance of the conducting polymer. It was proposed that methanol and water molecules first adsorb on the surface of gold nanoparticles. The Au nanoparticles may be slightly positively charged due to covalent bonding to the sulfur atom in the polymer. The positively charged gold atom may interact with the slightly negatively charged oxygen atom owing to hydrogen bonding among the analyte molecules, thereby leading to an increase in the Fermi level of the gold nanoparticles. As a result, the potential barrier between the metal nanoparticle and the polymer increased, leading to the overall increment in resistance.

Recently, an aptamer-functionalized multidimensional conducting-polymer (3-carboxylate polypyrrole) nanoparticle (M_CPPyNPs) was reported to detect bisphenol (BPA), a compound associated with cardiovascular disease, diabetes and cancer [173]. Owing to large surface area of the films, the interaction between the aptamer and polymer was enhanced (Figure 1.12). The aptamer-conjugated polymer immobilized on FET sensor exhibited outstanding stability against environmental perturbation as well as ultra-high sensitivity (~1 fM) to BPA. The reusability properties of the sensor were attributed to the covalent bonding used during an immobilization process.

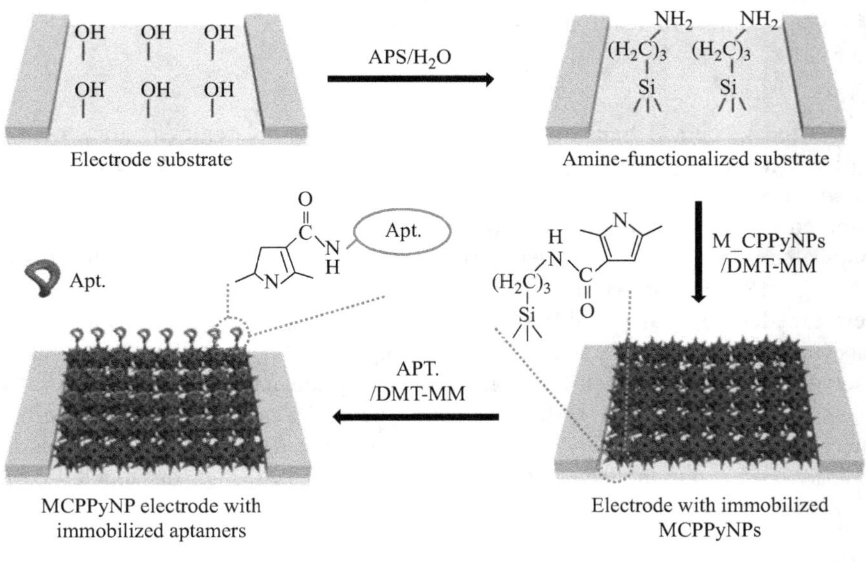

Figure 1.12 Schematic diagram of the immobilization process used to produce the sensor electrode based on aptamer-conjugated M_CPPyNPs (A_M_CPPyNPs) on the IDA electrode substrate. Apt. represents aptamer © 2013 IEEE. Reprinted, with permission, from Reference [173]

A single component ink system for thermochromic-conjugated polymers was also fabricated by introducing oligoethylene oxide moieties and bisurea groups into DA monomers [174]. The hydrogen bonding interaction between bisurea groups of the polymerized ink solution was found to increase as the number of the ethylene oxide unit increases without disturbing the closely packed self-assembled structures. The DA-printed paper resulted in blue-colored conjugated polydiacetylenes (PDAs), which displayed completely reversible thermochromic transitions in the temperature range 30–200 °C. Thermally as well as pH-stimulated reversible PDA Langmuir–Schaefer (LS) films consisting of a novel single-chain diacetylene derivative were also reported [175]. Upon repeating thermal cycles (25–100 °C), the color of the LS films switched between blue and purple. On the other hand, the film underwent a colorimetric transition by changing the pH from 13.7 to 5.5. The reversible colorimetric transition against both temperature and pH stimuli was attributed to double hydrogen bonding among head groups of the molecules which provide its supramolecular structure.

1.4 Optical sensors

As discussed earlier, the interaction of the analyte with CP may bring about changes in the band gap and therefore the optical properties. The sensing by using optical properties may be classified into three main categories as discussed in the following sections in detail and summarized in Table 1.5.

1.4.1 Fluorescence quenching

In addition to optical transitions involving π–π^* and polaronic–π^* transition, a third optical absorption is also possible for transition between the two doping energy levels. Due to these inter-gap transitions, an intense color change is observed from the pristine to the doped state of the conjugated polymer. Several conjugated polymers and their derivatives such as polyfluorene, poly(*p*-phenylene-ethynylene) (PPE), PPV, PTh and PDA were explored as a fluorescent probe to tap their excellent optical properties in the sensing of biologically and chemically important analytes [191,192,213,214]. The sensing is achieved by (i) fluorescence resonance energy transfer (FRET), (ii) fluorescence quenching and (iii) ratiometric response upon interaction with the analyte. Several sensors were developed to detect fluoride anions [193,194,215,216], anionic quenchers [217], nitroaromatic explosives in solution and in the vapor phase [210,218] as discussed in the following sections.

1.4.1.1 Poly(*p*-phenylene vinylene)

PPVs, such as polyacetylenes, are also strong luminescent-conjugated polymers with high permeabilities to nitroaromatics. Luminescence quenching on exposure to TNT and DNT vapor occurs at low concentration in the ppm level [206]. It was observed that MEH-PPV polymer shows greater quenching efficiency than

Table 1.5 Sensors based on optical properties of conducting polymers to several analytes

Analyte	Deposition technique	Sensing material	Range	Response/ recovery time	Sensitivity	Selectivity against	Transducer	Ref.
Ethylene	Spin coat	PPE derivative	1,000 ppm	–	K_{SV} ~2×10^3 M^{-1}	Ethanol, acetaldehyde, water	Fluorescence	[176]
Chloroform	Inkjet printing	PDA	–	Blue to red yellow/green fluorescence	–	THF, MC	Colorimetric	[177]
BPA	Dip coating (self-assembly)	3-Carboxylate PPy	1 fM–10 pM	–	–	Bisphenol B, 4,4′-bisphenol, 4,4-bis(4-hydroxyphenyl)valeric acid and 4,4′-(hexa-fluoroisopropylidene) diphenol	FET	[173]
VOCs	Electrospinning	PDA	–	–	–	Ethyl acetate, *n*-hexane	Colorimetric	[178]
VOCs	Inkjet	PDA	–	3 s	–	PCA analysis	Colorimetric	[177]
Hydrocarbons in kerosene	Solution cast	PDA	0%–100%	–	–	–	Colorimetric	[179]
CO_2	Electrospinning	PDA	400 ppm	–	–	–	Optical	[180]
CO_2	Solution	PTh	Dynamic at a flow rate of 60 cm^3/min^{-1}	–	–	–	Colorimetric	[181]
NH_3	Drop cast	Polyaniline	92–4,618 ppm	14 min	–	–	Absorbance	[120]
NH_3	Spin coat	Polyaniline	10–500 ppm	1 min/10 min for 100 ppm	–	–	Optical Transmittance	[182]
NH_3	*In situ* PZ	PANI	5 ppb-25 ppm		–	Light green to blue	Optical	[183]
NH_3	DC magnetron	Polyaniline	80–512 ppm	2 min/2–4 min	–	–	SPR	[118]
NH_3	Drop Cast	PANI	–	32–992 ppm	6–7 min/120 min for 192 ppm	–	SPR	[184]
NH_3	DC magnetron Sputtering	PANI	25–100 ppm	–	–	–	SPR	[118]
NH_3/HCl	Drop cast	PANI	–	5–100 ppm	–	–	Optical stop band	[185]
HCl	Spin coat	PDA	Real time	–	–	–	Colorimetric	[186]

(Continues)

Table 1.5 (Continued)

Analyte	Deposition technique	Sensing material	Range	Response/ recovery time	Sensitivity	Selectivity against	Transducer	Ref.
HCl	Electrospinning	PDA	2 ppm	1 s	–	–	Optical/ Fluorescence	[187]
HCl vapor	Drop cast	Poly(thiophene-3-propionic acid)	Yellow to purple	–	–	–	Colorimetric	[188]
Alkali earth metals			Red to purple, magenta and brown	–	–	–		
pH/heat	Spin coat	PDA	5.5–13.7 25–100 °C	–	–	–	Colorimetric	[175]
Hydrazine	Spin cast	PANI	0.05–5 ppm	–	–	–	Absorbance	[125]
Anionic surfactants	Spin coat	PDA	1 mM	% 90 RGB	–	Cationic surfactants	Colorimetric	[189]
Cationic surfactants	Solution	PDA	20–200 μM	–	–	Anionic surfactants and ammonium salt	Colorimetric	[190]
Li^+/Na^+	Solution	PDA	0.05–1.2 mM	–	–	–	Fluorescence	[191]
Na^+	Solution	Poly(phenylene bithiophene)	0–0.72 mM	–	–	Li^+, K^+, Ca^{2+}	Fluorescence	[192]
Pyrophosphate (PPi) anion	Solution	P1, P2, P3	1×10^{-5}–3.6×10^{-5} M	–	3.3×10^5 M^{-1}	F^-, Cl^-, Br^-, I^-, NO_3^-, HSO_4^-, AcO^-, HCO_3^-, $H_2PO_4^-$, CO_3^{--}, HPO_4^{--}, PO^{3-}	Fluorescence	[193]
F^-	Solution	Polyquinoline	20–200 μM	–	–	F^-, Cl^-, Br^-, $H_2PO_4^-$	Fluorescence	[194]
Amine	Spin coat	Polyfluorene/Pd tetraphenyl porphyrin	Saturated vapors	–	–	–	Fluorescence	[195]
Melamine	Solution	PDA	1–40 ppm	30 s	–	–	Colorimetric	[196]
Hg^{2+}	Solution	PFP	0.2–200 nM	–	–	Na^+, K^+, Mg^{2+}, Fe^{3+}, Cu^{2+}, Zn^{2+}, Pb^{2+}, Cd^{2+}, Ag^+, Au^{3+}, Ca^{2+}	FRET	[197]

Muscle Contraction Sensor	Spin coat	PTB7	Real time	–	–	–	Optical	[198]
Temperature	Inkjet print	PDA	30–200 °C	–	–	–	Colorimetric	[174]
γ-RAY	Spin cast	Poly[9-hexyl-9-(2-ethylhexyl)-fluorene-2,7-diyl] (HEH-PF) BiI	–	–	–	–	Photolumine scence	[199]
γ-Ray	Spin cast	Poly(HEH-PF)/1,3-diiodobenzene	5 Gray	–	–	–	Photolumine scence	[200]
γ-Ray		PANI-ES	1–10 kGy	–	–	–	Absorbance	[201]
γ-Ray	Spin cast	MEH-PPV	1–5 kGy	–	–	–	Absorbance	[202]
γ-Ray	Spin cast	MEH-PPV	0–152 Gy	–	–	–	Absorbance	[203]
Nitro compounds	Spin cast	Polyacetylene derivative	Vapors	–	–	–	Fluorescence	[204]
DNT	Solution	MI-PPV	Saturated vapors	–	–	–	Fluorescence	[205]
TNT	Solution	MEH-PPV, DP10-PPV, poly(dipheny-lace tylene)	Vapors	–	–	–	Fluorescence	[206]
TNT	Solution	P1, P2	1–5 mM	–	K_{SV} ~5	–	Fluorescence	[207]
TNT/DNT	Spin cast	PPV derivative	Vapors	–	–	–	Fluorescence	[208]
TNT/DNT	Solution cast	PPE derivative	10–60 μM		3.65×10^{-4} M^{-1}	Inorganic salts, seawater, acid, alkali	Fluorescence	[209]
TNT/picric	Spin cast	P1, P3	0–0.3 mM	–	1.2×10^{-5} M^{-1}	–	Fluorescence	[210]
Picric acid	Solution	Polyfluorene nanoparticle network	0–0.5 μM		1.12×10^{-8} M^{-1}	Other nitro compounds	Fluorescence	[211]
UV	Solution cast	P-PPV	–	–	–	–	FRET	[212]

DP10-PPV. This was attributed to the dialkoxy-substituted aromatic ring of MEH-PPV, which is more electron rich than the diphenyl-substituted aromatic ring of DP10-PPV, thus allowing for stronger columbic interactions between the electron-donating MEH-PPV and the electron-accepting nitroaromatics. Other than this planar backbone of MEH-PPV allows for efficient exciton migration as compared with twisted DP10-PPV. An enhanced sensitivity to nitroaromatic detection was achieved through the lasing action of poly(*p*-phenylene vinylene) DADP-PPV [208]. The pendant phenyl groups with branched alkoxy substituents encapsulate the polymer backbone and therefore protect it from self-quenching and photo-oxidation. Upon exposure to DNT vapor, luminescence quenching is observed in polymer thin films for both the spontaneous emission at 500 nm and the stimulated laser emission at 535 nm. A 2-min exposure to saturated vapor showed twofold and tenfold enhanced quenching in the 500-nm and 535-nm peak, respectively. It was demonstrated that the polymer coated on the surface of an optical fiber is sufficient to cause attenuation in the lasing emission upon exposure to a low concentration of TNT vapors (5 ppb). A new approach of a cross-linked molecularly imprinted fluorescent CP material was also developed for the detection of DNT and related nitroaromatic compounds [205].

In addition to this, the FRET process was also utilized for development of radiation sensors as in case of conventional plastic scintillators. In this process, the emission spectrum can be tuned and self-absorption is suppressed. Direct detection of X-ray-induced photocurrents was demonstrated in MEH-PPV and poly(9,9-dioctylfluorenyl-2,7-diyl) (PFO) [58]. The linear behavior was observed in PFO devices up to 18.5 mGy/s under different biases. Similarly, low-dose dosimetry was demonstrated for MEH-PPV for X-rays [203]. It was observed that MEH-PPV in solution is sensitive to very low doses of ionizing radiation. Low-dose ionizing gamma radiation was also demonstrated in MEH-PPV [202]. The absorption peak of MEH-PPV in chloroform solution shifted linearly with radiation dose below 10 Gy. It was suggested that Cl radicals were generated upon radiation, which together with water and oxygen in solution lead to a reduction of conjugation. A conjugated polymer composite suitable for gamma-ray detection with high sensitivity and response linearity was also demonstrated [212]. A transparent polymer, polymerized from trimethylolpropane trimethylacrylate, served as the matrix for a PPV derivative and high-Z compounds. Upon gamma-ray irradiation, excitons are generated in the high-Z compound phase with high probability, and subsequently energy is transferred to PPV via the FRET process and finally fluorescence is exhibited. Further, it was suggested that a ternary composite made up of an iodine compound, conjugated polymer and fluorescent compound can be used for gamma-rays. The iodine compound absorbs gamma-ray photo-sensitizer owing to high atomic number and the transfer energy to the polymer, the fluorescent material then receives energy from the polymer via the FRET process and results in color change [200]. In addition, composites of CP MEH-PPV with high-Z compounds, Quantum dot (QD) of CdSe–ZnSe spin-casted onto a sapphire substrate were also demonstrated for radiation scintillation [219].

1.4.1.2 Poly(*p*-phenylene-ethynylene)

A very high fluorescence property of conjugated polymer, PPE, was utilized for detection of nitroaromatic compounds at a low level [220]. Thin films of the PPE show rapid quenching of their blue luminescence (465 nm) when exposed to TNT vapor at sub-part-per-trillion (ppt) levels. The exceptional sensitivity was the result of long polymer chain lengths ($M_n = 56{,}000$), which provides exciton delocalization along the chain. Additionally, two bulky pentiptycene moieties present on each alternating phenyl unit of the backbone result in a porous packing arrangement in the solid state, which prevents interchain p-stacking and self-quenching of luminescence. The pentiptycene-conjugated polymer has also been prepared into a semiselective nitroaromatic sensor array for the tip of an optical fiber bundle [221]. The sensor was demonstrated to detect the more volatile DNT vapor from landmines and a short (200 ms) sampling time was required to detect explosives. Nomadics Inc. has fabricated and commercialized a landmine detector using this technology (www. nomadics.com.). Further, copolymers of poly(pyrene-co-phenylene-ethynylene)s having different compositions (PyPE-1 and PyPE-2) were synthesized and demonstrated to have TNT-sensitive fluorescence emissions in the aqueous phase, promising supersensitive detection of TNT in groundwater or seawater [209]. Further experiments demonstrated reversible sensing which was insensitive to interference from commonly found compounds, including other nitroaromatics.

Sensors for other analytes as well as fluoride ions were also developed. By usage of the copper(I) complex of fluorinated tris(pyrazolyl)borate ligand in PPEs thin films, a sensor was developed to sense ethylene [176]. copper(I) binds to the triple bonds in the conducting polymer film resulting in a significant fluorescence quenching. When this film is exposed to ethylene gas, the copper complex binds to the ethylene molecules, causing fluorescence. Using this method, the concentration as low as 1,000 ppm of ethylene could be detected.

A supramolecular cross-linked network of PPE was demonstrated as a multiple fluorescent sensor [222]. As compared with PPE, the network has weak fluorescence due to the aggregation of polymer chains. The fluorescence intensity of the supramolecular system could be enhanced by potassium cation, chloride anion, pH increase and heating owing to host–guest interactions and therefore shows the ability of the network structure as cation/anion sensor, pH as well as temperature sensor.

Thin films of the polymer network were also used to demonstrate ammonia sensing. A series of new conjugated polymers based on poly[*p*-(phenylene-ethynylene)-alt-(thienyleneethynylene)] (PPETE) with oligopyridine pendant groups as receptors for transition metals were synthesized as fluorescent probes [223]. All polymers showed good fluorescence and good selectivity for Ni^{2+} concentration as against other transition metal ions. For the terpyridine–receptor polymer, the signal could be detected for as low as 4 nM concentration.

1.4.1.3 Others

Owing to high fluorescence efficiency of PDA, TNT can easily produce intramolecular charge transfer from the photoinduced excited state of the polymer to the analyte, and thereby resulting in fluorescence quenching and therefore a very low detection limit of the analyte. Polyacetylenes having a high fractional free volume of 0.26 was investigated as potential nitroaromatic sensors [204]. The luminescence quenching was observed with even PPb concentrations of nitroaromatic analytes. The quenching was attributed to the formation of a charge-transfer complex between the nitroaromatic analyte and the electron-rich polyacetylene chain.

Besides aforementioned polymers, conjugated polymers were also synthesized to induce specificity for the analyte such as ions and TNT compounds [197,207,211]. The exciton migration increases the probability of interaction with a bound quencher and therefore, contributes to enhanced detection sensitivity even in the ppb levels. A combination of oligonucleotides, DNA intercalators (YOYO-1) and poly(9,9-bis(6-*N*,*N*,*N*-triethyl-ammonium)-hexyl-fluorenephenylene) (PFP) was used to produce a sensor for the detection of mercury ions in aqueous media at room temperature [197]. Upon addition of Hg^{2+}, thymine–Hg^{2+}–thymine (T–Hg^{2+}–T) complex formation induces a change in the poly-T conformation from a random coil to a folded double-stranded DNA (dsDNA) structure. The presence of PFP causes fluorescence enhancement of the energy acceptor YOYO-1 via FRET, thus enabling detection of Hg^{2+} at 0.27 nM. Recently, conjugated polymer nanoparticles of poly(3,3′-((2-phenyl-9*H*-fluorene-9,9-diyl)bis(hexane-6,1-diyl)) bis(1-methyl-1*H*-imidazol-3-ium)bromide) (PFMI) exhibited remarkable fluorescence response toward nitroexplosive picric acid (PA) in multiple environments including 100% aqueous media, solid support using portable paper strips and vapor-phase detection via two terminal devices [211]. Highest quenching constant value (1.12×10^8 M^{-1}) as well as a very low detection limit (30.9 pM/7.07 ppt) was obtained in 100% aqueous environment. The ultra-sensitivity was assigned due to the "molecular-wire effect", electrostatic interaction, photoinduced electron transfer, and possible resonance energy transfer.

1.4.2 Colorimetric

Different doping levels in the CP give rise to intermediate energy levels in the energy gap of CP and may lead to color changes depending upon the doping level, e.g. the protonation of the PANI backbone by hydrochloride forms ES which is green in color transforms into blue emeraldine base upon deprotonation. This color change is enough to be seen by the naked eye, thereby paving the pathway for the on-site and real-time colorimetric detection as discussed in the following sections.

1.4.2.1 Polydiacetylene

Other than PANI, PDAs and its substituents were utilized for the fabrication of color sensor due to its unique property of blue-to-red phase transition. PDA has A_g symmetry in blue form that shows no fluorescence due to the dipole-forbidden transition during the emission from a singlet excited state. A fluorescence and blue-to-red phase transition, owing to the lowest excited B_u state, take place in the presence of external stimuli, which arise due to a change in the preferred backbone conformation. Several derivatives of PDAs were used as sensing element in a different manner such as encapsulation of PDA vesicles in the host matrix, immobilization of PDAs on solid substrates and electrospun fiber and microfluidic system as discussed herein.

PDA was utilized extensively to develop a sensor for different VOCs. A litmus-type colorimetric and fluorometric sensor system was developed for the detection of VOCs based on inkjet-printed PDAs on paper [177]. The dot image on the paper-based sensor strip yielded blue-colored PDAs by UV-induced photopolymerization. The chloroform vapor-exposed image on the paper sensor strip undergoes a blue-to-red-to-yellow color transition accompanied by the generation of green fluorescence. A similar kind of sensor for VOCs was also fabricated by using the electrospinning technique, which provided large surface area [178]. A sensor was also fabricated using an inkjet-printable microemulsion system of the DA monomer on paper which was then photopolymerized [187]. Upon exposure different VOCs, different color changes were observed from original blue color, e.g. to purple color for methanol, red for ethanol, acetone, toluene, ethyl acetate (EA) or ether, dark for THF or MC. In addition, a reversible yellow-to-red colorimetric transition was observed for chloroform vapors, which was attributed to the exceptional solubility of the polymer in chloroform. Using organic solvent-dependent color change property of the PDA, a straightforward "color pattern" procedure was demonstrated to differentiate between several common organic solvents, such as chloroform, hexane, EA and THF, based on the color patterns. Further, a combinatorial approach for the colorimetric differentiation of organic solvents was developed [224]. Using arrays of PDA-embedded microfibers, unique color patterns were developed upon exposure to common organic solvents. This could be detected by the naked eye and showed the possibility to differentiate between a more diverse set of organic solvents by using computer-aided color analysis program. By making use of mechanochromic properties of PDA and hydrocarbon-induced swelling properties of PDMS, a colorimetric sensor was developed to differentiate a number of aliphatic hydrocarbons (C5–C14) [179]. The naked-eye differentiation between *n*-pentane and *n*-heptane could be possible with these devices. A significant expansion and tearing in PDA microcrystal in the film upon a blue-to-red colorimetric transition during the swelling process provides direct evidence for the mechanism for the mechanochromic behavior of PDA. Recently, solvatochromic PDA, which contains hygroscopic elements in its headgroup, was demonstrated to respond colorimetrically to water contents in organic solvent [225]. The hydrochromic PDA was found to undergo a remarkable fast structural change in the presence of water that brings about a blue-to-red color transition. The polymer was further utilized for sweat pore-based hydrochromic fingerprint detection system for latent fingerprint analysis.

Sensors based on PDA were also fabricated to detect gases and ions. A sensor for CO_2 was also fabricated using PDA; primary amines in PDA reacts with CO_2 to form carbamoate anions [180]. This resulted in neutralization of the positive charge of imidazolium cations of PDA, causing a change in color as well as fluorescence. Electrospun fiber of PDA incorporated in poly(ethylene oxide) (PEO) matrix were also used for the detection of HCl gas by simply observing changes in color and fluorescence [186]. Even at a low concentration of 2 ppm, a color transition from blue to red was observed by the naked eye. A pH colorimetric sensor was also developed which was reversibly switched with pH 4–10 [181]. At high pH, the polymer has a random-coiled conformation with a yellow color, while at low pH, the color of the solution changed to pink. This halochromism was attributed to the conformational stretching of the PTh backbone and formation of highly ordered aggregations. At pH 9.8, this polymer also exhibited distinct colorimetric transition from yellow to red-orange upon exposure to CO_2 gas.

The pH chromism of PDAs was also investigated with respect to the molecular size and acidity of acid analytes, along with the alkyl spacer length of primary-amine-functionalized diacetylene lipid [226]. The degree of the chromatic transition of PDA liposomes was found to be dependent on both the molecular size of the acid analytes and their pH levels, upon addition of various acids. Compared to the small-molecular-weight acids such as HF, HCl, HBr, HNO_3, H_2SO_4, H_3PO_4, diethyl phosphate induced the largest change from the blue to the red phase. In addition, the longer alkyl spacer produced a weaker chromatic transition of PDA liposomes, indicating that the flexible alkyl spacers produced an unfavorable force on the conjugated backbone.

A spin-coated PDA film on transparent PMMA substrates was also used as optical sensor which could detect distinctive colorimetric and spectroscopic transformations, corresponding to purple, red and yellow PDA phases specifically induced by different surfactant classes [189]. The spectroscopic transformation was quantified by %RGB which measures the blue–red–yellow transformations, having values corresponding to yellow color and lower %RGB corresponding to less pronounced, blue–purple and blue–red, color transitions. The %RGB analysis showed high %RGB upon addition of cationic surfactants (1-hexadecyltrimethylammonium bromide, CTAB and cetylpyridinium chloride) and low %RGB upon exposure to non-ionic surfactants (4-*O*-lauryl-1,6-anhydroglucopyranose, C12 LG and Triton X-100). More specifically, the PDA-bearing boronic acid group was synthesized which showed diverse colorimetric changes, such as yellow, orange and red, upon the addition of cationic surfactants, cetyltrimethylammonium chloride, at pH 7.4 [190]. The colorimetric changes with cationic surfactants were attributed to the charge interaction between ammonium groups in the cationic surfactants and boronate group in polymer at pH 7.4. In addition, it was also possible to differentiate between different head groups and alkyl chains via the penetration process.

Sensors for bacterial growth and melamine were also fabricated using PDA. A sensor for bacterial growth was also developed using PDA-based sensor [227]. The interaction of imidazolium and imidazole-derived PDA with bacteria such as MRSA (methicillin-resistant *Staphylococcus aureus*) and ESBL-EC (extended-spectrum β-lactamase-producing *Escherichia coli*) results in distinct blue-to-red colorimetric change of the solution. An analysis of Zeta potential suggests that the interaction results from an electrostatic interaction between the negatively charged bacterial cell surface and the positively charged polymers leading to a color change. Cyanuric acid-carrying PDA monomers were synthesized and examined for melamine detection [196]. Owing to a stable melamine–cyanurate complex formed through hydrogen bonding between imide and amino-pyridine moieties at a neutral pH, colorimetric sensor to detect low concentration of melamine (0.5 ppm) was developed.

1.4.2.2 Polyaniline and polythiophene

A few colorimetric sensors have been developed suing PANI and PTh and the sensitivity of the sensors was found to be limited due to nanostructured morphologies. A filter paper coated with PANI film for the colorimetric detection of gaseous and aqueous ammonia was fabricated; however, morphological hierarchy and microscopic porosity of filter posed difficulty in precise processing of PANI films [24]. An optical dosimeter for gamma radiation (^{60}Co) based on the color change of PANI-ES was also demonstrated [201]. The deep blue color films become green as the film is irradiated. A linear dependence of absorption coefficient with logarithm of the irradiation dose from 1 to 10 kGy was demonstrated in these films. Regioregular PThs with pendant carboxylic acid functionality, poly(thiophene-3-propionic acid) (PTPA, 3) and poly(thiophene-3-octanoic acid) (PTOA, 4), studied for chemosensory response [188]. PTPA showed a different CR to each of the alkali earth metals whereas the longer chain PTOA was only sensitive to ions larger than Et4N+. Distinct color changes were observed in studies of divalent cations known to selectively bind carboxylates, such as Zn^{2+}, Mn^{2+} and Cd^{2+}, and were attributed to relaxation of polymer chains from twisted conformation to a violet π-stacked conformation. A simple sol–gel process was employed to form nanowires PDA having lamellar mesostructure [228]. The nanowire papers fabricated through filtration of nanowire solutions could be freely folded and cut like regular printer papers. These papers were shown to have significant sensitivity to different ions, solvents and UV irradiation, as well as reversible blue-to-red chromatic switching induced by UV irradiation and solvatochromic transition.

1.4.3 Surface plasmon resonance (SPR)

Other than fluorescence and color, sensors were also developed using surface plasmon resonance technique. Using SPR, a very low concentration of gas/analyte molecule could be detected with a good stability, and re-producibility, e.g. using SPR changes in refractive index of approximately 10^{-5} can be detected. PANI doped with CSA was used to sense NH_3 by way of SPR spectroscopy [184]. The optical constants (n and k) as determined with spectroscopic ellipsometry confirmed the wavelength-dependent response observed via SPR. The analytical analyses showed a limit of detection (LOD) of 32 ppm NH_3 using this technique, which is comparable to other established sensing method. The ammonia sensor was also fabricated using DC magnetron sputtered PANI films doped with hydrochloric acid [118]. Using SPR spectroscopy, sensing behavior of these films towards of NH_3 investigated. A reversible behavior upon exposure to NH_3 was observed with LOD ~0.4 ppm in dry air. In addition to nanostructures, highly ordered inverse opals nanostructures of PANI have been used with respect to ethanol vapor, hydrogen chloride and ammonia. The reflectance spectra of PANI changed significantly in the presence of hydrogen chloride and ammonia altered the reflectance spectra significantly [185]. The optical stop band (the reflectance peak at 945 nm) arising due to the periodic structures was used to observe the changes upon interaction with the analyte. A significant shift of more than 100 nm in the stop band was observed for 75 ppm hydrogen chloride. On the other hand, only 8 nm shift was observed for ethanol exposure. Further, PANI-infiltrated TiO_2 inverse opal photonic crystal film was applied as sensor for NH_3 and HCl vapors [229]. A reversible color change by the naked eye could be seen film is exposed from HCl to NH_3 vapors. A significant shift of 132 nm in the stop band of film was observed, corresponding to a clear color change from green to red, in a saturated NH_3 environment. The stop band undergoes a blue shift when exposed to HCl vapor, and the color turns back to green. The results were attributed to both doping/dedoping of PANI by acid/base reaction as well as effective refractive index of the film. Further, sensor for heavy metal ions such as, mercury, lead were also developed based upon SPR measurements. A Cu^{2+} sensor based on PPy and chitosan composite was reported [230]. Different corrosion levels classified as class 1A according to the standard copper strip test could be recognized in the sample using this optical sensor. Further, an SPR sensor for Hg(II) detection was reported using PPy and 2-mercaptobenzothiazole (2-MBT) [231]. Owing to chelating action of 2-MBT, the sensor was specific to Hg^{2+} down to 0.01 ppm concentration. The plasmon coupling between nanoparticles results in large electric field enhancements, which has the ability to amplify the fluorescence signal in CPs [232]. A change in the optical properties of PANI has also been utilized for low ammonia detection (10–4,000 ppm) [182]. A decrease in the polaron density inside band gap occurred upon interaction of ammonia with ES of PANI, resulting in band gap changes.

1.5 Conclusions

In summary, the synergetic adoption of sensors with specific recognition capacities, physical/chemical operational conditions of devices and the miniaturization of transducers such as microelectrodes, sensor arrays, FETs hold promise for the development of sensors as per demand. Recent advances in supramolecular chemistry in terms of nanostructured morphology control and molecular recognition and/or chiral discrimination induced by intermolecular interactions promise to lead the pathway for good sensing material synthesis. In addition, software development in terms of better recognition algorithm and judicial design of advanced circuits will be the decisive factor for developing future applications relying on sensors based on CPs. Over the past decade, considerable scientific efforts have been deployed toward sensors in terms of (i) sophisticated working principles to provide fundamental knowledge for the sensing mechanism, (ii) the development of state-of-the-art sensor for high performance and (iii) the development of advanced sensory arrays to fabricate e-noses, e-tongue, touch sensors, wearable sensor, etc. Nevertheless, persistent efforts are required to further improve the overall performance of sensors, including sensitivity, specificity, selectivity, response time, stability, reproducibility, LOD, capability of digital and intelligent readout, and real-time workability for meeting the commercial requirements. In our opinion, the following aspects addressed here will be major subjects of research in the field:

1. **Molecular engineering**
   The main problem of the analytical sensors based on CPs is selectivity. In general, high conductive polymers based on simple polyaromatic structures do not have specificity to analytes owing to the absence of reactive sites. To this extent choice of conducting polymers as the materials for developing sensors poses a disadvantage since sensors are needed for high sensitivity as well as specificity or selectivity. Molecular engineering in terms of design of specific CPs with specific functionalities, for creating reactive sensing sites via covalent or non-covalent bonds with the analyte is the most widely adopted strategy for addressing this drawback. A lot of research has been devoted on this subject and success has been achieved mainly with the incorporation of function groups on CPs. Further development requires the optimized strategy for meeting selective and sensitive responses to target analytes since introducing functional sensing sites might induce changes to morphology of, solubility, and other characteristics of the CPs and have an effect on the overall device performances. More efforts are required to develop new host–guest link type of molecules without compromising the sensitivity and electrical/optical properties. An interesting future research field would be molecularly imprinted polymers (MIPs) combined with fluorescent CPs for the development of radiation sensors, explosives sensors and other chemical sensors. A relatively unexplored approach for developing sensors based on the FRET mechanism also holds immense promise.
2. **Blends or composite**
   In addition to molecular engineering, the requirement of sensors with specificity was partly addressed through the usage of blends of CPs with high electrical performances and sensing molecules of specific recognition capacity, e.g. enzyme entrapment, composites of CPs with metal oxide or other inorganic semiconductors. These blends or composites of CPs with metals, metal oxide and other semiconducting polymers hold promise for development of sensors for specific analyte in an efficient manner. However, uniform distribution, synergetic percolation for transport continue to pose challenges that need to be addressed for further developments in this class of sensors.
3. **Film preparation**
   Preparation of films of CPs is another important aspects which needs attention for development of devices at large scales. The deposition technique is important in the sense it affects the packing of the molecules, π-stacking of molecules on the substrates as well as the surface area.
4. **Transducers, Circuits and algorithms.**
   In addition to the above, the miniaturization of sensors as well as read-out devices is an important consideration for developing futuristic sensors especially wearable sensors. The integration of signal processing units with sensors are expected to pave the path for this upcoming technology. More complex algorithms as well as

appropriate softwares for analyzing the data are also needed to develop sensor for identification of components in a mixture, e.g. e-nose, e-tongue, touch sensors and wearable sensors.

Apart from these a better understanding of the underlying sensing mechanism of the CPs will also enable the researchers to fabricate high-quality sensors. For CPs, other information such as charge transport mechanism, doping and trapping studies will play an important role in designing new molecules with better sensor performance. In spite of large progress made in fabricating transducer circuit and towards developing software capable of performing pattern recognition algorithms, fabrication of sensors array at a large scale still poses a challenge. Combined efforts for development of hardware and software technology will be helpful to meet these challenges. The potential and hurdles of the upcoming technologies require interdisciplinary research efforts from physics, chemistry and engineering.

References

1. (a) Seyedin, S.; Razal, J. M.; Innis, P. C.; Jeiranikhameneh, A.; Beirne, S.; Wallace, G. G., Knitted strain sensor textiles of highly conductive all-polymeric fibers. *ACS Applied Materials & Interfaces* 2015, *7* (38), 21150–21158; (b) Tai, Y.-L.; Yang, Z.-G., Flexible, transparent, thickness-controllable SWCNT/PEDOT:PSS hybrid films based on coffee-ring lithography for functional noncontact sensing device. *Langmuir* 2015, *31* (48), 13257–13264.
2. Li, Z.-F.; Blum, F. D.; Bertino, M. F.; Kim, C.-S., Understanding the response of nanostructured polyaniline gas sensors. *Sensors and Actuators B: Chemical* 2013, *183*, 419–427.
3. Fang, Q.; Chetwynd, D. G.; Covington, J. A.; Toh, C. S.; Gardner, J. W., Micro-gas-sensor with conducting polymers. *Sensors and Actuators B: Chemical* 2002, *84* (1), 66–71.
4. (a) Charlesworth, J. M.; Partridge, A. C.; Garrard, N., Mechanistic studies on the interactions between poly(pyrrole) and organic vapors. *The Journal of Physical Chemistry* 1993, *97* (20), 5418–5423; (b) Gozde, O.-I.; Anna Maria, C.; Karen, K. G., CVD of polymeric thin films: applications in sensors, biotechnology, microelectronics/organic electronics, microfluidics, MEMS, composites and membranes. *Reports on Progress in Physics* 2012, *75* (1), 016501.
5. Rella, R.; Siciliano, P.; Quaranta, F.; Primo, T.; Valli, L.; Schenetti, L., Poly [3-(butylthio)thiophene] Langmuir–Blodgett films as selective solid state chemiresistors for nitrogen dioxide. *Colloids and Surfaces A: Physicochemical and Engineering Aspects* 2002, *198–200*, 829–833.
6. Stamenov, P.; Madathil, R.; Coey, J. M. D., Dynamic response of ammonia sensors constructed from polyaniline nanofibre films with varying morphology. *Sensors and Actuators B: Chemical* 2012, *161* (1), 989–999.

7. Gardner, J. W.; Bartlett, P. N.; Pratt, K. F. E., Modelling of gas-sensitive conducting polymer devices. *IEE Proceedings – Circuits, Devices and Systems* 1995, *142* (5), 321–333.
8. Zhang, Y.; Kim, J. J.; Chen, D.; Tuller, H. L.; Rutledge, G. C., Electrospun polyaniline fibers as highly sensitive room temperature chemiresistive sensors for ammonia and nitrogen dioxide gases. *Advanced Functional Materials* 2014, *24* (25), 4005–4014.
9. (a) Kofler, J.; Nau, S.; List-Kratochvil, E. J. W., A paper based, all organic, reference-electrode-free ion sensing platform. *Journal of Materials Chemistry B* 2015, *3* (25), 5095–5102; (b) Michalska, A.; Dumańska, J.; Maksymiuk, K., Lowering the detection limit of ion-selective plastic membrane electrodes with conducting polymer solid contact and conducting polymer potentiometric sensors. *Analytical Chemistry* 2003, *75* (19), 4964–4974.
10. Bandodkar, A. J.; Hung, V. W. S.; Jia, W., *et al.*, Tattoo-based potentiometric ion-selective sensors for epidermal pH monitoring. *Analyst* 2013, *138* (1), 123–128.
11. Dong, S.; Sun, Z.; Lu, Z., A new kind of chemical sensor based on a conducting polymer film. *Journal of the Chemical Society, Chemical Communications* 1988, (15), 993–995.
12. Qu, Y.; Zhang, X.; Wu, Y.; Li, F.; Hua, J., Fluorescent conjugated polymers based on thiocarbonyl quinacridone for sensing mercury ion and bioimaging. *Polymer Chemistry* 2014, *5* (10), 3396–3403.
13. Mahanta, D.; Munichandraiah, N.; Radhakrishnan, S.; Madras, G.; Patil, S., Polyaniline modified electrodes for detection of dyes. *Synthetic Metals* 2011, *161* (9–10), 659–664.
14. Migdalski, J.; Błaż, T.; Lewenstam, A., Conducting polymers – mechanisms of cationic sensitivity and the methods of inducing thereof. *Electrochimica Acta* 2014, *133*, 316–324.
15. Janata, J.; Josowicz, M., A fresh look at some old principles: the Kelvin probe and the Nernst equation. *Analytical Chemistry* 1997, *69* (9), 293A–296A.
16. Zhan, C.-G.; Nichols, J. A.; Dixon, D. A., Ionization potential, electron affinity, electronegativity, hardness, and electron excitation energy: molecular properties from density functional theory orbital energies. *The Journal of Physical Chemistry A* 2003, *107* (20), 4184–4195.
17. Barman, T.; Hussain, A. A.; Sharma, B.; Pal, A. R., Plasmonic hot hole generation by interband transition in gold-polyaniline. *Scientific Reports* 2015, *5*, 18276.
18. Dong, B.; Yang, M.; Ge, S.; Cao, Y.; Li, B.; Lu, Y., Synthesis and photoluminescence modulating of polypyrrole fluorescent nano-spheres/dots. *RSC Advances* 2016, *6* (28), 23737–23745.
19. Dong, H.; Zhu, H.; Meng, Q.; Gong, X.; Hu, W., Organic photoresponse materials and devices. *Chemical Society Reviews* 2012, *41* (5), 1754–1808.
20. Yang, Y.; Nakamichi, T.; Yoshioka, H., *et al.*, Spectral-resolving capable and integratable multilayered conductive films via an inkjet method. *Journal of Materials Chemistry C* 2013, *1* (9), 1739–1744.

21. Parker, I. D., Carrier tunneling and device characteristics in polymer light-emitting diodes. *Journal of Applied Physics* 1994, *75* (3), 1656–1666.
22. Giridharagopal, R.; Kelly, K. F., Substrate-dependent properties of poly-diacetylene nanowires on graphite and MoS_2. *ACS Nano* 2008, *2* (8), 1571–1580.
23. Li, G.; Josowicz, M.; Janata, J.; Semancik, S., Effect of thermal excitation on intermolecular charge transfer efficiency in conducting polyaniline. *Applied Physics Letters* 2004, *85* (7), 1187–1189.
24. Jaczewska, J.; Raptis, I.; Budkowski, A., *et al.*, Swelling of poly(3-alkyl-thiophene) films exposed to solvent vapors and humidity: evaluation of solubility parameters. *Synthetic Metals* 2007, *157* (18–20), 726–732.
25. Blackwood, D.; Josowicz, M., Work function and spectroscopic studies of interactions between conducting polymers and organic vapors. *The Journal of Physical Chemistry* 1991, *95* (1), 493–502.
26. Pedrosa, V. A.; Luo, X.; Burdick, J.; Wang, J., "Nanofingers" based on binary gold–polypyrrole nanowires. *Small* 2008, *4* (6), 738–741.
27. Yan, X. B.; Han, Z. J.; Yang, Y.; Tay, B. K., NO_2 gas sensing with polyaniline nanofibers synthesized by a facile aqueous/organic interfacial polymerization. *Sensors and Actuators B: Chemical* 2007, *123* (1), 107–113.
28. Xie, D.; Jiang, Y.; Pan, W.; Li, D.; Wu, Z.; Li, Y., Fabrication and characterization of polyaniline-based gas sensor by ultra-thin film technology. *Sensors and Actuators B: Chemical* 2002, *81* (2–3), 158–164.
29. Lange, U.; Mirsky, V. M., Integrated electrochemical transistor as a fast recoverable gas sensor. *Analytica Chimica Acta* 2011, *687* (1), 7–11.
30. Srinives, S.; Sarkar, T.; Mulchandani, A., Nanothin polyaniline film for highly sensitive chemiresistive gas sensing. *Electroanalysis* 2013, *25* (6), 1439–1445.
31. Navale, S. T.; Mane, A. T.; Chougule, M. A.; Sakhare, R. D.; Nalage, S. R.; Patil, V. B., Highly selective and sensitive room temperature NO_2 gas sensor based on polypyrrole thin films. *Synthetic Metals* 2014, *189*, 94–99.
32. Yang, Y.; Li, S.; Yang, W.; Yuan, W.; Xu, J.; Jiang, Y., In situ polymerization deposition of porous conducting polymer on reduced graphene oxide for gas sensor. *ACS Applied Materials & Interfaces* 2014, *6* (16), 13807–13814.
33. Saxena, V.; Choudhury, S.; Gadkari, S. C.; Gupta, S. K.; Yakhmi, J. V., Room temperature operated ammonia gas sensor using polycarbazole Langmuir–Blodgett film. *Sensors and Actuators B: Chemical* 2005, *107* (1), 277–282.
34. Arsat, R.; Yu, X. F.; Li, Y. X.; Wlodarski, W.; Kalantar-zadeh, K., Hydrogen gas sensor based on highly ordered polyaniline nanofibers. *Sensors and Actuators B: Chemical* 2009, *137* (2), 529–532.
35. Fowler, J. D.; Virji, S.; Kaner, R. B.; Weiller, B. H., Hydrogen detection by polyaniline nanofibers on gold and platinum electrodes. *The Journal of Physical Chemistry C* 2009, *113* (16), 6444–6449.
36. Sadek, A. Z.; Wlodarski, W.; Kalantar-Zadeh, K.; Baker, C.; Kaner, R. B., Doped and dedoped polyaniline nanofiber based conductometric hydrogen gas sensors. *Sensors and Actuators A: Physical* 2007, *139* (1–2), 53–57.

37. Li, W.; Jang, D. M.; An, S. Y.; Kim, D.; Hong, S.-K.; Kim, H., Polyaniline–chitosan nanocomposite: high performance hydrogen sensor from new principle. *Sensors and Actuators B: Chemical* 2011, *160* (1), 1020–1025.
38. Virji, S.; Kojima, R.; Fowler, J. D.; Kaner, R. B.; Weiller, B. H., Polyaniline nanofiber–metal salt composite materials for arsine detection. *Chemistry of Materials* 2009, *21* (14), 3056–3061.
39. Agbor, N. E.; Petty, M. C.; Monkman, A. P., Polyaniline thin films for gas sensing. *Sensors and Actuators B: Chemical* 1995, *28* (3), 173–179.
40. Arulraj, A. D.; Vijayan, M.; Vasantha, V. S., Highly selective and sensitive simple sensor based on electrochemically treated nano polypyrrole-sodium dodecyl sulphate film for the detection of para-nitrophenol. *Analytica Chimica Acta* 2015, *899*, 66–74.
41. Singh, V. V.; Gupta, G.; Sharma, R., *et al.*, Detection of chemical warfare agent nitrogen mustard-1 based on conducting polymer phthalocyanine nanorod modified electrode. *Synthetic Metals* 2009, *159* (19–20), 1960–1967.
42. Brady, S.; Diamond, D.; Lau, K.-T., Inherently conducting polymer modified polyurethane smart foam for pressure sensing. *Sensors and Actuators A: Physical* 2005, *119* (2), 398–404.
43. Pan, L.; Chortos, A.; Yu, G., *et al.*, An ultra-sensitive resistive pressure sensor based on hollow-sphere microstructure induced elasticity in conducting polymer film. *Nature Communications* 2014, *5*, 1–8.
44. Li, Y.; Leung, M. Y.; Tao, X. M.; Cheng, X. Y.; Tsang, J.; Yuen, M. C. W., Polypyrrole-coated conductive fabrics as a candidate for strain sensors. *Journal of Materials Science* 2005, *40* (15), 4093–4095.
45. Ding, Y.; Invernale, M. A.; Sotzing, G. A., Conductivity trends of PEDOT-PSS impregnated fabric and the effect of conductivity on electrochromic textile. *ACS Applied Materials & Interfaces* 2010, *2* (6), 1588–1593.
46. Wang, Q.; Wu, Y.; Li, T.; Zhang, D.; Miao, M.; Zhang, A., High performance two-ply carbon nanocomposite yarn supercapacitors enhanced with a platinum filament and in situ polymerized polyaniline nanowires. *Journal of Materials Chemistry A* 2016, *4*, 3828–3834.
47. Enzo Pasquale, S.; Lorussi, F.; Mazzoldi, A.; Rossi, D. D., Strain-sensing fabrics for wearable kinaesthetic-like systems. *IEEE Sensors Journal* 2003, *3* (4), 460–467.
48. Fan, Q.; Zhang, X.; Qin, Z., Preparation of polyaniline/polyurethane fibers and their piezoresistive property. *Journal of Macromolecular Science, Part B* 2012, *51* (4), 736–746.
49. Liu, N.; Fang, G.; Wan, J.; Zhou, H.; Long, H.; Zhao, X., Electrospun PEDOT:PSS-PVA nanofiber based ultrahigh-strain sensors with controllable electrical conductivity. *Journal of Materials Chemistry* 2011, *21* (47), 18962–18966.
50. Lipomi, D. J.; Lee, J. A.; Vosgueritchian, M.; Tee, B. C. K.; Bolander, J. A.; Bao, Z., Electronic properties of transparent conductive films of PEDOT:PSS on stretchable substrates. *Chemistry of Materials* 2012, *24* (2), 373–382.

51. Chen, M.; Duan, S.; Zhang, L.; Wang, Z.; Li, C., Three-dimensional porous stretchable and conductive polymer composites based on graphene networks grown by chemical vapour deposition and PEDOT:PSS coating. *Chemical Communications* 2015, *51* (15), 3169–3172.
52. Daoud, W. A.; Xin, J. H.; Szeto, Y. S., Polyethylenedioxythiophene coatings for humidity, temperature and strain sensing polyamide fibers. *Sensors and Actuators B: Chemical* 2005, *109* (2), 329–333.
53. Seo, J.; Park, S.; Nam, S.; Kim, H.; Kim, Y., Liquid crystal-on-organic field-effect transistor sensory devices for perceptive sensing of ultralow intensity gas flow touch. *Scientific Reports* 2013, *3*, 2452.
54. Kwon, O. S.; Park, S. J.; Lee, J. S., *et al.*, Multidimensional conducting polymer nanotubes for ultrasensitive chemical nerve agent sensing. *Nano Letters* 2012, *12* (6), 2797–2802.
55. Bang, D.; Chang, Y. W.; Park, J., *et al.*, Fabrication of a near-infrared sensor using a polyaniline conducting polymer thin film. *Thin Solid Films* 2012, *520* (22), 6818–6821.
56. Yoshino, K.; Hayashi, S.; Kaneto, K., *et al.*, Radiation effect in conducting polymers. *Molecular Crystals and Liquid Crystals* 1985, *121* (1–4), 255–258.
57. Graham, S. C.; Friend, R. H.; Fung, S.; Moratti, S. C., The effect of X-ray irradiation on poly(*p*-phenylene vinylene) and derivatives. *Synthetic Metals* 1997, *84* (1–3), 903–904.
58. Boroumand, F. A.; Zhu, M.; Dalton, A. B.; Keddie, J. L.; Sellin, P. J.; Gutierrez, J. J., Direct x-ray detection with conjugated polymer devices. *Applied Physics Letters* 2007, *91* (3), 033509.
59. Qian, R.; Pei, Q.; Li, Y., Proton doping of reduced polypyrrole. *Synthetic Metals* 1993, *61* (3), 275–278.
60. Kunzo, P.; Lobotka, P.; Micusik, M.; Kovacova, E., Palladium-free hydrogen sensor based on oxygen-plasma-treated polyaniline thin film. *Sensors and Actuators B: Chemical* 2012, *171–172*, 838–845.
61. Chowdhury, D., Ni-coated polyaniline nanowire as chemical sensing material for cigarette smoke. *The Journal of Physical Chemistry C* 2011, *115* (28), 13554–13559.
62. Li, G.; Josowicz, M.; Janata, J., Electrochemical modulation of the electronic properties in conducting poly(phenylenesulfidephenyleneamine). *Synthetic Metals* 2001, *125* (3), 273–278.
63. Potje-Kamloth, K.; Polk, B. J.; Josowicz, M.; Janata, J., Doping of polyaniline in the solid state with photogenerated triflic acid. *Chemistry of Materials* 2002, *14* (6), 2782–2787.
64. Domanský, K.; Baldwin, D. L.; Grate, J. W., *et al.*, Development and calibration of field-effect transistor-based sensor array for measurement of hydrogen and ammonia gas mixtures in humid air. *Analytical Chemistry* 1998, *70* (3), 473–481.
65. Hanawa, T.; Yoneyama, H., Gas sensitiveness of polypyrrole films doped chemically in the gas phase. *Synthetic Metals* 1989, *30* (3), 341–350.

66. Saxena, V.; Aswal, D. K.; Kaur, M., *et al.*, Enhanced NO_2 selectivity of hybrid poly(3-hexylthiophene): ZnO-nanowire thin films. *Applied Physics Letters* 2007, *90* (4), 043516.
67. Bildirir, H.; Osken, I.; Ozturk, T.; Thomas, A., Reversible doping of a dithienothiophene-based conjugated microporous polymer. *Chemistry – A European Journal* 2015, *21* (26), 9306–9311.
68. Chen, Q.; Hajagos, T.; Pei, Q., Conjugated polymers for radiation detection. *Annual Reports Section "C" (Physical Chemistry)* 2011, *107* (0), 298–318.
69. Davenas, J.; Stevenson, I.; Celette, N., *et al.*, Stability of polymers under ionising radiation: the many faces of radiation interactions with polymers. *Nuclear Instruments and Methods in Physics Research Section B: Beam Interactions with Materials and Atoms* 2002, *191* (1–4), 653–661.
70. Hatchett, D. W.; Josowicz, M.; Janata, J., Acid doping of polyaniline: spectroscopic and electrochemical studies. *The Journal of Physical Chemistry B* 1999, *103* (50), 10992–10998.
71. Brie, M.; Turcu, R.; Neamtu, C.; Pruneanu, S., The effect of initial conductivity and doping anions on gas sensitivity of conducting polypyrrole films to NH_3. *Sensors and Actuators B: Chemical* 1996, *37* (3), 119–122.
72. Huang, J.; Virji, S.; Weiller, B. H.; Kaner, R. B., Polyaniline nanofibers: facile synthesis and chemical sensors. *Journal of the American Chemical Society* 2003, *125* (2), 314–315.
73. Ma, H.-y.; Li, Y.-w.; Yang, S.-x.; Cao, F.; Gong, J.; Deng, Y.-l., Effects of solvent and doping acid on the morphology of polyaniline prepared with the ice-templating method. *The Journal of Physical Chemistry C* 2010, *114* (20), 9264–9269.
74. Bai, S.; Sun, C.; Wan, P., *et al.*, Transparent conducting films of hierarchically nanostructured polyaniline networks on flexible substrates for high-performance gas sensors. *Small* 2015, *11* (3), 306–310.
75. Basak, S. P.; Kanjilal, B.; Sarkar, P.; Turner, A. P. F., Application of electrical impedance spectroscopy and amperometry in polyaniline modified ammonia gas sensor. *Synthetic Metals* 2013, *175*, 127–133.
76. Liu, H.; Kameoka, J.; Czaplewski, D. A.; Craighead, H. G., Polymeric nanowire chemical sensor. *Nano Letters* 2004, *4* (4), 671–675.
77. Wang, J.; Chan, S.; Carlson, R. R., *et al.*, Electrochemically fabricated polyaniline nanoframework electrode junctions that function as resistive sensors. *Nano Letters* 2004, *4* (9), 1693–1697.
78. Tuan, C. V.; Tuan, M. A.; Hieu, N. V.; Trung, T., Electrochemical synthesis of polyaniline nanowires on Pt interdigitated microelectrode for room temperature NH_3 gas sensor application. *Current Applied Physics* 2012, *12* (4), 1011–1016.
79. Reemts, J.; Parisi, J.; Schlettwein, D., Electrochemical growth of gas-sensitive polyaniline thin films across an insulating gap. *Thin Solid Films* 2004, *466* (1–2), 320–325.
80. Shirsat, M. D.; Bangar, M. A.; Deshusses, M. A.; Myung, N. V.; Mulchandani, A., Polyaniline nanowires-gold nanoparticles hybrid network

based chemiresistive hydrogen sulfide sensor. *Applied Physics Letters* 2009, *94* (8), 083502.

81. Carquigny, S.; Sanchez, J.-B.; Berger, F.; Lakard, B.; Lallemand, F., Ammonia gas sensor based on electrosynthesized polypyrrole films. *Talanta* 2009, *78* (1), 199–206.
82. Hernandez, S. C.; Chaudhuri, D.; Chen, W.; Myung, N. V.; Mulchandani, A., Single polypyrrole nanowire ammonia gas sensor. *Electroanalysis* 2007, *19* (19–20), 2125–2130.
83. Hangarter, C. M.; Bangar, M.; Hernandez, S. C., *et al.*, Maskless electrodeposited contact for conducting polymer nanowires. *Applied Physics Letters* 2008, *92* (7), 073104.
84. Lee, J. S.; Jun, J.; Shin, D. H.; Jang, J., Urchin-like polypyrrole nanoparticles for highly sensitive and selective chemiresistive sensor application. *Nanoscale* 2014, *6* (8), 4188–4194.
85. Chen, J.; Yang, J.; Yan, X.; Xue, Q., NH_3 and HCl sensing characteristics of polyaniline nanofibers deposited on commercial ceramic substrates using interfacial polymerization. *Synthetic Metals* 2010, *160* (23–24), 2452–2458.
86. Klug, A.; Denk, M.; Bauer, T., *et al.*, Organic field-effect transistor based sensors with sensitive gate dielectrics used for low-concentration ammonia detection. *Organic Electronics* 2013, *14* (2), 500–504.
87. Liu, C.; Hayashi, K.; Toko, K., Template-free deposition of polyaniline nanostructures on solid substrates with horizontal orientation. *Macromolecules* 2011, *44* (7), 2212–2219.
88. Talwar, V.; Singh, O.; Singh, R. C., ZnO assisted polyaniline nanofibers and its application as ammonia gas sensor. *Sensors and Actuators B: Chemical* 2014, *191*, 276–282.
89. Zou, W.; Quan, B.; Wang, K.; Xia, L.; Yao, J.; Wei, Z., Patterned growth of polyaniline nanowire arrays on a flexible substrate for high-performance gas sensing. *Small* 2011, *7* (23), 3287–3291.
90. Deshpande, N. G.; Gudage, Y. G.; Sharma, R.; Vyas, J. C.; Kim, J. B.; Lee, Y. P., Studies on tin oxide-intercalated polyaniline nanocomposite for ammonia gas sensing applications. *Sensors and Actuators B: Chemical* 2009, *138* (1), 76–84.
91. Tai, H.; Jiang, Y.; Xie, G.; Yu, J., Preparation, characterization and comparative NH_3-sensing characteristic studies of PANI/inorganic oxides nanocomposite thin films. *Journal of Materials Science & Technology* 2010, *26* (7), 605–613.
92. Gong, J.; Li, Y.; Hu, Z.; Zhou, Z.; Deng, Y., Ultrasensitive NH_3 gas sensor from polyaniline nanograin enchased TiO_2 fibers. *The Journal of Physical Chemistry C* 2010, *114* (21), 9970–9974.
93. Penza, M.; Milella, E.; Alba, M. B.; Quirini, A.; Vasanelli, L., Selective NH_3 gas sensor based on Langmuir–Blodgett polypyrrole film. *Sensors and Actuators B: Chemical* 1997, *40* (2–3), 205–209.
94. Bai, H.; Chen, Q.; Li, C.; Lu, C.; Shi, G., Electrosynthesis of polypyrrole/sulfonated polyaniline composite films and their applications for ammonia gas sensing. *Polymer* 2007, *48* (14), 4015–4020.

95. Jeong, J. W.; Lee, Y. D.; Kim, Y. M., *et al.*, The response characteristics of a gas sensor based on poly-3-hexylithiophene thin-film transistors. *Sensors and Actuators B: Chemical* 2010, *146* (1), 40–45.
96. Jung, H. S.; Shimanouchi, T.; Morita, S.; Kuboi, R., Novel fabrication of conductive polymers contained micro-array gas sensitive films using a micromanipulation method. *Electroanalysis* 2003, *15* (18), 1453–1459.
97. Bangar, M. A.; Hangarter, C. M.; Yoo, B., *et al.*, Magnetically assembled multisegmented nanowires and their applications. *Electroanalysis* 2009, *21* (1), 61–67.
98. Paul, S.; Chavan, N. N.; Radhakrishnan, S., Polypyrrole functionalized with ferrocenyl derivative as a rapid carbon monoxide sensor. *Synthetic Metals* 2009, *159* (5–6), 415–418.
99. Watcharaphalakorn, S.; Ruangchuay, L.; Chotpattananont, D.; Sirivat, A.; Schwank, J., Polyaniline/polyimide blends as gas sensors and electrical conductivity response to CO–N_2 mixtures. *Polymer International* 2005, *54* (8), 1126–1133.
100. Liu, C.; Noda, Z.; Sasaki, K.; Hayashi, K., Development of a polyaniline nanofiber-based carbon monoxide sensor for hydrogen fuel cell application. *International Journal of Hydrogen Energy* 2012, *37* (18), 13529–13535.
101. Ogura, K.; Shiigi, H.; Oho, T.; Tonosaki, T., CO_2 sensor with polymer composites operating at ordinary temperature. *Journal of the Electrochemical Society* 2000, *147* (11), 4351–4355.
102. Neethirajan, S.; Freund, M. S.; Jayas, D. S.; Shafai, C.; Thomson, D. J.; White, N. D. G., Development of carbon dioxide (CO_2) sensor for grain quality monitoring. *Biosystems Engineering* 2010, *106* (4), 395–404.
103. Oho, T.; Tonosaki, T.; Isomura, K.; Ogura, K., A CO_2 sensor operating under high humidity. *Journal of Electroanalytical Chemistry* 2002, *522* (2), 173–178.
104. Radhakrishnan, S.; Paul, S., Conducting polypyrrole modified with ferrocene for applications in carbon monoxide sensors. *Sensors and Actuators B: Chemical* 2007, *125* (1), 60–65.
105. Waghuley, S. A.; Yenorkar, S. M.; Yawale, S. S.; Yawale, S. P., Application of chemically synthesized conducting polymer-polypyrrole as a carbon dioxide gas sensor. *Sensors and Actuators B: Chemical* 2008, *128* (2), 366–373.
106. Zeng, F.-W.; Liu, X.-X.; Diamond, D.; Lau, K. T., Humidity sensors based on polyaniline nanofibres. *Sensors and Actuators B: Chemical* 2010, *143* (2), 530–534.
107. Lin, Q.; Li, Y.; Yang, M., Investigations on the sensing mechanism of humidity sensors based on electrospun polymer nanofibers. *Sensors and Actuators B: Chemical* 2012, *171–172*, 309–314.
108. Cao, Y.; Kovalev, A. E.; Xiao, R.; Kim, J.; Mayer, T. S.; Mallouk, T. E., Electrical transport and chemical sensing properties of individual conducting polymer nanowires. *Nano Letters* 2008, *8* (12), 4653–4658.
109. Vieira, N. C. S.; Fernandes, E. G. R.; Faceto, A. D.; Zucolotto, V.; Guimarães, F. E. G., Nanostructured polyaniline thin films as pH sensing

membranes in FET-based devices. *Sensors and Actuators B: Chemical* 2011, *160* (1), 312–317.

110. Karyakin, A. A.; Bobrova, O. A.; Lukachova, L. V.; Karyakina, E. E., Potentiometric biosensors based on polyaniline semiconductor films. *Sensors and Actuators, B: Chemical* 1996, *33* (1–3), 34–38.
111. Ameen, S.; Akhtar, M. S.; Shin, H. S., Hydrazine chemical sensing by modified electrode based on in situ electrochemically synthesized polyaniline/graphene composite thin film. *Sensors and Actuators B: Chemical* 2012, *173*, 177–183.
112. Virji, S.; Kaner, R. B.; Weiller, B. H., Hydrazine detection by polyaniline using fluorinated alcohol additives. *Chemistry of Materials* 2005, *17* (5), 1256–1260.
113. Ellis, D. L.; Zakin, M. R.; Bernstein, L. S.; Rubner, M. F., Conductive polymer films as ultrasensitive chemical sensors for hydrazine and monomethylhydrazine vapor. *Analytical Chemistry* 1996, *68* (5), 817–822.
114. Nicolas-Debarnot, D.; Poncin-Epaillard, F., Polyaniline as a new sensitive layer for gas sensors. *Analytica Chimica Acta* 2003, *475* (1–2), 1–15.
115. Vilkman, M.; Lehtinen, K.; Mäkelä, T.; Rannou, P.; Ikkala, O., Poly(aniline) doped with 5-formyl-2-furansulfonic acid: a humidity memory. *Organic Electronics* 2010, *11* (3), 472–478.
116. Shabnam, V.; Richard, B. K.; Bruce, H. W., Detection of toxic chemicals for homeland security using polyaniline nanofibers. In: John G. Reynolds, Glenn E. Lawson, Carolyn Jean Koester, eds. *Antiterrorism and Homeland Defense*, American Chemical Society: 2007; Vol. 980, *pp 101–115*.
117. Agbor, N. E.; Petty, M. C.; Monkman, A. P.; Harris, M., Proceedings of the international conference on science and technology of synthetic metals Langmuir–Blodgett films of polyaniline. *Synthetic Metals* 1993, *57* (1), 3789–3794.
118. Menegazzo, N.; Boyne, D.; Bui, H.; Beebe, T. P.; Booksh, K. S., DC magnetron sputtered polyaniline–HCl thin films for chemical sensing applications. *Analytical Chemistry* 2012, *84* (13), 5770–5777.
119. Liu, C.; Hayashi, K.; Toko, K., Electrochemical deposition of nanostructured polyaniline on an insulating substrate. *Electrochemistry Communications* 2010, *12* (1), 36–39.
120. Kebiche, H.; Debarnot, D.; Merzouki, A.; Poncin-Epaillard, F.; Haddaoui, N., Relationship between ammonia sensing properties of polyaniline nanostructures and their deposition and synthesis methods. *Analytica Chimica Acta* 2012, *737*, 64–71.
121. Ma, Y.; Zhang, J.; Zhang, G.; He, H., Polyaniline nanowires on Si surfaces fabricated with DNA templates. *Journal of the American Chemical Society* 2004, *126* (22), 7097–7101.
122. Chen, X.; Yuan, C. A.; Wong, C. K. Y.; Ye, H.; Leung, S. Y. Y.; Zhang, G., Molecular modeling of protonic acid doping of emeraldine base polyaniline for chemical sensors. *Sensors and Actuators B: Chemical* 2012, *174*, 210–216.

123. Ullah, H.; Shah, A.-u.-H. A.; Bilal, S.; Ayub, K., DFT study of polyaniline NH_3, CO_2, and CO gas sensors: comparison with recent experimental data. *The Journal of Physical Chemistry C* 2013, *117* (45), 23701–23711.
124. Karyakin, A. A.; Vuki, M.; Lukachova, L. V., *et al.*, Processible polyaniline as an advanced potentiometric pH transducer. Application to biosensors. *Analytical Chemistry* 1999, *71* (13), 2534–2540.
125. Sambasevam, K. P.; Mohamad, S.; Phang, S.-W., Enhancement of polyaniline properties by different polymerization temperatures in hydrazine detection. *Journal of Applied Polymer Science* 2015, *132* (13), 41746.
126. Pei, Q.; Qian, R., Protonation and deprotonation of polypyrrole chain in aqueous solutions. *Synthetic Metals* 1991, *45* (1), 35–48.
127. Kemp, N. T.; Kaiser, A. B.; Trodahl, H. J., *et al.*, Effect of ammonia on the temperature-dependent conductivity and thermopower of polypyrrole. *Journal of Polymer Science Part B: Polymer Physics* 2006, *44* (9), 1331–1338.
128. Zotti, G.; Schiavon, G.; Zecchin, S.; D'Aprano, G., The reversible hydroxide doping of polypyrrole. Delay of polaron and mobile bipolaron injection in polypyrrole by H bonding of hydroxide anions. *Synthetic Metals* 1996, *80* (1), 35–39.
129. Ullah, H.; Ayub, K.; Ullah, Z., *et al.*, Theoretical insight of polypyrrole ammonia gas sensor. *Synthetic Metals* 2013, *172*, 14–20.
130. Geng, L.; Wang, S.; Zhao, Y., *et al.*, Study of the primary sensitivity of polypyrrole/r-Fe_2O_3 to toxic gases. *Materials Chemistry and Physics* 2006, *99* (1), 15–19.
131. Kriván, E.; Visy, C.; Dobay, R.; Harsányi, G.; Berkesi, O., Irregular response of the polypyrrole films to H_2S. *Electroanalysis* 2000, *12* (15), 1195–1200.
132. Li, M.; Li, H.; Zhong, W.; Zhao, Q.; Wang, D., Stretchable conductive polypyrrole/polyurethane (PPy/PU) strain sensor with netlike microcracks for human breath detection. *ACS Applied Materials & Interfaces* 2014, *6* (2), 1313–1319.
133. Hwang, B.-U.; Lee, J.-H.; Trung, T. Q., *et al.*, Transparent stretchable self-powered patchable sensor platform with ultrasensitive recognition of human activities. *ACS Nano* 2015, *9* (9), 8801–8810.
134. Glidle, A.; Pearson, P. E.; Smith, E. L., *et al.*, Determining compositional profiles within conducting polymer films following reaction with vapor phase reagents. *The Journal of Physical Chemistry B* 2007, *111* (16), 4043–4053.
135. Svetlicic, V.; Schmidt, A. J.; Miller, L. L., Conductometric sensors based on the hypersensitive response of plasticized polyaniline films to organic vapors. *Chemistry of Materials* 1998, *10* (11), 3305–3307.
136. Ruangchuay, L.; Sirivat, A.; Schwank, J., Electrical conductivity response of polypyrrole to acetone vapor: effect of dopant anions and interaction mechanisms. *Synthetic Metals* 2004, *140* (1), 15–21.
137. Athawale, A. A.; Kulkarni, M. V., Polyaniline and its substituted derivatives as sensor for aliphatic alcohols. *Sensors and Actuators B: Chemical* 2000, *67* (1–2), 173–177.

138. Steffens, C.; Corazza, M. L.; Franceschi, E., *et al.*, Development of gas sensors coatings by polyaniline using pressurized fluid. *Sensors and Actuators B: Chemical* 2012, *171–172*, 627–633.
139. Weerakoon, K. A.; Shu, J. H.; Park, M.-K.; Auad, M. A.; Chin, B. A., Detecting insect infestation using a polymer based sensor array. *Sensors and Actuators B: Chemical* 2012, *174*, 506–512.
140. Torsi, L.; Tafuri, A.; Cioffi, N., *et al.*, Regioregular polythiophene field-effect transistors employed as chemical sensors. *Sensors and Actuators B: Chemical* 2003, *93* (1–3), 257–262.
141. Zhang, X.; Qin, Z.; Liu, X., *et al.*, Flexible sensing fibers based on polyaniline-coated polyurethane for chloroform vapor detection. *Journal of Materials Chemistry A* 2013, *1* (35), 10327–10333.
142. Anitha, G.; Subramanian, E., Recognition and exposition of intermolecular interaction between CH_2Cl_2 and $CHCl_3$ by conducting polyaniline materials. *Sensors and Actuators B: Chemical* 2005, *107* (2), 605–615.
143. Anitha, G.; Subramanian, E., Dopant induced specificity in sensor behaviour of conducting polyaniline materials with organic solvents. *Sensors and Actuators B: Chemical* 2003, *92* (1–2), 49–59.
144. Li, B.; Santhanam, S.; Schultz, L., *et al.*, Inkjet printed chemical sensor array based on polythiophene conductive polymers. *Sensors and Actuators B: Chemical* 2007, *123* (2), 651–660.
145. Wang, X.; Hou, S.; Goktas, H., *et al.*, Small-area, resistive volatile organic compound (VOC) sensors using metal–polymer hybrid film based on oxidative chemical vapor deposition (oCVD). *ACS Applied Materials & Interfaces* 2015, *7* (30), 16213–16222.
146. Qi, J.; Xu, X.; Liu, X.; Lau, K. T., Fabrication of textile based conductometric polyaniline gas sensor. *Sensors and Actuators B: Chemical* 2014, *202*, 732–740.
147. Barisci, J. N.; Wallace, G. G.; Andrews, M. K.; Partridge, A. C.; Harris, P. D., Conducting polymer sensors for monitoring aromatic hydrocarbons using an electronic nose. *Sensors and Actuators B: Chemical* 2002, *84* (2–3), 252–257.
148. Lin, C. W.; Liu, Y. L.; Thangamuthu, R., Investigation of the relationship between surface thermodynamics of the chemically synthesized polypyrrole films and their gas-sensing responses to BTEX compounds. *Sensors and Actuators B: Chemical* 2003, *94* (1), 36–45.
149. Jun, H.-K.; Hoh, Y.-S.; Lee, B.-S., *et al.*, Electrical properties of polypyrrole gas sensors fabricated under various pretreatment conditions. *Sensors and Actuators B: Chemical* 2003, *96* (3), 576–581.
150. de Souza, J. E. G.; dos Santos, F. L.; Neto, B. B.; dos Santos, C. G.; dos Santos, M. V. B.; de Melo, C. P., Free-grown polypyrrole thin films as aroma sensors. *Sensors and Actuators B: Chemical* 2003, *88* (3), 246–259.
151. Sakurai, Y.; Jung, H.-S.; Shimanouchi, T., *et al.*, Novel array-type gas sensors using conducting polymers, and their performance for gas identification. *Sensors and Actuators B: Chemical* 2002, *83* (1–3), 270–275.

152. Yu, J.-B.; Byun, H.-G.; So, M.-S.; Huh, J.-S., Analysis of diabetic patient's breath with conducting polymer sensor array. *Sensors and Actuators B: Chemical* 2005, *108* (1–2), 305–308.
153. Torsi, L.; Tanese, M. C.; Cioffi, N., *et al.*, , Side-chain role in chemically sensing conducting polymer field-effect transistors. *The Journal of Physical Chemistry B* 2003, *107* (31), 7589–7594.
154. Pramanik, S.; Das, G.; Karak, N., Facile preparation of polyaniline nanofibers modified bentonite nanohybrid for gas sensor application. *RSC Advances* 2013, *3* (14), 4574–4581.
155. English, J. T.; Deore, B. A.; Freund, M. S., Biogenic amine vapour detection using poly(anilineboronic acid) films. *Sensors and Actuators B: Chemical* 2006, *115* (2), 666–671.
156. Li, G.; Martinez, C.; Semancik, S., Controlled electrophoretic patterning of polyaniline from a colloidal suspension. *Journal of the American Chemical Society* 2005, *127* (13), 4903–4909.
157. Liu, C.; Hayashi, K.; Toko, K., Au nanoparticles decorated polyaniline nanofiber sensor for detecting volatile sulfur compounds in expired breath. *Sensors and Actuators B: Chemical* 2012, *161* (1), 504–509.
158. Ayad, M. M.; Torad, N. L., Quartz crystal microbalance sensor for detection of aliphatic amines vapours. *Sensors and Actuators B: Chemical* 2010, *147* (2), 481–487.
159. Khot, L. R.; Panigrahi, S.; Lin, D., Development and evaluation of piezoelectric-polymer thin film sensors for low concentration detection of volatile organic compounds related to food safety applications. *Sensors and Actuators B: Chemical* 2011, *153* (1), 1–10.
160. Gannot, Y.; Hertzog-Ronen, C.; Tessler, N.; Eichen, Y., Sensing of alkylating agents using organic field-effect transistors. *Advanced Functional Materials* 2010, *20* (1), 105–110.
161. Do, J.-S.; Wang, S.-H., On the sensitivity of conductimetric acetone gas sensor based on polypyrrole and polyaniline conducting polymers. *Sensors and Actuators B: Chemical* 2013, *185*, 39–46.
162. Wang, Y.; Ding, P.; Hu, R., *et al.*, A dibutyl phthalate sensor based on a nanofiber polyaniline coated quartz crystal monitor. *Sensors* 2013, *13* (3), 3765.
163. Wu, R.-J.; Huang, Y.-C.; Chavali, M.; Lin, T. H.; Hung, S.-L.; Luk, H.-N., New sensing technology for detection of the common inhalational anesthetic agent sevoflurane using conducting polypyrrole films. *Sensors and Actuators B: Chemical* 2007, *126* (2), 387–393.
164. Virji, S.; Huang, J.; Kaner, R. B.; Weiller, B. H., Polyaniline nanofiber gas sensors: examination of response mechanisms. *Nano Letters* 2004, *4* (3), 491–496.
165. Tan, C. K.; Blackwood, D. J., Interactions between polyaniline and methanol vapour. *Sensors and Actuators B: Chemical* 2000, *71* (3), 184–191.
166. Liu, S.-S.; Bian, L.-J.; Luan, F.; Sun, M.-T.; Liu, X.-X., Theoretical study on polyaniline gas sensors: examinations of response mechanism for alcohol. *Synthetic Metals* 2012, *162* (9–10), 862–867.

167. Yang, L.-Y.; Liau, W.-B., Environmental responses of nanostructured polyaniline films based on polystyrene–polyaniline core–shell particles. *Materials Chemistry and Physics* 2009, *115* (1), 28–32.
168. Li, W.; Hoa, N. D.; Cho, Y.; Kim, D.; Kim, J.-S., Nanofibers of conducting polyaniline for aromatic organic compound sensor. *Sensors and Actuators B: Chemical* 2009, *143* (1), 132–138.
169. Freund, M. S.; Lewis, N. S., A chemically diverse conducting polymer-based "electronic nose". *Proceedings of the National Academy of Sciences of the United States of America* 1995, *92* (7), 2652–2656.
170. Li, B.; Lambeth, D. N., Chemical sensing using nanostructured polythiophene transistors. *Nano Letters* 2008, *8* (11), 3563–3567.
171. Chang, J. B.; Liu, V.; Subramanian, V., *et al.*, Printable polythiophene gas sensor array for low-cost electronic noses. *Journal of Applied Physics* 2006, *100* (1), 014506.
172. Da-Jeng, Y.; Yong-Ruei, Y. *Bio-chemical sensor using polyaniline nanofibers for sensing amino-group type of gas*. 2005 IEEE International Conference on Robotics and Biomimetics (ROBIO), 2005; IEEE, 1708843, pp 763–767.
173. Lee, J. S.; Kim, S. G.; Jun, J.; Shin, D. H.; Jang, J., Aptamer-functionalized multidimensional conducting-polymer nanoparticles for an ultrasensitive and selective field-effect-transistor endocrine-disruptor sensors. *Advanced Functional Materials* 2014, *24* (39), 6145–6153.
174. Yoon, B.; Shin, H.; Kang, E.-M., *et al.*, Inkjet-compatible single-component polydiacetylene precursors for thermochromic paper sensors. *ACS Applied Materials & Interfaces* 2013, *5* (11), 4527–4535.
175. Ahn, D. J.; Chae, E.-H.; Lee, G. S., *et al.*, Colorimetric reversibility of polydiacetylene supramolecules having enhanced hydrogen-bonding under thermal and pH stimuli. *Journal of the American Chemical Society* 2003, *125* (30), 8976–8977.
176. Esser, B.; Swager, T. M., Detection of ethylene gas by fluorescence turn-on of a conjugated polymer. *Angewandte Chemie International Edition* 2010, *49* (47), 8872–8875.
177. Yoon, B.; Park, I. S.; Shin, H.; Park, H. J.; Lee, C. W.; Kim, J.-M., A litmus-type colorimetric and fluorometric volatile organic compound sensor based on inkjet-printed polydiacetylenes on paper substrates. *Macromolecular Rapid Communications* 2013, *34* (9), 731–735.
178. Yoon, J.; Chae, S. K.; Kim, J.-M., Colorimetric sensors for volatile organic compounds (VOCs) based on conjugated polymer-embedded electrospun fibers. *Journal of the American Chemical Society* 2007, *129* (11), 3038–3039.
179. Park, D.-H.; Hong, J.; Park, I. S.; Lee, C. W.; Kim, J.-M., A colorimetric hydrocarbon sensor employing a swelling-induced mechanochromic polydiacetylene. *Advanced Functional Materials* 2014, *24* (33), 5186–5193.
180. Xu, Q.; Lee, S.; Cho, Y.; Kim, M. H.; Bouffard, J.; Yoon, J., Polydiacetylene-based colorimetric and fluorescent chemosensor for the detection of carbon dioxide. *Journal of the American Chemical Society* 2013, *135* (47), 17751–17754.

181. Yao, Z.; Hu, X.; Huang, B., *et al.*, Halochromism of a polythiophene derivative induced by conformational changes and its sensing application of carbon dioxide. *ACS Applied Materials & Interfaces* 2013, *5* (12), 5783–5787.
182. Nicho, M. E.; Trejo, M.; García-Valenzuela, A.; Saniger, J. M.; Palacios, J.; Hu, H., Polyaniline composite coatings interrogated by a nulling optical-transmittance bridge for sensing low concentrations of ammonia gas. *Sensors and Actuators B: Chemical* 2001, *76* (1–3), 18–24.
183. Gu, Y.; Huang, J., Colorimetric detection of gaseous ammonia by polyaniline nanocoating of natural cellulose substances. *Colloids and Surfaces A: Physicochemical and Engineering Aspects* 2013, *433*, 166–172.
184. Menegazzo, N.; Herbert, B.; Banerji, S.; Booksh, K. S., Discourse on the utilization of polyaniline coatings for surface plasmon resonance sensing of ammonia vapor. *Talanta* 2011, *85* (3), 1369–1375.
185. Yang, L.-Y.; Liau, W.-B., Optical responses of polyaniline inverse opals to chemicals. *Synthetic Metals* 2010, *160* (17–18), 1809–1814.
186. Jeon, H.; Lee, J.; Kim, M. H.; Yoon, J., Polydiacetylene-based electrospun fibers for detection of HCl gas. *Macromolecular Rapid Communications* 2012, *33* (11), 972–976.
187. Yoon, B.; Shin, H.; Yarimaga, O., *et al.*, An inkjet-printable microemulsion system for colorimetric polydiacetylene supramolecules on paper substrates. *Journal of Materials Chemistry* 2012, *22* (17), 8680–8686.
188. Ewbank, P. C.; Loewe, R. S.; Zhai, L.; Reddinger, J.; Sauvé, G.; McCullough, R. D., Regioregular poly(thiophene-3-alkanoic acid)s: water soluble conducting polymers suitable for chromatic chemosensing in solution and solid state. *Tetrahedron* 2004, *60* (49), 11269–11275.
189. Parambath Kootery, K.; Jiang, H.; Kolusheva, S., *et al.*, Poly(methyl methacrylate)-supported polydiacetylene films: unique chromatic transitions and molecular sensing. *ACS Applied Materials & Interfaces* 2014, *6* (11), 8613–8620.
190. Lee, S.; Lee, K. M.; Lee, M.; Yoon, J., Polydiacetylenes bearing boronic acid groups as colorimetric and fluorescence sensors for cationic surfactants. *ACS Applied Materials & Interfaces* 2013, *5* (11), 4521–4526.
191. Chen, Z.; Xue, C.; Shi, W., *et al.*, Selective and sensitive fluorescent sensors for metal ions based on manipulation of side-chain compositions of poly(*p*-phenyleneethynylene)s. *Analytical Chemistry* 2004, *76* (21), 6513–6518.
192. Crawford, K. B.; Goldfinger, M. B.; Swager, T. M., Na^+ specific emission changes in an ionophoric conjugated polymer. *Journal of the American Chemical Society* 1998, *120* (21), 5187–5192.
193. Bao, Y.; Wang, H.; Li, Q., *et al.*, 2,2′-Biimidazole-based conjugated polymers as a novel fluorescent sensing platform for pyrophosphate anion. *Macromolecules* 2012, *45* (8), 3394–3401.
194. Tong, H.; Wang, L.; Jing, X.; Wang, F., "Turn-On" conjugated polymer fluorescent chemosensor for fluoride ion. *Macromolecules* 2003, *36* (8), 2584–2586.

195. Dong, L.; Deng, C.; He, C., *et al.*, Highly sensitive vapor detection of amines with fluorescent conjugated polymer: a novel lasing turn-on sensory mechanism. *Sensors and Actuators B: Chemical* 2013, *180*, 28–34.
196. Lee, J.; Jeong Jeong, E.; Kim, J., Selective and sensitive detection of melamine by intra/inter liposomal interaction of polydiacetylene liposomes. *Chemical Communications* 2011, *47* (1), 358–360.
197. Ren, X.; Xu, Q.-H., Highly sensitive and selective detection of mercury ions by using oligonucleotides, DNA intercalators, and conjugated polymers. *Langmuir* 2009, *25* (1), 29–31.
198. Bansal, A. K.; Hou, S.; Kulyk, O.; Bowman, E. M.; Samuel, I. D. W., Wearable organic optoelectronic sensors for medicine. *Advanced Materials* 2015, *27* (46), 7638–7644.
199. Haizheng, Z.; YongSheng, Z.; Yongfang, L.; Qibing, P., Photoluminescence quenching of conjugated polymer nanocomposites for gamma ray detection. *Nanotechnology* 2008, *19* (50), 505503.
200. Zhao, Y. S.; Zhong, H.; Pei, Q., Fluorescence resonance energy transfer in conjugated polymer composites for radiation detection. *Physical Chemistry Chemical Physics* 2008, *10* (14), 1848–1851.
201. Laranjeira, J. M. G.; Khoury, H. J.; de Azevedo, W. M.; de Vasconcelos, E. A.; da Silva Jr, E. F., Polyaniline nanofilms as a monitoring label and dosimetric device for gamma radiation. *Materials Characterization* 2003, *50* (2–3), 127–130.
202. Bazani, D. L. M.; Lima, J. P. H.; Andrade, A. M. d., MEH-PPV thin films for radiation sensor applications. *IEEE Sensors Journal* 2009, *9* (7), 748–751.
203. Silva, E. A. B.; Borin, J. F.; Nicolucci, P.; Graeff, C. F. O.; Netto, T. G.; Bianchi, R. F., Low dose ionizing radiation detection using conjugated polymers. *Applied Physics Letters* 2005, *86* (13), 131902.
204. Liu, Y.; Mills, R. C.; Boncella, J. M.; Schanze, K. S., Fluorescent polyacetylene thin film sensor for nitroaromatics. *Langmuir* 2001, *17* (24), 7452–7455.
205. Li, J.; Kendig, C. E.; Nesterov, E. E., Chemosensory performance of molecularly imprinted fluorescent conjugated polymer materials. *Journal of the American Chemical Society* 2007, *129* (51), 15911–15918.
206. Chang, C.-P.; Chao, C.-Y.; Huang, J. H., *et al.*, Fluorescent conjugated polymer films as TNT chemosensors. *Synthetic Metals* 2004, *144* (3), 297–301.
207. Narayanan, A.; Varnavski, O. P.; Swager, T. M.; Goodson, T., Multiphoton fluorescence quenching of conjugated polymers for TNT detection. *The Journal of Physical Chemistry C* 2008, *112* (4), 881–884.
208. Rose, A.; Zhu, Z.; Madigan, C. F.; Swager, T. M.; Bulovic, V., Sensitivity gains in chemosensing by lasing action in organic polymers. *Nature* 2005, *434* (7035), 876–879.
209. He, G.; Yan, N.; Yang, J., *et al.*, Pyrene-containing conjugated polymer-based fluorescent films for highly sensitive and selective sensing of TNT in aqueous medium. *Macromolecules* 2011, *44* (12), 4759–4766.

210. Xu, B.; Wu, X.; Li, H.; Tong, H.; Wang, L., Selective detection of TNT and picric acid by conjugated polymer film sensors with donor–acceptor architecture. *Macromolecules* 2011, *44* (13), 5089–5092.
211. Malik, A. H.; Hussain, S.; Kalita, A.; Iyer, P. K., Conjugated polymer nanoparticles for the amplified detection of nitro-explosive picric acid on multiple platforms. *ACS Applied Materials & Interfaces* 2015, *7* (48), 26968–26976.
212. Zhao, Y. S.; Yu, Z.; Douraghy, A.; Chatziioannou, A. F.; Mo, Y.; Pei, Q., A facile route to bulk high-Zpolymer composites for gamma ray scintillation. *Chemical Communications* 2008, (*45*), 6008–6010.
213. Wang, B.; Wasielewski, M. R., Design and synthesis of metal ion-recognition-induced conjugated polymers: an approach to metal ion sensory materials. *Journal of the American Chemical Society* 1997, *119* (1), 12–21.
214. Yoon, B.; Lee, S.; Kim, J.-M., Recent conceptual and technological advances in polydiacetylene-based supramolecular chemosensors. *Chemical Society Reviews* 2009, *38* (7), 1958–1968.
215. Zhou, G.; Cheng, Y.; Wang, L.; Jing, X.; Wang, F., Novel polyphenylenes containing phenol-substituted oxadiazole moieties as fluorescent chemosensors for fluoride ion. *Macromolecules* 2005, *38* (6), 2148–2153.
216. Kim, T.-H.; Swager, T. M., A fluorescent self-amplifying wavelength-responsive sensory polymer for fluoride ions. *Angewandte Chemie* 2003, *115* (39), 4951–4954.
217. Harrison, B. S.; Ramey, M. B.; Reynolds, J. R.; Schanze, K. S., Amplified fluorescence quenching in a poly(*p*-phenylene)-based cationic polyelectrolyte. *Journal of the American Chemical Society* 2000, *122* (35), 8561–8562.
218. Chen, S.; Zhang, Q.; Zhang, J.; Gu, J.; Zhang, L., Synthesis of two conjugated polymers as TNT chemosensor materials. *Sensors and Actuators B: Chemical* 2010, *149* (1), 155–160.
219. Campbell, I. H.; Crone, B. K., Quantum-dot/organic semiconductor composites for radiation detection. *Advanced Materials* 2006, *18* (1), 77–79.
220. Yang, J.-S.; Swager, T. M., Fluorescent porous polymer films as TNT chemosensors: electronic and structural effects. *Journal of the American Chemical Society* 1998, *120* (46), 11864–11873.
221. Albert, K. J.; Myrick, M. L.; Brown, S. B.; James, D. L.; Milanovich, F. P.; Walt, D. R., Field-deployable sniffer for 2,4-dinitrotoluene detection. *Environmental Science & Technology* 2001, *35* (15), 3193–3200.
222. Ji, X.; Yao, Y.; Li, J.; Yan, X.; Huang, F., A supramolecular cross-linked conjugated polymer network for multiple fluorescent sensing. *Journal of the American Chemical Society* 2013, *135* (1), 74–77.
223. Zhang, Y.; Murphy, C. B.; Jones, W. E., Poly[*p*-(phenyleneethynylene)-alt-(thienyleneethynylene)] polymers with oligopyridine pendant groups: highly sensitive chemosensors for transition metal ions. *Macromolecules* 2002, *35* (3), 630–636.
224. Yoon, J.; Jung, Y.-S.; Kim, J.-M., A combinatorial approach for colorimetric differentiation of organic solvents based on conjugated polymer-embedded electrospun fibers. *Advanced Functional Materials* 2009, *19* (2), 209–214.

225. Lee, J.; Pyo, M.; Lee, S.-h., *et al.*, Hydrochromic conjugated polymers for human sweat pore mapping. *Nature Communications* 2014, 5.
226. Seo, D.; Kim, J., Effect of the molecular size of analytes on polydiacetylene chromism. *Advanced Functional Materials* 2010, *20* (9), 1397–1403.
227. Lee, S.; Cheng, H.; Chi, M., *et al.*, Sensing and antibacterial activity of imidazolium-based conjugated polydiacetylenes. *Biosensors and Bioelectronics* 2016, *77*, 1016–1019.
228. Bai, F.; Sun, Z.; Lu, P.; Fan, H., Smart polydiacetylene nanowire paper with tunable colorimetric response. *Journal of Materials Chemistry* 2012, *22* (30), 14839–14842.
229. Liu, C.; Gao, G.; Zhang, Y.; Wang, L.; Wang, J.; Song, Y., The naked-eye detection of NH_3–HCl by polyaniline-infiltrated TiO_2 inverse opal photonic crystals. *Macromolecular Rapid Communications* 2012, *33* (5), 380–385.
230. Sadrolhosseini, A. R.; Moksin, M. M.; Yunus, W. M. M.; Talib, Z. A.; Abdi, M. M., Surface plasmon resonance detection of copper corrosion in biodiesel using polypyrrole-chitosan layer sensor. *Optical Review* 2011, *18* (4), 331–337.
231. Yu, J. C. C.; Lai, E. P. C.; Sadeghi, S., Surface plasmon resonance sensor for Hg(II) detection by binding interactions with polypyrrole and 2-mercaptobenzothiazole. *Sensors and Actuators B: Chemical* 2004, *101* (1–2), 236–241.
232. Leong, K.; Zin, M. T.; Ma, H.; Sarikaya, M.; Huang, F.; Jen, A. K. Y., Surface plasmon enhanced fluorescence of cationic conjugated polymer on periodic nanoarrays. *ACS Applied Materials & Interfaces* 2010, *2* (11), 3153–3159.

Chapter 2

Electrolyte-gated organic transistors for biosensing applications

Fabrizio Torricelli[1], *Kyriaki Manoli*[2], *Eleonora Macchia*[2], *Luisa Torsi*[2] *and Maria Magliulo*[2]

Abstract

The interfacing of biomaterials to electronic devices is one of the most challenging research fields that has relevance to both fundamental studies and the development of highly performing biosensors. Important aspects connected to the field of biosensors based on electrolyte-gated organic transistors are discussed. The main features of biomolecules used as recognition elements along with the strategies reported so far for their deposition on the biosensors transducer surface are presented. An introduction of the involving surface interactions such as covalent binding, physical adsorption, self-assembly and bio-affinity binding strategy is given. Besides, the most relevant surface materials for electronic biosensors development are analysed. In the last part, different electronic biosensors based on electrolyte-gated organic transistors are presented. Particular attention is paid to the biosensors operation mechanism and to the analytical figure of merits. Specific applications such as DNA and proteins detection are also discussed.

2.1 Electrolyte-gated transistor: theory and operation

Organic transistors are expected to open up a range of new application opportunities including active-matrix displays, smart electronic and photonic surfaces, imperceptible and wearable electronics, robotic skin and sensors, biological and medical devices [1–7]. In particular, organic field-effect transistors (OFETs) have already been proven to be suitable candidates as transducers for various sensor applications [7–11]. Owing to their intrinsic amplification, OFET-based sensors typically show a high sensitivity, viz. a small variation of the (input) physical, chemical or biological quantity results in a large variation of the (output) electronic signal. In conjunction

[1]Department of Information Engineering, University of Brescia, Brescia 25123, Italy
[2]Department of Chemistry, University of Bari Aldo Moro, Bari 70126, Italy

with the outstanding features of organic electronic devices, such as economic production, integration on flexible substrates and biocompatibility, low-cost disposable sensor assemblies are not a future vision anymore. Especially, the emerging fields of medical diagnostics and environmental monitoring require sensor platforms for in situ sensing of ions and biological substances in appropriate aqueous media. There is an increasing interest in low cost, time-saving yet reliable, point-of-care assays. Direct electronic, label-free transduction of biorecognition events represents a compelling alternative offering miniaturization, easy data handling and processing. Low costs and versatility can be provided if organic electronic devices such as OFETs are used as transducers [12,13]. Interestingly, OFET-based sensors – proven to be extremely sensitive [13–15] – can be endowed with selectivity by using biomolecules as receptors for recognizing a target molecule. Because an aqueous electrolyte solution is the natural environment for biological receptors, the current challenge is to develop OFET biosensors that, in addition to integrating biomolecules, can also operate in an aqueous electrolyte.

At this aim, electrolyte-gated organic transistors (EGOTs) are the transducers of choice. Originally developed in the 1980s, EGOTs are re-emerging as a promising platform for biosensing [16,17]. Figure 2.1(a) shows a schematic cross-section of an EGOT. It comprises an organic semiconductor (OS) directly contacted with two metal electrodes named source and drain. A portion of the semiconductor (typically the top surface) is in ionic contact with a gate electrode via an electrolyte. The ionically conducting and electronically insulating electrolyte serves as a gate insulator. The OS can be made of oligomers or polymers that are deposited as films (a few tens of nanometers thick at most) by solution casting, spin coating or sublimation. Other deposition methods can be also used, including printing techniques. Solution processed OSs are generally polycrystalline films composed of contiguous grains with linear dimensions of a few hundred nanometers. Source and drain contacts to the OS can be easily defined either by thermal evaporation through a shadow-mask or again by printing. Gold is the most convenient contact metal because its work function is the closest to that of most p-type organic materials. The region of the organic semiconducting film delimited by the source and drain electrodes defines the transistor channel, whose geometry is characterized by the inter-electrode distance (channel length, L) and the electrode width (channel width, W).

For hole-transport Oss, the transistor (p-type) is operated by negatively biasing the drain (D) and the gate (G) contacts applying the V_D and V_G potentials with respect to the grounded source (S). This is called common source configuration. The current flowing between S and D (I_D) is modulated by V_G. Application of a voltage to the gate results in polarization of the electrolyte and an increase in the conductivity of the semiconductor channel. There are two basic operation mechanisms for EGOTs. In the case where the OS is impermeable to ions of the electrolyte (Figure 2.1(b)), application of a voltage to the gate electrode causes migration and accumulation of ions at the gate–electrolyte and semiconductor–electrolyte interfaces. The ions at these interfaces screen the charges in the gate metal and cause accumulation of charge carriers in the semiconductor channel. The result is that "electrical double layers" (EDLs) are formed at each interface.

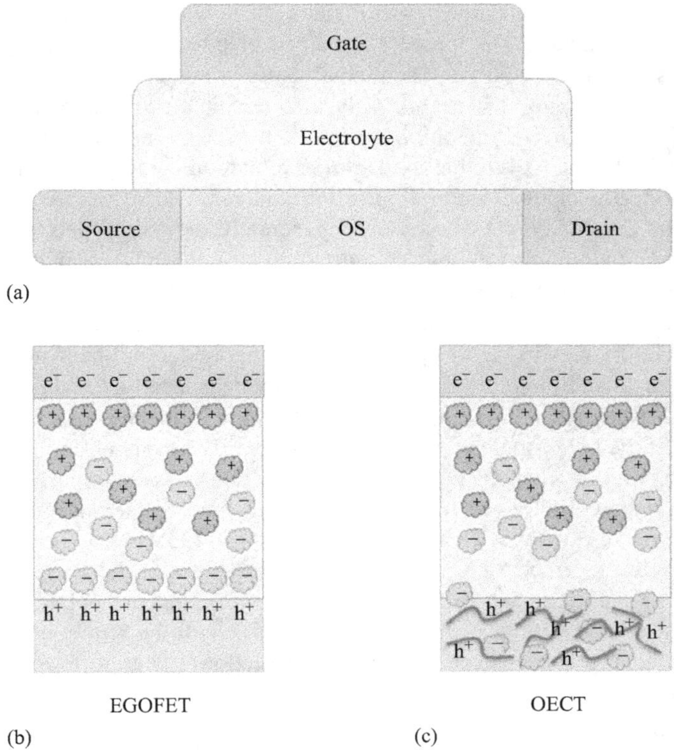

Figure 2.1 (a) Cross-section of an electrolyte-gated organic transistor. The gate voltage is biased at a lower potential with respect to the source and drain. (b) Zoom on the central region. The semiconductor is impermeable. Two electric double layers are formed. Cations are accumulated at the gate–electrolyte interface while anions are accumulated at the electrolyte–semiconductor interface. Anions lead to hole accumulation in the OS and a channel extending from source to drain is formed. This transistor configuration is named electrolyte-gated organic field-effect transistor (EG-OFET). (c) Zoom on the central region. The semiconductor is permeable. One electric double layer is formed. Cations are accumulated at the gate–electrolyte interface while anions diffuse into the OS and compensate induced charge carriers. This process is termed electrochemical doping. This transistor configuration is named organic electro-chemical transistor (OECT)

At steady-state conditions, nearly all the applied gate potential drops across the double layers and there is little potential drop in the bulk of the electrolyte [18,19].

On the other hand, in the case of ion-permeable semiconductors (Figure 2.1(c)), an EDL may be formed at the gate–electrolyte interface upon application of a gate voltage, while ions on the semiconductor side diffuse into the OSC and compensate induced charge carriers. This process is named electrochemical doping, and it is very well established for polymer semiconductors in contact with electrolyte [20]. Transistors that employ permeable OSs are named electrochemical transistors (OECTs) [21,22]. OECT operation proceeds by the reversible electrochemical doping and de-doping (reversible oxidation or reduction, i.e. Faradaic processes) of the semiconductor channel upon application and removal of a gate bias, respectively. In the case of OECTs based on degenerately doped p-type OSs (e.g. poly (3,4-ethylenedioxythiophene):poly(styrene sulfonate), PEDOT:PSS), application of positive gate bias leads to cation migration into the film and de-doping of the semiconductor. This results in a decrease of channel current, turning the normally ON device OFF. This mode of operation has generated significant interest in the area of chemical and biological sensing, primarily because PEDOT:PSS is stable in contact with water [22]. Although understanding the influences of the ions on the magnitude of the hole (or electron) conductivity in ECTs remains an important area of research, in this chapter we focus on the EGOTs with impermeable semiconductors, where EDLs are formed at both the gate metal and OS interfaces. A comprehensive review of OECTs can be found in [23,24].

The double-layer charging at flat, impermeable semiconductor interfaces is similar to the gating action in field-effect transistors, i.e. the polarization of the electrolyte produces a large electric field at the surface of the semiconductor that in turn causes charge carrier accumulation (or depletion). Therefore, in electrolyte-gated organic transistors employing impermeable semiconductors, named electrolyte gate-organic field-effect transistors (EG-OFETs), the EDL at the interface can be considered as a nanometer-thick capacitor. At steady state, the ionic current from the gate to source electrodes is nearly zero because once the double layers form, the driving force for ion migration in the bulk is eliminated. It is worth noting that strictly speaking this will never be the case due to impurities, for example, and hence a gate leakage current is always observed.

Figure 2.1(b) illustrates a p-type EG-OFET in case of a negative applied gate voltage, leading to the formation of an EDL at the gate–electrolyte and semiconductor–electrolyte interfaces via migration of cations and anions, respectively. In detail, the EDL at the gate side consists of electrons within the metal and cations in the electrolyte, while at the semiconductor side holes and anions form the EDL. Within the electrolyte, the ion layer is composed of two layers, the Helmholtz layer and the diffusion layer (Figure 2.2 shows the OS side). The Helmholtz layer accounts for the capacitance due to the minimum distance between the solvated ions and the semiconductor surface (or metal). The diffusion layer, on the other hand, describes the layer of free ions, which migrate under electrostatic interaction and thermal motion. The excess of these ions decreases with increasing distance from the interface. Their distribution can be calculated by solving the Nernst–Planck equation in equilibrium conditions (i.e. no ionic current flow).

The specific capacitance C of each double layer is given by the series of the Helmholtz double layer and the Gouy–Chapman diffuse layer. The former is $C_H = \varepsilon_r \varepsilon_0 d^{-1}$, where ε_0 is the vacuum permittivity, ε_r is the relative permittivity of the electrolyte and d is the dimension of the solvated ion. With a sufficiently large

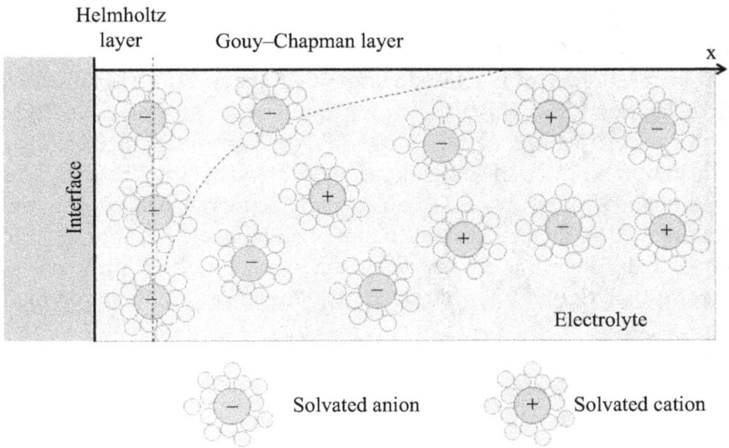

Figure 2.2 Formation of an electric double layer. When a positive voltage is applied, solvated anions are accumulated at the interface. The minimum distance between the interface and the anions is given by the solvated anion radius, typically on the order of few Angstrom. This first layer is called the Helmholtz layer. In the case of large voltages and/or high ionic strength, other anions are accumulated and a diffuse layer is formed. The latter is called the Gouy-Chapman layer

gate electrode, the applied voltage fully drops across the Helmholtz double layer generating a specific capacitance as high as $C_H \approx 10\ \mu F/cm^2$. This strictly holds only in the presence of high ion concentrations, as, for example, in ionic liquids and solid polyelectrolytes. In conventional electrolyte solutions, only a fraction of the charges are confined in the Helmholtz double layer; the residual aliquot forms Gouy–Chapman diffuse double layers with a capacitance that is inversely proportional to Debye's screening length, defined as follows:

$$\lambda = \sqrt{\frac{\epsilon_0 \epsilon_r k_B T}{e^2 \sum_i z_i^2 c_i}} \tag{2.1}$$

where k_B is the Boltzmann constant, T is the temperature, e is the elementary charge, z_i and c_i are the valence and the concentration of the ion specie i, respectively. Assuming low-surface potential, the Gouy–Chapman capacitance per unit area can be estimated as follows:

$$C_{GC} \cong \frac{\epsilon_0 \epsilon_r}{\lambda} = \sqrt{\frac{e^2 \epsilon_0 \epsilon_r i_s}{k_B T}} \tag{2.2}$$

where i_s is the ionic strength of the solution. Since the total capacitance of the electric double layer can be described, at first instance, as the Helmholtz layer in series with the Gouy–Chapman layer, the overall capacitance is dominated by the smaller one, namely C_{GC}.

Typical values of λ are of the order of 1 nm, and with an electrolyte relative permittivity $\varepsilon_r \sim 10$, a C_{GC} value of the order of 1 $\mu F/cm^2$ is commonly attainable, which is at least an order of magnitude larger than the C values obtained with conventional dielectrics [25]. This enables the low-voltage operation of EG-OFETs. Owing to this exceptional property, EG-OFETs were first recognized and subject of intense research in the field of low-voltage and low-power applications, where they are good alternatives for OFETs based on solid dielectrics such as conventional oxide gate insulators, high-k dielectrics or ultrathin organic layers [17].

Figure 2.3 displays the energy band diagrams for three principal states of operation for p-type EG-OFETs: flat-band, depletion and accumulation modes. The

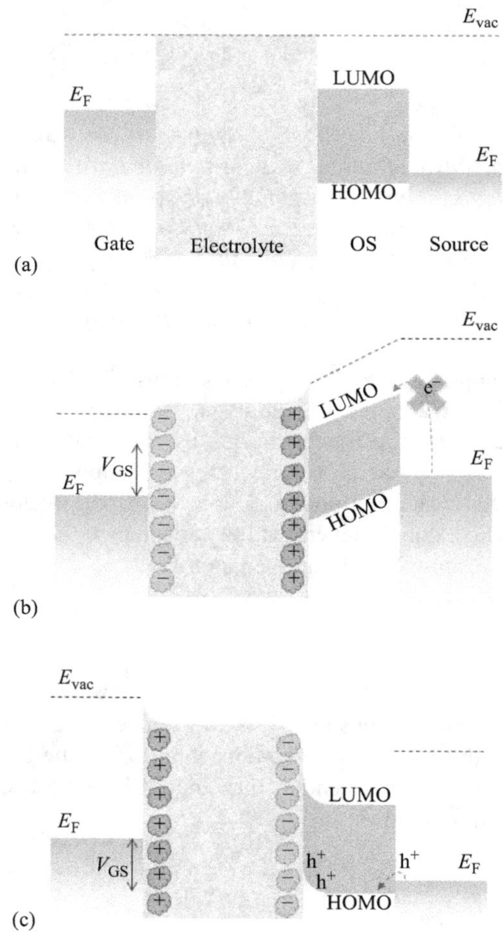

Figure 2.3 Energy band diagrams of gate/electrolyte/organic semiconductor interfaces in p-type EGOFETs: (a) flat-band condition, (b) depletion mode and (c) accumulation mode

three states are displayed in Figure 2.3(a)–(c), respectively. In each diagram, the locations of the electric double layers are shown. The source voltage is the reference voltage, viz. $V_S = 0$ V. When $V_G > V_{FB}$, a large energy barrier at the source/semiconductor interface prevents injection of electrons (depletion mode). The EG-OFET is operated in the OFF state and no electric current can flow from source to drain. When $V_G = V_{FB}$, the EDLs are not formed and both the electrolyte and the OS are in equilibrium (Figure 2.3(a)). On the other hand, when $V_G < V_{FB}$, holes can be injected into the highest occupied molecular orbital energy states and accumulate into the OS at the electrolyte–semiconductor interface (Figure 2.3(c)), as is typical in any p-type FET. Under this condition, a high-conductivity channel connecting the source and drain electrodes is formed and the EG-OFET is operated in the ON state. By applying negative drain voltage, holes are injected by the source, transport through the channel and are extracted at the drain. Figure 2.4(a) shows

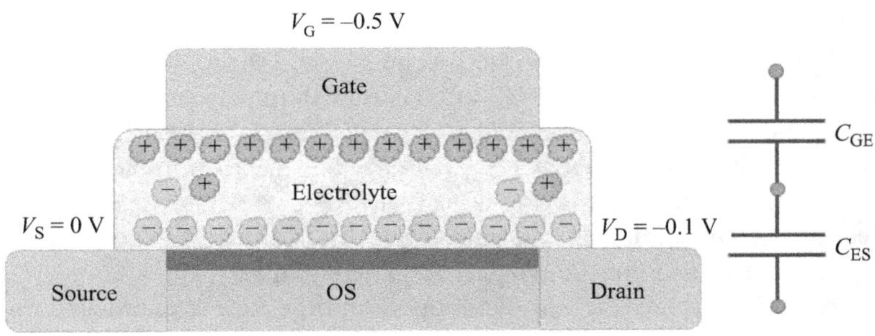

(a)

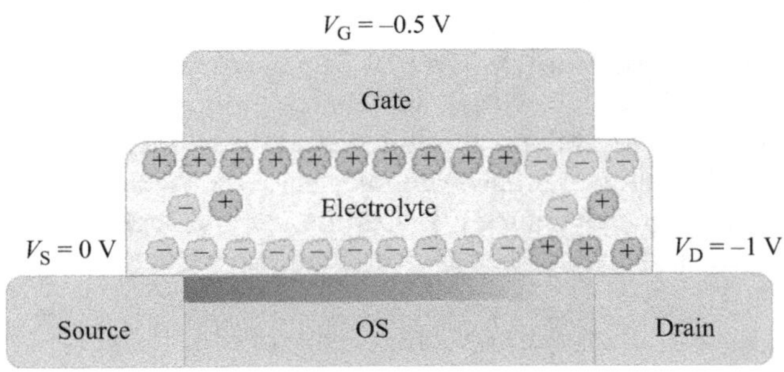

(b)

Figure 2.4 EGOFET operation (p-type). (a) Linear regime, holes are accumulated at the semiconductor/electrolyte forming a continuous high-conductivity channel from source to drain. Two EDL are formed and the corresponding capacitances are C_{GE} and C_{ES}. (b) Saturation regime, holes are accumulated at the semiconductor/electrolyte until $|V_G| > |V_{CH}|$, where V_{CH} is the potential along the channel. When $V_G = V_{CH}$ the channel is pinched-off. In the region of the OS where $|V_G| < |V_{CH}|$ the semiconductor is depleted

that at small drain voltage, $|V_D| < |V_G - V_{FB}|$, the accumulated channel extends from the source to the drain electrode and the EG-OFET operates in a linear regime. The source–drain current reads as:

$$I_{D(lin)} = \frac{W}{L}\beta C_i\mu_h\left[(V_G - V_{FB})V_D - \frac{V_D^2}{2}\right] \cong \frac{W}{L}\beta C_i\mu_h(V_G - V_{FB})V_D \qquad (2.3)$$

where μ_h is the hole mobility of the OS, and $\beta = C_{GE}/(C_{GE} + C_{ES})$ accounts for the voltage partition across the series capacitors C_{GE} and C_{ES}. As depicted in Figure 2.4(a), the total capacitance (C_{TOT}) of the gate–electrolyte–semiconductor stack is given by the series of the gate–electrolyte capacitance (C_{GE}) and electrolyte–semiconductor capacitance (C_{ES}):

$$\frac{1}{C_{TOT}} = \frac{1}{C_{GE}} + \frac{1}{C_{ES}} \qquad (2.4)$$

where $C_{GE} = C_{gate} \times A_{gate}$, $C_{ES} = C_i \times A_{osc}$, C_{gate} and C_i are the gate–electrolyte and the electrolyte–OS capacitance per unit area, respectively, while A_{gate} and A_{osc} are the area of gate electrode and OS in contact with the electrolyte, respectively. Both C_{gate} and C_{osc} can be estimated to be equal to C_{GC} given by (2.2). If the two EDL capacitances are equal ($C_{GE} = C_{ES} = C$), the total capacitance of the gate–electrolyte–semiconductor stack is $C_{TOT} = C/2$. In many cases, however, the electrolyte–semiconductor interface typically exhibits smaller capacitance values per unit area and thus dominates, viz. $C_{TOT} \approx C_{ES}$.

Figure 2.4(b) depicts the ion distribution and the hole accumulated channel when the EG-OFET is operated in saturation. Since $|V_D| > |V_G - V_{FB}|$ the channel does not extend from the source to the drain and it is pinched-off when $|V_G| = |V_{CH}|$, where V_{CH} is the pseudo-Fermi potential at the OS–electrolyte interface. This can be explained as follows. The electrolyte–OS EDL is composed by anions from the source to the pinch-off where $|V_G| < |V_{CH}|$ and cations from the pinch-off to the drain where $|V_G| > |V_{CH}|$. The cations accumulated close to the drain electrode deplete the OS. Hence, in saturation holes are injected at the source, drift through the channel, diffuse at the channel pinch-off, drift through the depleted region and are eventually collected at the drain. The source–drain current is given by:

$$I_{D(sat)} = \frac{W}{2L}\beta C_i\mu_h(V_G - V_{FB})^2 \qquad (2.5)$$

Equations (2.3) and (2.5) can be extended accounting for the morphological and energetic disorder of the OS. The charge transport in the OS can be described by the variable range hopping theory, i.e. a carrier may either hop over a small distance with a high activation energy or hop over a long distance with a low activation energy. Assuming a single-exponential approximation of the density of states (DOS), the drain current reads:

$$I_D = \frac{W}{L}\beta C_i\mu_h\Gamma_t\left[H(V_G - V_{FB} - V_S)^{2Tt/T} - H(V_G - V_{FB} - V_D)^{2Tt/T}\right] \qquad (2.6)$$

where $H(x)$ return x if $x > 0$ and 0 otherwise, T_t is the DOS energy width and Γ_t is a function of the OS parameters, namely T_t, the total DOS N_t, and the localization radius α. It is worth noting that (2.6) is valid in both linear and saturation operation, and it reduces to (2.3) when $|V_D| \ll |V_G - V_{FB}|$ or to (2.5) when $|V_D| > |V_G - V_{FB}|$, provided $\Gamma_t = 1$ and $T_t = T$. The detailed model derivation and the theoretical background are reported in [26,27].

When used in biosensing applications, the EG-OFETs can offer the unique combination of high-sensitivity FET amplification, operation in aqueous environment, and can be endowed with selectivity by using biomolecules as receptors for recognizing a target molecule. A layer of receptors can be linked to one of the electronic active interfaces of the transistor, and the interactions with a target molecule affect the output current of the device. In an EG-OFET, the interfaces between the gate electrode and electrolyte [28] and between the electrolyte and the OS [29] can be modified with the deposition of bioreceptors. The ligand binding to the receptor changes the receptor conformation and/or the charge distribution at the bio-modified interface, eventually causing a shift in the threshold voltage of the transistor, and/or impacting the field-effect mobility, and/or the capacitance. In particular, in the EG-OFET the bioreceptors can be integrated on the gate electrode or on the OS because both the interfaces are in contact with the electrolyte. Equations (2.3) and (2.5) show that the sensitivity of the EG-OFET can be easily enhanced by the proper design of the transistor geometries. For example, if bioreceptors are integrated on the gate electrode, the biorecognition event may result in a variation of the gate capacitance and hence the maximum sensitivity is achieved by making $C_{GE} \ll C_{ES}$, which gives $\beta \cong C_{GE}/C_{ES}$ and:

$$I_{D(\text{lin})} \cong \frac{W}{L}\frac{A_{\text{gate}}}{A_{\text{osc}}} C_{\text{gate}}\, \mu_h (V_G - V_{FB}) V_D \tag{2.7}$$

$$I_{D(\text{sat})} = \frac{W}{2L}\left(\frac{A_{\text{gate}}}{A_{\text{osc}}}\right)^2 \frac{C_{\text{gate}}{}^2}{C_i} \mu_h (V_G - V_{FB})^2 \tag{2.8}$$

This concept has recently been proposed in [30] where a minute change in protein layer capacitance is shown to efficiently modulate a water-gated OFET, allowing for chiral differential detection with large enantiomeric discrimination factor and accurate estimation of interaction energies as low as few kilojoules per mole. This provides a unique tool with general applicability that allows neutral ligand detection in the picomolar concentration range, thus opening a new pathway for the study of ultra-weak molecular interactions in surface-bound protein–ligand complexes.

By shifting the molecular "capture surface" from the gate electrode to the semiconductor, the EG-OFET sensitivity can be optimized following analogous considerations and assuring $C_{ES} \ll C_{GE}$. Therefore, the EG-OFET architecture provides two principal avenues for the device optimization. Basing on these considerations, a high-sensitivity EG-OFET biosensor for stable and reliable label-free electronic sensing in electrolyte solution with physiologically relevant ionic strength has been recently demonstrated [29]. This was achieved by anchoring, just at the OS surface, a capturing bio-layer capable of controlling ionic diffusion into poly(3-hexylthiophene) (P3HT). It is worth noting that the latter is one of the most widely used OSs in EG-OFETs [17,24].

The aforementioned idea has been further investigated in a recent work proposed by Palazzo *et al.* [31], where it has been proven that the EG-OFET overcomes the detection limit imposed by Debye's length. A bio-EG-OFET sensing platform comprising a biological layer at the interface between the OS and the electrolyte was used (Figure 2.5(a)). The biological layer was composed of a phospholipid (PL) bilayer covalently anchored to the OS surface through a plasma-deposited (–COOH)-functionalized thin coating. Some of the anchored PLs were endowed with a biotin moiety, having an incomparably high-binding affinity for streptavidin (SA) or avidin (AV) proteins. Both proteins hold four binding sites and are similar in size ($\approx$5 nm), but they have opposite charges: SA is negative and AV is positive at the operating pH of 7.4. The binding of the negatively charged SA or positively charged AV to the biotinylated PL results in the PL/SA or PL/AV stacked layers, respectively [29]. For the sake of clarity, the types of stacked layers are referred to as PL/SA(AV). After binding to the PL bilayer, SA and AV have empty binding sites available to bind a biotinylated antibody (Ab), leading to the PL/SA(AV)/Ab multilayers (Figure 2.5(a), zoom on the right). Here, the antibody against the C-reactive protein (CRP) was used; it is negatively charged and has a size of $\approx$15 nm. The Ab selectively binds the CRP, a large ($\approx$10 nm) negatively charged protein that is stimulating a great deal of interest as a marker in cardiovascular diseases as well as a prognostic indicator in gastroesophageal cancer. The binding of CRP to the PL/SA(AV)/Ab gives rise to a three-stacked protein system, labeled as PL/SA(AV)/Ab/CRP (Figure 2.5(a), zoom). Such biological multilayers were conceived to allow the electronic probing of a binding event occurring at different distances from the surface of the PLs; that is, 20 nm for the Ab binding to PL/SA(AV), and 30 nm for the CRP binding to PL/SA(AV)/Ab. Rather surprisingly, the fractional change of the drain current as a function of the antibody concentration and CRP concentration revealed that (i) bio-EG-OFETs can measure a response associated with a binding event taking place at a distance of 20–30 nm (30λ–40λ) from the OSC–electrolyte interface where the electronic channel is formed; and (ii) the electronic transduction is dominated by the outermost protein in the stacked layer and not by those residing underneath, even though they are closer to the device electronic channel. Evaluation of the fractional change of the capacitance as a function of the ionic strength i_s showed that the charged moieties of the bound proteins along with the counter-ions form a layer that is similar to an ionic gel. The fixed polyelectrolyte ions generate an electric field that confines the mobile counter-ions in the region of the fixed charges. Eventually, Donnan's equilibrium [32,33] is reached where an electrical double layer is formed at the boundary of the fixed-ion region as sketched in Figure 2.5(b). The bio-EG-OFET gating system capacitance can therefore be modeled by the following elements in

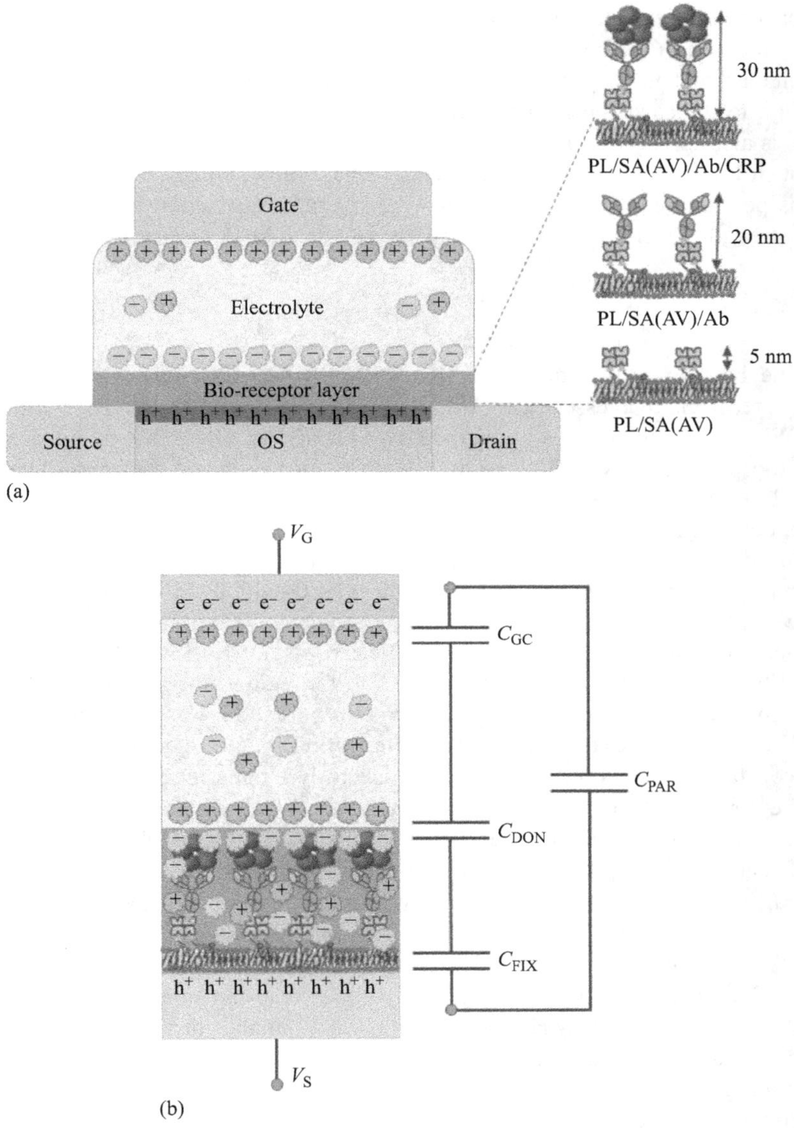

Figure 2.5 *(a) Schematic cross-section of the bio EG-OFET proposed in [31] with different multilayers acting as the bioreceptors layer shown on the right. (b) Schematic of the charge arrangements in the PL/SA/Ab/CRP multiplayer. The equivalent circuit of the capacitance involved is also shown. Adapted from [31]*

series: a fixed capacitance, C_{FIX}, accounting for the HDLs and the PL bilayer capacitance, the $C_{GC} \infty \sqrt{i_s}$ and, in the presence of the protein layer, Donnan's capacitance that is proportional to the ionic strength ($C_{DON} \infty i_s$) [33]. A contribution to the capacitance that is in parallel to the series of described elements, C_{PAR}, is also foreseen to fully account for the stacked proteins (see the equivalent circuit in Figure 2.5(b)). This is expected to be a rather small term (a few micro-Farads per square centimeter), which will govern the detection only at very low ionic strength. Overall, the gating capacitance C can be described as:

$$C = \left(\frac{1}{C_{FIX}} + \frac{1}{C_{GC}} + \frac{1}{C_{DON}} \right)^{-1} + C_{par} \tag{2.9}$$

The binding adds a new layer that changes the capacitance impacting the $1/C_{DON}$ term and to a lesser extent the C_{PAR} term. According to this model, the bio-EG-OFET response is indeed governed by a capacitive effect that is independent from the position at which Donnan's equilibrium is set and therefore does not depend on Debye's length. The latter dominates in electrostatic, rather than capacitive, detections.

In addition to the DC parameters defining the quasi-static operation, a key factor to consider concerning the operation of EG-OFETs is the switching time between OFF and ON states. Normally, transistor switching times (t) depend on the channel length and the mobility of the charge carriers, i.e. $t = $ L2/μVD. However, in EG-OFETs, the polarization speed or the EDL formation time of the electrolyte is also important, and it can dominate the switching frequency of the transistor.

The figure of merit describing the polarization time of electrolytes is the RC time constant, which is the product of the electrolyte ionic resistance (R in Ohms) and the double-layer capacitance (C in Farads). A small $\tau = R\ C$ time constant is desirable for device operation because it means that high C double layers can be established quickly. In terms of intrinsic parameters of an electrolyte, $\tau = l_{el}\ C_{GC}\ \sigma^{-1} = l_{el}\ \varepsilon_r\ \varepsilon_0\ \sigma^{-1}\ \lambda^{-1}$, where σ is the ionic conductivity, l_{el} is the electrolyte layer thickness. For common electrolytes σ is 10^{-5}–10^{-2} S/cm and C_{GC} is 1–10 μF/cm^2 [34]. With a film thickness of ~1 μm, the polarization time can be as small as 0.1 μs. This, in turn, implies that, in principle, one can operate an EG-OFET potentially as fast as 10 MHz. Whether this is attainable in any specific device depends on additional factors such as the source–drain channel length, source–gate and drain–gate area overlap, the charge carrier mobility of the OS, and whether the semiconductor is permeable or impermeable to ions. Permeability to ions leads to increased channel formation times. Nevertheless, it is not unreasonable to anticipate EG-OFET operation in the MHz regime, which is very relevant to explore the frequency domain. For example, in biosensing application the EG-OFET transducers could be used in a ring-oscillator circuit and the bio-event can be readily converted in a frequency response. This approach dramatically simplifies the transducer conditioning as the frequency response can be straightforwardly digitalized and transmitted. This shifts the approach from the single-device level to the circuit level, thus paving the way to electrolyte-gated organic bio-electronics.

2.2 Electrolyte-gated organic transistor: integration of recognition elements

As already discussed earlier, EGOTs have been recently proposed for the development of printable label-free electronic biosensors [11]. These devices offer interesting features such as operation at very low voltages (below 1 V) [35], ability to work directly in biological fluids [31], selectivity and sensitivity that compete with that of conventional bioanalytical techniques [29]. The selectivity towards the detection of a specific analyte is conferred by the integration in the electronic device of proper recognition elements. Both the gate and the OS of EGOTs can serve as surfaces for (bio)molecules deposition and different immobilization strategies have been explored so far [29,36]. In this section, an overview of the molecules that can be employed as recognition elements in EGOTs, along with a discussion on the strategies that have been proposed for their successful integration in the electronic platforms, will be presented.

2.2.1 Recognition elements

It is worth to mention that biosensors can be used for detecting a wide range of target analytes in clinical, foods, environmental and pharmaceutical samples, by just coupling the suitable recognition elements to the transducer surface. Many different molecules have been proposed as recognition elements and they can be grouped into two main categories: (1) biomolecules such as proteins or nucleic acids that are isolated from living systems and (2) synthetic molecules such as aptamers or peptide nucleic acids (PNA), with affinity or catalytic properties similar to biological components, that are chemically designed in laboratory [37]. Among biomolecules, enzymes are the most exploited recognition elements of commercially available biosensors and have been employed since the development of the first glucose biosensor by Clark and Lyons in 1962 [38,39]. The reason for this is that they are extremely stable, easy to obtain and cheap. Among the enzyme biosensors, the glucose biosensor is the most widely studied example. In this device, the glucose oxidase (GOx) enzyme is deposited on the surface of a working electrode and the glucose concentration in the sample is electrochemically detected [40]. EGOT devices, and in particular OECTs, have been widely explored for glucose enzyme biosensors realization and several milestones have been achieved using these novel biosensing platforms [41]. Malliaras *et al.* have developed OECT biosensors capable to measure glucose concentrations down to few micromoles per liter, which are sensitive enough for detecting glucose in noninvasive human fluids such as saliva [42,43]. Lately, Yan *et al.* have improved the detection limit of glucose OECT biosensor of three orders of magnitude by modifying the gate electrodes with nanomaterials such as multi-walled carbon nanotubes or platinum nanoparticles. The improvement of the device performance is due to the excellent electrocatalytic properties of the nanomaterials and more efficient immobilization of enzyme on the gate electrodes. Apart from GOx, many other enzymes have been integrated in EGOT biosensors and metabolites such as lactate [44], penicillin [45], glutamate and acetylcholine [46] have been detected.

Another type of biomolecules that have been successfully integrated in EGOT devices are antibodies [31]. This type of recognition elements have been employed for detection purposes since 1959 [47] and the so-called immunosensors, that make use of the highly selective antibody–antigen interaction, are still considered among the most powerful diagnostic tools with broad applications in several analytical fields including clinical diagnostics [48]. The major advantage of immunosensors is that the target analyte need not be purified prior to detection. The selectivity of the antibodies allows, in fact, these biomolecules to be employed with high efficiency even with complex biological matrices, such as blood, plasma or urine. Furthermore, a high sensitivity can be achieved using antibodies as recognition elements. Magliulo *et al.* have reported label-free EG-OFET immunosensors that are able to detect CRP, a specific biomarker of inflammatory and infection diseases, in a wide dynamic range and with good sensitivity [49]. Antibodies against interleukin-4 (IL4), an anti-inflammatory cytokine relevant in several pathologies, have been also incorporated in the EG-OFET structure for the development of a label-free immunosensor [50].

EGOTs genosensors, based on DNA biomolecules as recognition elements, have also been employed. These biosensors have attracted much interest due to their applications in gene expression monitoring, viral and bacterial identification, bioterrorism agents detection and clinical medicine. Recently, an OECT fabricated on flexible polyethylene terephthalate substrates and integrated in a microfluidic channel formed by poly(dimethylsiloxane) has been proposed for label-free DNA sensing [51]. In this device, single-stranded DNA molecules were immobilized on the surface of gold gate electrodes and complementary DNA targets at concentrations as low as 1 nM were detected. DNA strands have been directly anchored on the surface of the transistors channel as well, for sensing DNA hybridization using an EG-OFET [52]. The immobilization of the DNA probes on the surface of the OS was accomplished through the carboxyl groups present at the side chains of the polymer.

In addition to biomolecules, many other synthetic recognition elements can be integrated in EGOTs. Of particular interest are molecules such as aptamers or PNA that exhibit many advantages as recognition elements in biosensing when compared to traditional biomolecules. These molecules are small in size, chemically stable and cost effective. Besides, they can offer high sensitivity and selectivity [53, 54].

2.2.2 Strategies for recognition elements immobilization in EGOTs

The integration of recognition elements in an electronic device is a key step in the fabrication process of the biosensor. The adopted immobilization method is determined by (i) the properties of the involved recognition elements and the surface of the supporting material, (ii) the operation principle of the sensor and (iii) the application for which the sensor is designed. Certain requirements should be fulfilled in order the selected recognition element to maintain its integrity and full functionality, when immobilized onto a surface or entrapped into a matrix. The conditions of the liquid phase during the fabrication and operation state, such as the temperature, ionic strength and pH, are important parameters that directly affect the stability of the biomolecules Moreover, the immobilized biomolecules are essential to be sensitive and highly selective during the recognition process as part of the sensing device.

Ideally, biomolecules need to be immobilized onto surfaces with high density, limited non-specific adsorption and in a manner so as to provide efficient binding between the immobilized probe and the ligand. Therefore, the study of the surface chemistry of different materials, including noble metals, metal oxides, OS, conductive polymers, hydrogels etc., has become critical for the optimization of the incorporation of biomaterials in miniaturized electronic sensors. Particularly, highly oriented recognition layers on solid support surfaces are required to attain selectivity as well as sensitivity and reproducibility for biosensing. Also in order to enhance sensitivity, the distance between the transducer surface and the immobilized biomolecules layer should be minimized.

Recognition elements may be immobilized on the sensors surface via physical adsorption, covalent attachment or electrostatic entrapment in a polymer matrix [55,56]. This can be achieved using a variety of substances including nafion, sol–gels, lipid membranes, conducting polymers or self-assembled monolayers (SAMs). SAMs are among the most popular ones, due to their ability to form ultrathin, ordered and stable monolayers, and the wide variety of SAMs with different functional groups. Also, SAMs can provide a membrane-like micro-environment for the biomolecules and only require minimal amounts of what are often expensive biomolecules. A variety of SAMs have been used to immobilize biomolecules [57–59]. Another alternative method is to employ a conducting polymer as a matrix to immobilize biocomponents. After the pioneering work of Foulds and Lowe [60], the immobilization of biomolecules such as enzymes, DNA, antibodies and even whole cells in conducting polymers has been widely studied to fabricate biosensors, including immunosensors. Among all of the conducting polymers studied up to date, polypyrrole (PPy) can be considered as one of the most attractive materials due to its excellent conductivity, stability and biocompatibility even in neutral pH medium [61]. Besides PPy and its derivative, the employment of conducting polymers such as poly(aniline), poly(3-hexylthiophene) and poly (*o*-aminobenzoic acid) has been investigated for the realization of biosensors.

The use of cell-membrane mimics such as liposome and lipid bilayers has also attracted the attention in biosensors development and several immobilization methods have been reported [62]. The intrinsically low bioactivity of supported membranes makes them interesting as an interface between the non-biological material on the surface of a sensor and biologically active fluids. Lipid membranes also provide a natural environment for the immobilization of biomolecules such as enzymes, antibodies or protein receptors. The natural environment for the active molecules offered by lipid membranes preserves their high sensitivity and selectivity by preventing denaturation, the low bioactivity of the supporting membrane reduces spurious signals and the localization of the active molecules at a surface aids in signal transduction. Lipid membranes may be deposited on solid supports by several ways. In particular, the sensors surface can be modified to provide specific binding sites for lipid molecules, thus allowing a stable immobilization of lipid membranes.

Different immobilization processes suitable for the integration of biomolecules in EGOTs electronic devices are depicted in Figure 2.6 and they are described in the following.

2.2.2.1 Physical adsorption

Physical adsorption (Figure 2.6(a)) is a simple and inexpensive technique that is commonly employed for biomolecules immobilization on several surfaces. Biomolecules can be adsorbed onto different surfaces via hydrophobic, electrostatic and/or ionic interactions and several physicochemical parameters affect the physisorption process [63]. For instance, proteins like antibodies can be directly adsorbed onto noble metals (Au, Ag, Pt), carbon materials (glassy carbon, carbon paste, nanotubes), metal oxides and conductive or semiconductive materials. This method has been also exploited for DNA and proteins immobilization on the OSs of electronic devices such as pentacene, P3HT and PEDOT [24,64]. The main advantage of the physical adsorption is that reagents for conjugation are not needed and the biomolecules structure is not modified. However, the resulting physically adsorbed bio-layer lacks of homogeneity and is usually weakly attached on the surfaces. Furthermore, it is difficult to obtain well-oriented structures and non-specific adsorption is likely to occur, leading to high background signal. In this respect, improvements, especially in terms of surface loading capacity, are obtained by using the layer–by-layer (LBL) assembly procedure. This method allows obtaining ordered multilayers formed by stepwise adsorption of oppositely charged materials [65]. Biomolecules are easily incorporated in LBL films since charge sites can be obtained on their structure by controlling the pH of the solution in which they are dissolved. The LBL deposition has been successfully employed to deposit fairly homogeneous poly(diallyl-dimethylammoniumchloride)/SA films on the surface of silicon oxide dielectric in electronic devices [66]. Feng *et al.* have also used a similar approach for gate functionalization of EGOTs [67].

2.2.2.2 Entrapment

Polymeric membranes and hydrogels have been widely utilized for immobilizing biomolecules. Due to the high surface area of these materials, higher sorption capacity is expected compared to planar, two-dimensional films [55]. Furthermore, polymeric membranes and hydrogels offer a more amenable environment for the biomolecules. Biomolecules can maintain their 3D structure since they diffuse inside the whole matrix. Due to their hydrophilicity, these materials are able to capture a controlled amount of aqueous solution, thus preventing the biomolecules from denaturing. The biomolecules can be entrapped within the matrix by both physical adsorption and chemical anchoring. The advantage is that entrapment strategies are easy to perform and permit to deposit more components (enzyme,

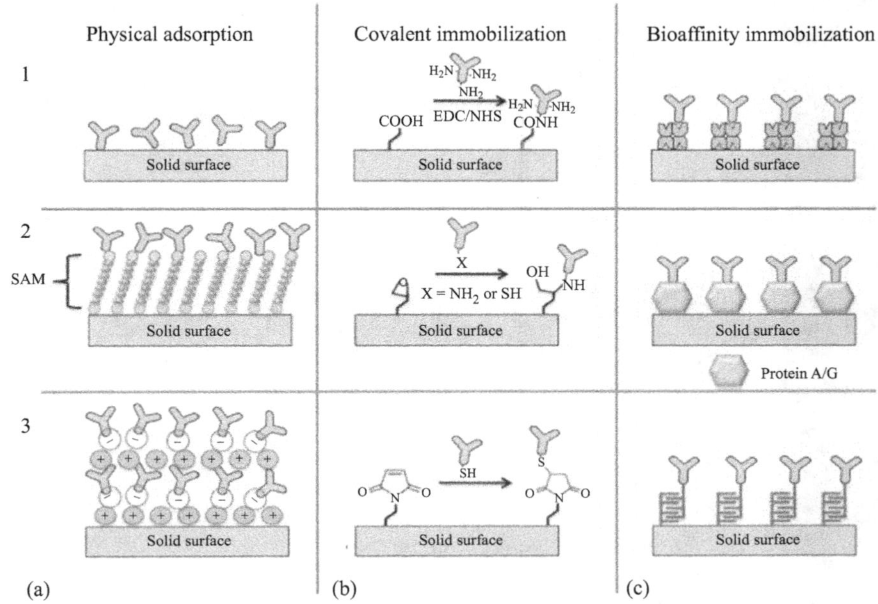

Figure 2.6 *Strategies for the integration of recognition elements in EGOTs. (a) Physical adsorption. Recognition elements are adsorbed on an unmodified surface (a1) or, to obtain a higher density, on a surface modified with a self-assembled monolayer (a2). Layer–by-layer adsorption (a3) through stepwise deposition of biomolecules having opposite charge is another efficient strategy to deposit recognition elements on the surface. (b) Covalent attachment. Recognition elements can be covalently anchored on a solid surface using: the reaction between carboxylic groups on the surface and the ammine groups of the lysine residues of the biomolecules upon activation of the carboxylic functionalities by carbodiimide hydrochloride (EDC) – N-hydroxysuccinimide (NHS) chemistry (b1); the reaction between epoxides on the surface and nucleophilic residues such as amino or thiol groups of the biomolecules (b2); the reaction between maleimide groups on the surface and the thiol group of cysteine residues of the biomolecules (b3). (c) Bioaffinity binding. Biotinylated recognition elements can be deposited in an oriented manner through a streptavidin layer anchored on the surface (c1) and vice versa. Protein A or Protein G can be used to specifically orient antibodies molecules through the binding with their fragment crystallizable region, Fc (c2). Recognition elements can be modified with a proper tag that is selectively recognized by specific molecules on the surface (c3). Reprinted with permission from [11]. Copyright 2015, WILEY-VCH Verlag GmbH & Co. KGaA, Weinheim*

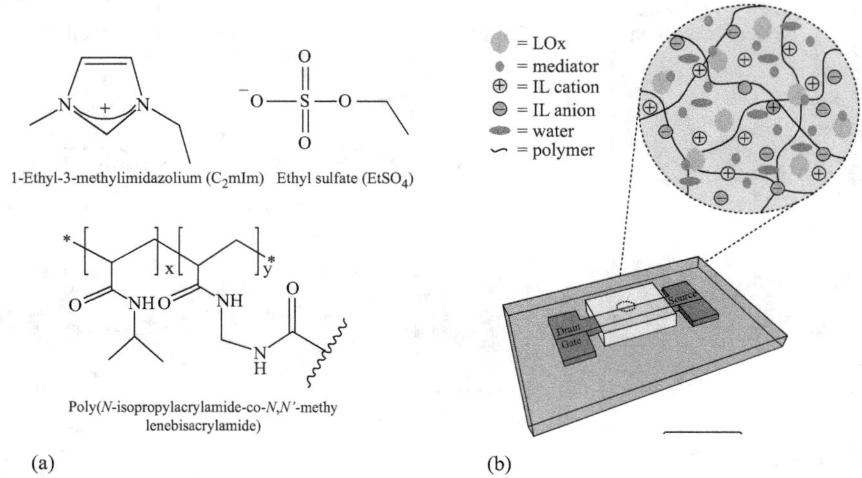

Figure 2.7 (a) Components of the ion gel, (b) scheme of the OECT device embedding the ionogel/enzyme mixture. Reprinted from [44] with permission of The Royal Society of Chemistry http://dx.doi.org/10.1039/C2JM15716K

mediators, additives, etc.) in the same sensing layer. Moreover, the functionality of the biomolecules is preserved, as they are not modified during the immobilization process. This technique is most often used for enzymes immobilization. For example, an ionogel consisting of two monomeric units: *N*-iso-propylacrylamide and *N,N*0-methylene-bis(acryla-mide) have been employed to immobilize lactate oxidase enzyme in EGOT devices [44]. As shown in Figure 2.7, the monomers were photo-polymerized within an ionic liquid matrix that is capable to dissolve the enzyme and to serve as a supporting electrolyte for EGOTs. An enhanced stability of the enzyme in the ionogel has been achieved as well as good electronic and biosensor analytical performance.

2.2.2.3 Covalent attachment

Biomolecules can be anchored on a solid surface by covalent immobilization through the formation of chemical bonds between complementary functional groups present on the biomolecules and on the support. Suitable functional groups are carboxylic –COOH, hydroxyl –OH, amine $-NH_2$ and thiol –SH and several chemical routes have been proposed (Figure 2.6(b)) [58]. While functional groups are naturally present on the biomolecules structure, usually they need to be introduced on the solid support by using intermediate compounds that act as linkers between the surface and the biomolecule.

Polymeric films or SAMs bearing functional groups are generally used at this aim. SAMs are monolayers formed on substrates through chemisorption of molecules able to self-assemble at the solid–liquid interface [68]. Such molecules should present two different head and tail functional groups where one end bounds to the sensor's surface and the other end anchors the biomolecules. For the formation of SAMs, the substrate is usually immersed into a dilute solution of the self-assembling molecules and a monolayer film forms progressively. The SAM formation can last few minutes up to hours. Parameters such as chain length, concentration of the solution and nature of the end groups affect the formation of SAMs. Recognition elements can be then attached to SAMs by chemical modification of the tail group.

This strategy has been successfully employed to anchor recognition elements to the gate electrode of EGOTs [30,36]. Particularly, thiolated SAM molecules bearing carboxylic tail groups, that are further used for biomolecules covalent attachment via carbodiimide chemistry, have been employed to functionalize gold gate electrodes. In this case, the amino groups of the biomolecules form amide bonds with the carboxyl groups of the SAM present on the gold electrode. Mulla *et al.* have used SAMs of 3-mercaptopropionic acid to covalently immobilize the porcine odorant-binding proteins (pOBPs) over the surface of gold gate electrode [30].

On the other hand, covalent attachment of bioreceptors on the surface of the active channel can be done by means of chemical tailoring of the OS. For instance, polythiophene-based semiconductors have been synthesized with functional groups in the side chains [52]. However, such an approach can impact dramatically on the electric properties of the semiconductor. Moreover, it is more beneficial if the functional groups are confined on the surface of the OS and not in the bulk. An effective method to obtain carboxylic functionality on the surface of the OS is through the deposition of ultrathin, hydrophilic polymeric coatings. Low-temperature plasma treatments are broadly used to modify the surface of polymers with specific chemical groups for biomolecules immobilization. Plasma-enhanced chemical vapor deposition (PE-CVD) of ethylene acrylic acid layer (pdEthAA) has been reported for the biofunctionalization of P3HT [69]. Alternatively, solution processed techniques, such as spin coating, can be used for depositing ultrathin layers. A relevant example is the deposition of polyacrylic acid on the surface of poly[2,5-bis(3-hexadecyllthiophen-2-yl)thieno[3,2*b*]thiophene] (PBTTT). In order to render the PAA film insoluble to water and other organic solvents, the film was cross-linked using UV irradiation [70]. In both cases, the —COOH groups were further used to anchor biotinylated phospholipids (B-PLs) molecules. Such immobilization processes have shown to have a negligible impact on the electrical performance of the EG-OFET devices.

2.2.2.4 Bio-affinity binding strategy

Bio-affinity binding approaches (Figure 2.6(c)) provide important advantages over other immobilization methods since biomolecules are precisely oriented on a surface and can be attached with high density. Among them, the (strep)avidin–biotin system is the most exploited to immobilize oriented biomolecules on a variety of surfaces. Specifically, the strong non-covalent (strep)avidin–biotin biointeraction (Kd 10^{-15} M) is used to bound biotinylated molecules on (strep)avidin-modified surfaces or vice versa. Biotinylated anti-bodies have been immobilized on the active channel of an EG-OFET device through a (strept)avidin bridge [31]. A synthesized co-polythiophene derivative with biotinylated side-chains able to capture (strept) avidin molecules has been proposed as OS in EG-OFET devices [71]. Recently, genetically engineered chimeric AV molecules containing C-terminal cysteine groups have been used as a general binding scaffold for biotinylated antibodies deposition on the gold surface [72]. A similar approach has also been employed to functionalize spin-coated films of P3HT [73]. Chimeric AVs showed remarkably increased binding characteristics compared with other AVs, such as wild-type AV, SA and bacterial AV, when merely physically adsorbed onto the polythiophene surface. The former gave the highest binding capacities, the highest affinity constant and the highest stability for biotinylated probe immobilization.

Another efficient method to properly orient antibodies on a solid surface is the use of antibody-binding proteins such as protein G and protein A that selectively bind the fragment crystallizable region (Fc) of the target antibody [74]. The main advantage of this strategy over the SA–biotin approach is that antibody modifications, i.e. conjugation with biotin, are not required, thus ensuring a full retention of the antibody binding capability. Casalini *et al.* [50] recently compared two different strategies to deposit the IL4 monoclonal antibody on the gate electrode of EG-OFETs (Figure 2.8). The first is based on a 6-aminohexanethiol (HSC_6NH_2) SAM monolayer activated by glutaraldehyde. This functional approach guarantees chemical binding between gold and the side chains of the lysine residues in the antibody. However, there is no control on the biomolecules orientation. The second

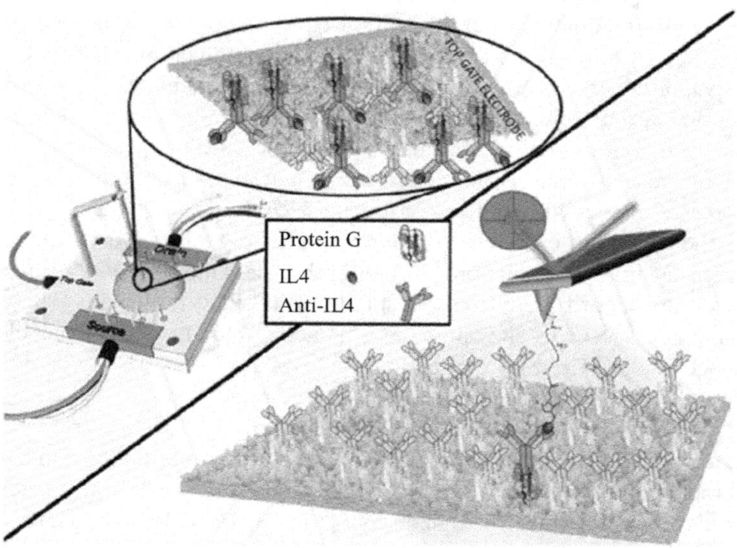

Figure 2.8 Scheme of the immobilization of IL4 antibodies on the gate electrode of an EG-OFET device. His-tagged proteins G are deposited on the gold surface to properly orient the antibodies on the surface. Reprinted with permission from [50]. Copyright (2015) American Chemical Society

strategy exploits the recombinant His-tagged protein G, whose N-terminus side is tailored by a 6-histidine chain (6-His-tag) capable to bind polycrystalline gold. His-tagged protein G forms an oriented layer which promotes antibody immobilization on the Au electrode. Substantial differences of sensitivity were found between the two immobilization strategies. A large probability of specific recognition events occurs on the protein G-functionalized electrodes with respect to the first strategy.

2.3 EG-OFET biosensors

Before discussing the recent advances in the field of label-free biosensors based on electrolyte-gated OFETs, let us recall the basic structure of this electronic device. An EG-OFET biosensor can be easily built up by depositing the OS on a substrate with pre-patterned the source and drain electrodes. Owing to the simplicity of the fabrication and the flexibility of organic materials, bio-EG-OFET sensors can be realized on different substrates among them flexible foils, textiles or biodegradable substrates, using mass production techniques [75]. As in the conventional FET, the drain current that passes through the semiconducting channel is modulated by the gate electrode that in this case can be a metal wire, a plate or even a conductive film deposited on the same plain with the S–D electrodes (Figure 2.9). The gate is capacitively coupled to the semiconductor channel through the electrolyte solution. The main approaches to incorporate biorecognition elements in an EG-OFET device are based on the triggering of the two interfaces of the electrolyte, and in particular, the semiconductor–electrolyte and the gate–electrolyte interface. The occurring changes in the active interfaces during the recognition event impact on the transistor electric properties, thus allowing label-free detection. The rest of this chapter is concerned with label-free sensing of biologically relevant molecules using electrolyte-gated OFETs.

In most of EG-OFET biosensors developed to date, bioreceptors are introduced in the semiconductor–electrolyte interface. Different protocols are reported in the literature so as to convert the hydrophobic surface of the OS into a surface able to host biomolecules. The first demonstration of such an EG-OFET biosensor was carried out by Kergoat *et al.* [52] a few years back. A polythiophene derivative with carboxyl side chains served as semiconductor and as surface for the attachment of bioprobes. DNA probes were covalently bound to the surface of the polymer via standard EDC/NHS chemistry. A shift of the threshold potential along with a decrease of the drain OFF current in the transfer *I–V* curve was seen after the immobilization of the DNA probes. The presence of the DNA probes on top of semiconductor acted as a barrier that reduced the penetration of ions in the polymers bulk. Due to the negative charge of the DNA backbone, the threshold voltage shifted to more negative values since more holes need to be injected for the transistor to switch on. Upon exposure to the complementary DNA strands, the device was able to sense the DNA hybridization when operated with pure water, while for non-complementary random target, no significant change was observed. In this case, a further decrease of the OFF current value was taken as measuring signal.

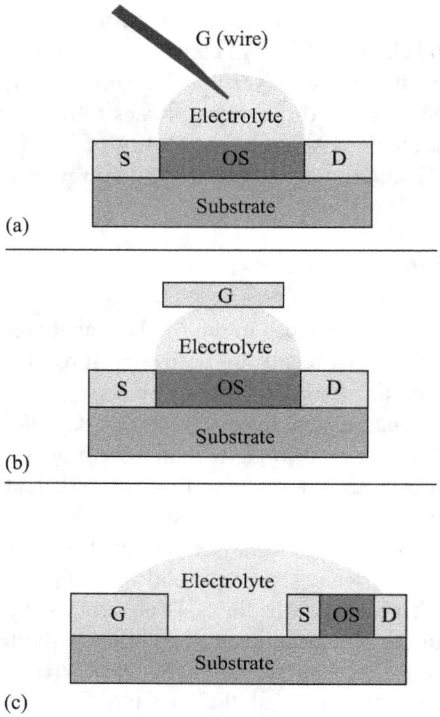

Figure 2.9 Schematic illustration of EG-OFETs with different gate contacts (a) wire, (b) plate, and (c) gate placed at the same plain with the transistor channel

When a high salt concentration buffer (i.e. PBS) was used to gate the device, the detection of DNA failed. The authors suggested that DNA hybridization could be effectively detected in water-gated OFET because the size of the DNA is below the Debye length in pure water, while for an electrolyte with high ionic strength, Debye's length is shorter than the DNA size, and therefore detection is not possible due to the charge screening effect. More recent studies, however, have shown that the detection of charged biomolecules with EGOFET is possible even if the recognition event takes place in a distance greater that the value of Debey's length. A more detailed discussion will be provided later on in the text.

Soon after, an EGOFET was employed as a bio-affinity sensor by Suspène *et al.* [71]. A copolythiophene bearing randomly located biotin molecules at the side chains was synthesized for selective detection of SA and AV in PBS. Tween 20 and 1-octanol have been investigated as blocking agents, with the latter showing less non-specific binding. After 1-octanol blocking step, a decrease of the drain current was observed after the binding of the protein for both negative charged SA and positive charged AV at pH 7.4. Almost unchanged was the transfer curve after incubation in HSA solution. The decrease in the current was attributed to a decrease in the capacitance of the double layer, whereas the opposite charges of the two proteins did not seem to play any significant role in the sensing mechanism.

Buth *et al.* [45] reported an enzyme-based sensor using the EG-OFET configuration. Penicillinase (PEN) was grafted on the surface of a-sexythiophene (a-6T) to specifically detect penicillin. The PEN catalyses the decomposition of penicillin into penicilloic acid and a proton, that consequently can be detected by changes in the pH of the electrolyte solution. PEN was both physically adsorbed on the untreated a-6T and covalently attached to a-6T surface using APTES and EDC/NHS. Transient current measurements were recorded, where the current decreases upon subsequent addition of increasing concentrations of penicillin. A shift towards more negative values in the threshold voltage was also observed. Selective detection of penicillin was demonstrated down to 5 μM for the device operated in PBS. The sensitivity was found similar for both physical adsorbed and covalently bound PEN. Furthermore, the devices were tested after rinsing and no significant changes in the signal were observed, indicating that in this case, there is no need of using complicated, expensive and multi-step biofunctionalization processes.

It is worth noting that in EG-OFET devices, the semiconductor is directly facing the electrolyte solution. This is believed to originate undesirable doping of the semiconductor due to ions penetration from the electrolyte into the bulk of the polymer, giving rise to hysteretic phenomena that eventually might lead to the degradation of the device. Nevertheless, by altering chemically the structure of the semiconductor in order to introduce anchoring groups, it is most likely that the electric properties of the material are also changed. As the bioreceptors immobilization is concerned, the bioprobes are distributed randomly in the bulk of the polymer layer and they are not confined or properly oriented at the surface, which is critical for ensuring effective binding of the target analyte.

To overcome these issues, several innovative approaches have been demonstrated in the past years. Such an example is the work reported by Magliulo *et al.* [29], where a pdEthAA layer with controlled surface density of carboxyl groups was deposited on the top of the OS, using PE-CVD. In order to render the surface of the active channel impermeable to ionic species and at the same time to obtain a surface that holds binding sites for selective detection of a target analyte, a bilayer of B-PLs was deposited on the top of the plasma-treated surface. Owing to their amphiphilic nature, the PL vesicles self-organized on the surface. The incorporated biotin molecules served as binding sites for SA. The proposed EG-OFET sensor was able to detect SA in high ionic strength PBS solution. Upon binding of SA, an increase in drain current was observed. This was attributed to the additional negative gating effect induced by the negative charge of SA when attached over a p-type semiconductor–electrolyte interface. More recently, the same group proposed a simpler and low-cost route for generating carboxyl groups on the surface of the semiconductor [70]. A thin PAA layer was deposited over a p-type OS, namely PBTTT, and crosslinked using the UV light ($\lambda = \sim 254$ nm). An illustration of two methodologies is depicted in Figure 2.10(a). Selective label-free detection of SA was demonstrated also in this case. The plasma-treated EG-OFET biosensors were able to detect SA in the 10 nM–1 μM range with an LOD equal to 10 nM (Figure 2.10(b)).

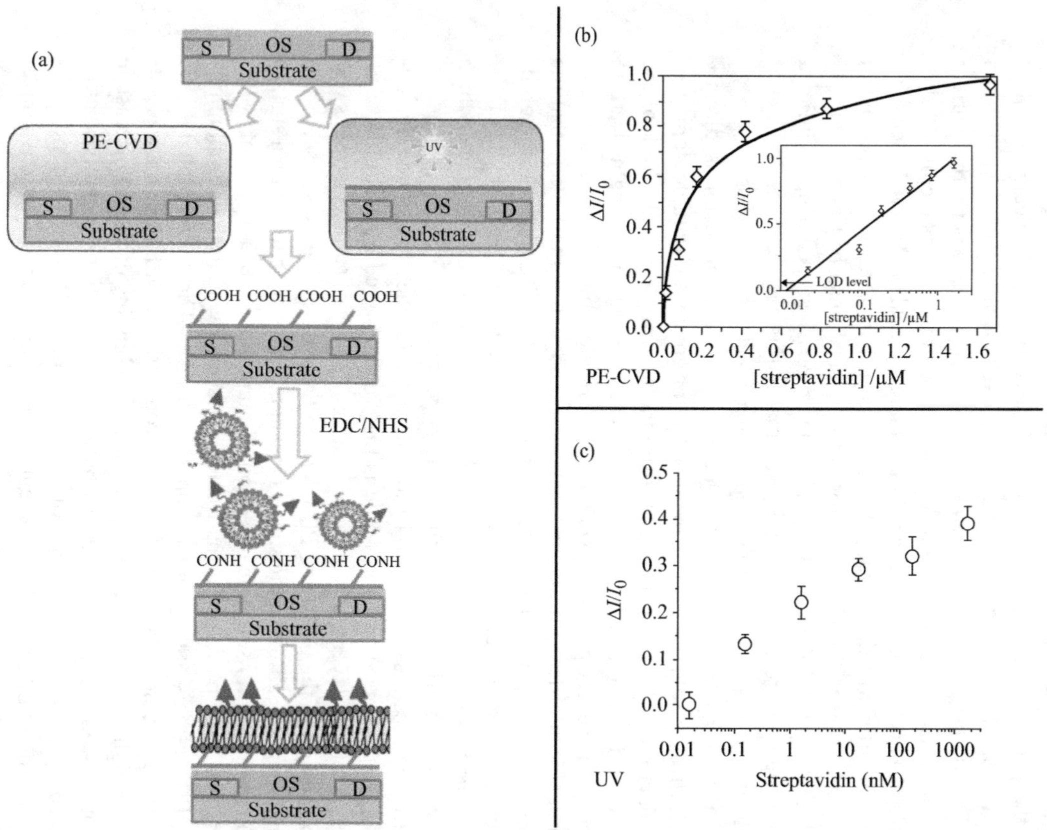

Figure 2.10 *(a) PE-CVD and UV treatment of the organic semiconductor surface in order to obtain a hydrophilic coating bearing carboxyl groups and covalent attachment of the biotin PL bilayer using the EDC/NHS chemistry. Calibration curves as obtained for (b) the plasma-treated EG-OFET, reprinted with permission from [29]. Copyright 2013, WILEY-VCH Verlag GmbH & Co. KGaA, Weinheim and (c) the UV crosslinked PAA-based EG-OFET biosensor, reprinted from [70] with permission of The Royal Society of Chemistry http://dx.doi.org/10.1039/C5TB00243E*

The UV crosslinked PAA EGOFET sensor detected SA down to 10 pM while the dynamic range was from 0.01 nM to 1.6 μM (Figure 2.10(c)). The performance of these EG-OFET biosensors competes the performances of traditional optical techniques.

The biotin PL/SA layer can be further functionalized with an antibody for detection of clinically relevant biomarkers. To this aim, the plasma-treated EG-OFET platform has been investigated as a biosensor for CRP detection [31]. The proposed EG-OFET immunosensor was operated in a phosphate buffered saline (PBS) (pH 7.4) with ionic strength, $i_s = 0.163\ _{\mathrm{M}}$, that corresponds to Debye's length, $\lambda = 0.7$ nm. The sensor was able to detect the biomarker, although the binding event takes place at more than 20 nm far from the transistor channel. The mechanism of sensing was related to changes in the capacitance of the gating system. Upon the binding of the CRP protein, an extra smaller capacitance in series is added to the gating system due to the formation of Donnan's equilibria within the protein layer. Donnan's capacitance, which is proportional to the ionic strength, is the lowest and therefore the dominant one. The authors systematically studied the response of the sensor as a function of Debye's length. It was shown that the EG-OFET biosensor can detect successfully the target analyte at a distance that exceeds the λ value. Moreover, the response was higher in the presence of high salt concentration. Sensing measurements in serum were also reported.

EG-OFET biosensors involving a gate functionalized with a biorecognition element have been also explored. Casalini *et al.* [36] pioneered such an approach of sensing using the EG-OFET structure to detect the dopamine (DA) neurotransmitter. In particular, the sensing element was confined at the Au gate–electrolyte solution interface, through a SAM composed of cysteamine (CA) and 4-formylphenyl boronic acid (BA). A gold wire was used as the gate electrode as shown in Figure 2.11(a). This gate modification allows selective and covalent binding of DA on the Au electrode, due to the esterification between BA and DA. Owing to the zwitterionic structure of BA and the amine group, a large surface dipole builds up after the adsorption process producing a modification of the device transconductance. The DA dipole formed on the surface of the gate changes the work function of the gate, increasing the potential drop at the gate–electrolyte interface and shifting the threshold voltage to more negative values. The adsorption of DA on the gate electrode also affects the capacitance of the double layer in the respective interface. These factors account for the sensor response, allowing the detection of DA down to the picomolar scale. A gain of three orders of magnitude in terms of sensitivity with respect to the traditional OECT has been reported. The device consisted of P3HT as the OS and was operated with PBS at pH 8.5, which is the optimal pH for both DA binding and SAM stability.

In order to obtain a calibration curve and determine the sensitivity of the proposed DA sensor, measurements at different concentrations of the analyte were realized. The functionalized gold wire was exposed to different DA solutions, with concentrations ranging from pM to mM. After rinsing, it is used as the gate electrode by immersing it in the drop of PBS solution that serves as the gate medium. Transfer characteristics were measured sweeping the gate voltage from +0.2 to

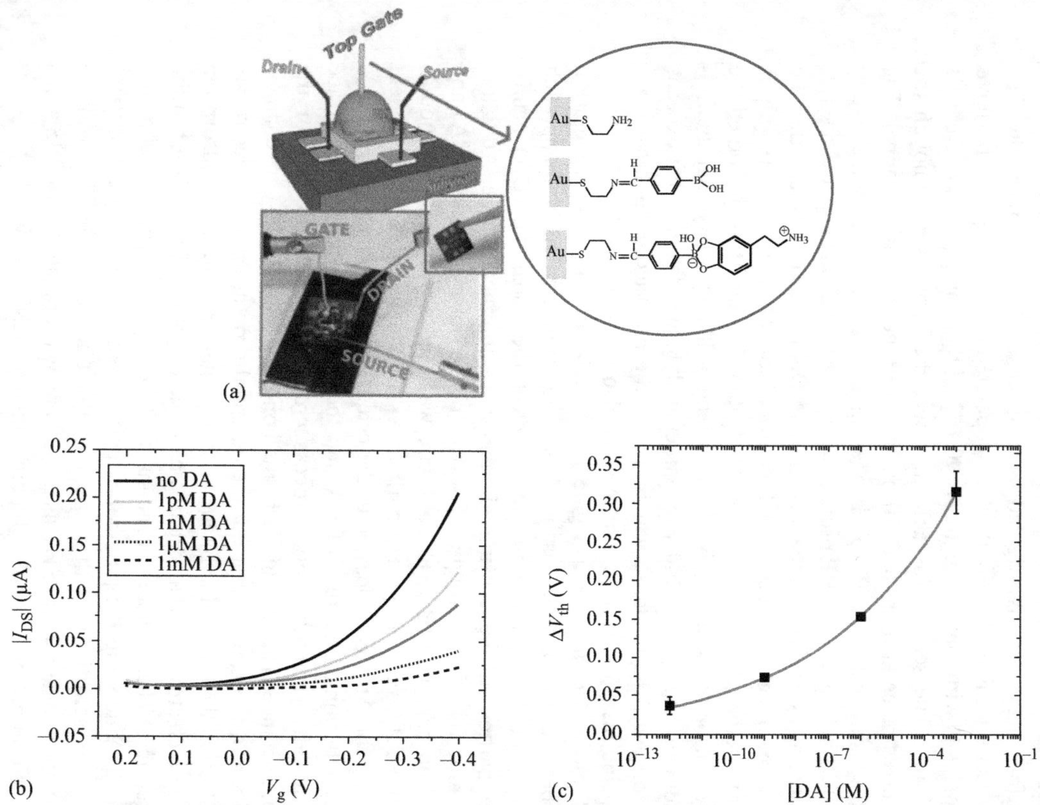

Figure 2.11 *(a) Schematic and real picture of the EG-OFET; (b) Transfer I–V characteristics, acquired in a linear regime ($V_{DS} = 0.1$ V), upon exposure of the functionalized gate to increasing concentrations of DA; (c) the change of the threshold voltage as a function of the DA concentration in the logarithmic scale. The solid line is a power law fitting. Reprinted from [36], Copyright (2013), with permission from Elsevier*

−0.4 V and applying a constant source–drain voltage of −0.1 V (linear regime). The data recorded are shown in Figure 2.11(b). At pM concentration, a clear decrease of the current in the transfer characteristics can be observed. With increasing DA concentration, the current shows a further decrease, and at 1 mM concentration I_{on} is reduced by more than one order of magnitude. A more profound dependence on the concentration was found for the threshold voltage. As reported in Figure 2.11(c), the threshold voltage shift upon exposure to 1 mM DA is *ca* 350 mV. The direction of the shift and its magnitude are in agreement with the zwitterionic structure of BA and the terminal amine.

Recently, sensitive and quantitative detection of the weak interactions associated with neutral enantiomers has been achieved by means of EG-OFET, using OBPs as biorecognition element. OBPs have been immobilized through an SAM to a gold plate (gate), as shown in Figure 2.12(a) [30]. The paper reports on the binding of (S)-(+)- and (R)-(−)-carvone enantiomers to a pOBP-mutant-F88W, detected by means of an EG-OFET, using water as gate medium (Figure 2.12(a)). The calibration curves for the (S)-(+)-carvone, (R)-(−)-carvone along with the control experiment with 2-phenylethanol ligands, that binds very weakly to pOBP, are reported in Figure 2.12(b) as $\Delta I/I$ versus the ligand concentration. The curves are distinguishable down to a few tens of picomolar concentration. The authors demonstrated enantiomeric differentiation at extremely low concentrations with an enantioselectivity factor (ESF) as high as 6.3. The ESF was taken as the ratio of the slopes in the linear region of the calibration curve for each enantiomer. The high enantioselectivity of the sensor was correlated to larger conformational changes upon pOBP-(S)-(+)-carvone interaction of the surface confined molecular complex

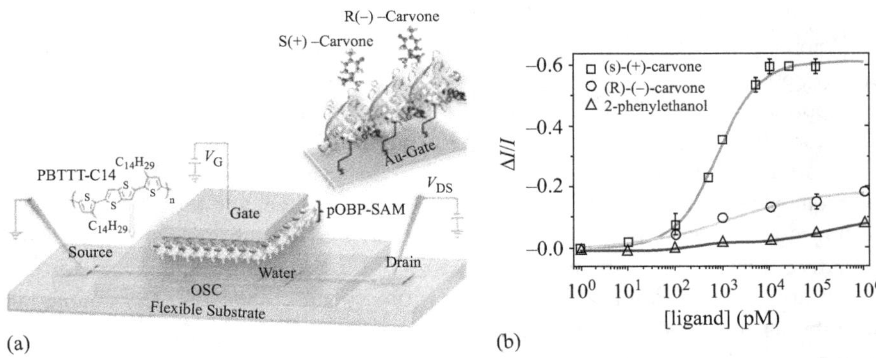

Figure 2.12 (a) Schematic illustration of the EG-OFET biosensor and of the biofunctionalized gold gate electrode. (b) The calibration curves as a variation in drain current upon exposure to the (R)-(−)- and (S)-(+)-carvone as well as the non-specific ligand 2-phenylethanol ligand in the concentration range from 1 to 10^6 pM. Reprinted from [30]

compared to the interaction with the (R)-(−)-carvone. The ability of the EG-OFET device to detect concentrations in the pM level was related to the changes in the capacitance of the protein layer [11]. As already discussed, in an EGOFET biosensor, when a bio-layer is introduced to one of the two active interfaces, a third and lower capacitance in series is added in the capacitance of gating system, that one of the bio-layers. Being the smallest, the capacitance of the bio-layer is modulating the transistor. Even minute changes in the protein layer capacitance due to the ligand binding are enough to give rise to a large variation in the signal.

It is worth mentioning that the use of EG-OFET as biosensor allowed reaching dissociation constants (K_{FET}) at least three orders of magnitude lower than those measured in solution (K_{Sol}). This has been explained observing that the binding at the pOBP-SAM is favored compared with pOBP in solution. The authors suggested that this occurrence reflects the lower entropy change associated with ligands binding to the proteins orderly confined on the gate surface.

The response of an EG-OFET operated as an inflammatory cytokine, IL4, immunosensor has been explored as well [50]. For this purpose, two strategies of gate functionalization have been assessed, one based on the use of amino-terminated SAMs, and another by means of His-tagged protein G (PG). The functionalized Au gate electrode was incubated in a reference solution of IL4 at a concentration of 5 nM. This study highlights the impact of different functionalization strategies of the Au gate electrode on the sensitivity of the biosensor. Specifically, EG-OFETs successfully detected IL4 down to 5 nM concentration when the gate electrode was functionalized with PG, while EG-OFETs functionalized using 6-aminohexanethiol showed negligible response upon exposure to the analyte. The absence of significant changes in the electrical properties EG-OFETs was correlated to the less ordered biofunctionalized film obtained in the latter case. In particular, the random orientation of the anti-IL4 limited the number of active sites for the target analyte to bind, thus leading to a poor signal.

Another interesting approach of EG-OFET with functionalized gate for biosensing application has been proposed by White *et al.* [76]. The authors described a new approach to label-free DNA hybridization detection using an electrolyte-gated transistor with a floating gate. Figure 2.13(a) shows the device architecture in detail. The floating gate connects two separate parts of the sensor, that is the right arm filled with aqueous buffer, where DNA hybridization takes place, and the left arm, where the transducer is placed. The transducer is an EG-OFET with printed P3HT as the OS. An ion gel electrolyte, compatible with P3HT, bridges the semiconductor and the left arm of the floating gate. A second, primary gate, is capacitively coupled to the right part of the floating gate through the aqueous electrolyte. This means that the application of a gate voltage to the primary gate equivalently biases the floating gate. The primary gate is electrostatically coupled with the transistor channel through the floating gate. Voltage is applied to the primary gate, while the floating gate is used as the surface for the immobilization of the DNA probes. The surface immobilization of ssDNA on the floating gold gate was achieved through thiol-modified DNA probes. A blocking layer 6-mercaptohexanol has been adsorbed to prevent non-specific absorption. The main advantage of the present configuration is

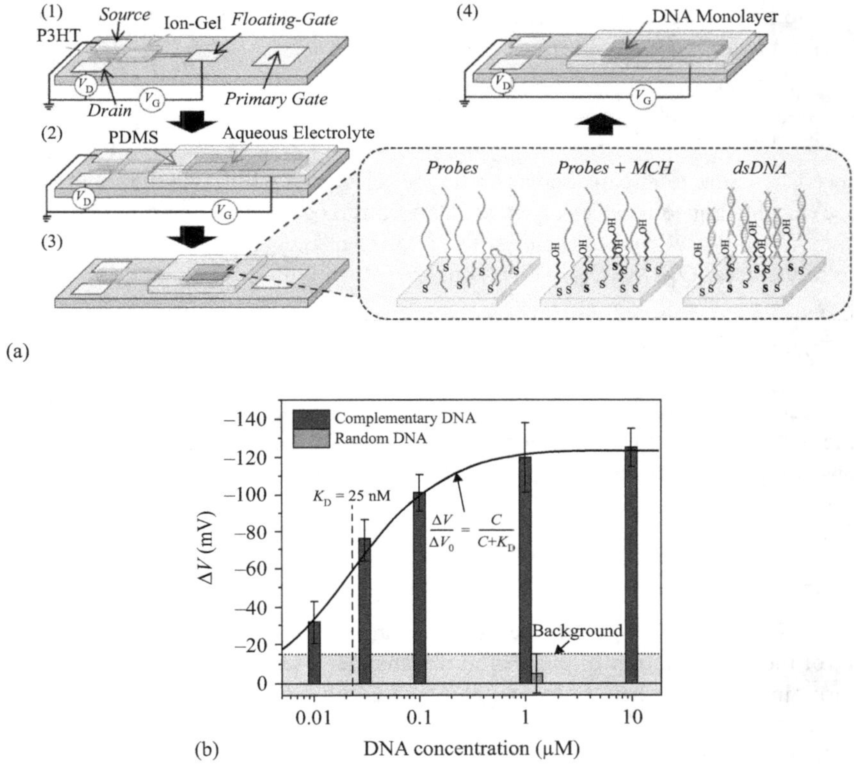

Figure 2.13 *(a) A schematic illustration of the EG-OFET floating gate sensing platform, (b) the calibration curve of the DNA hybridization as a change in the threshold voltage versus the complementary DNA stands concentration. The line is fitted using the Langmuir isotherm equation. Reprinted with permission from [76]. Copyright (2015) American Chemical Society*

the minimization of contamination and/or degradation of the device, since the recognition event occurs in a separate electrolyte from the transistor channel. However, geometrical parameters are critical to ensure the sensitivity of the proposed biosensing platform.

The operation of the floating gate transistor can be modeled as a network of capacitors in series. More specifically, there are two capacitors in the aqueous electrolyte area, one for the double layer at the floating gate–electrolyte interface and the other for the double layer at the primary gate–electrolyte interface. Another two capacitors are added from the ion gel electrolyte side, that is the double layer of the OS–gel electrolyte and that of the floating gate–gel electrolyte. The capacitors in the aqueous electrolyte represent the capacitive coupling between the floating and primary gate, whereas the capacitors in the gel electrolyte represent the coupling between the OS and floating gate. So, the total capacitance can be expressed as follows:

$$\frac{1}{C_{\mathrm{Tot}}} = \frac{1}{C_{\mathrm{SF}}} + \frac{1}{C_{\mathrm{FG}}} \tag{2.10}$$

where C_{SF} is the total capacitance in the gel electrolyte and C_{FG} the total capacitance in the aqueous electrolyte area. Assuming that the charge, $Q=CV$, is common to all involved capacitors, the total charge equals the charge in the gel electrolyte area, $Q_{\mathrm{TOT}}=Q_{\mathrm{SF}}$ which equals:

$$\frac{C_{\mathrm{SF}}C_{\mathrm{FG}}}{C_{\mathrm{SF}}+C_{\mathrm{FG}}}V_{\mathrm{G}} = C_{\mathrm{SF}}V_{\mathrm{gel}} \tag{2.11}$$

According to (2.11), if $C_{\mathrm{SF}} \ll C_{\mathrm{FG}}$, then $V_{\mathrm{gel}} \sim V_{\mathrm{G}}$, which is desirable to achieve the sharpest turn-on (steepest $I_{\mathrm{D}} - V_{\mathrm{G}}$ curve) because only a small voltage drops at the gate electrode. On the other hand, when $C_{\mathrm{SF}} \gg C_{\mathrm{FG}}$ (e.g. $C_{\mathrm{SF}}/C_{\mathrm{FG}}=20$) (2.11) results in:

$$V_{\mathrm{gel}} \cong V_{\mathrm{G}}\frac{C_{\mathrm{FG}}}{C_{\mathrm{SF}}} \tag{2.12}$$

and the occurring changes in the surface functionalized floating gate indirectly control the accumulation of charges in the channel. Hence, the latter case is particularly interesting when floating gate BIO-EGOFETs are used for biosensing applications.

Exposure of complementary DNA to the immobilized ssDNA resulted in the formation of dsDNA on the right part of the surface of floating gate. Upon DNA hybridization, negative shifts in the threshold voltage were recorded. The response of the sensor was associated with capacitance changes in the aqueous electrolyte area, as well as, with changes in the work function of the floating gate due to the analyte molecules adsorption. The authors suggested that DNA hybridization changes the effective double-layer potential at the floating gate–electrolyte interface, thus a more negative voltage is required on the primary gate for the transistor to switch on. In particular, the formation of dsDNA on the floating gate resulted in -120 mV shift in the $I-V$ transfer characteristics of the EG-OFET. The negative shift in the voltage was correlated to the decrease in the capacitance upon hybridization of DNA. The drop in the capacitance value is due to the increase in the bio-layer thickness upon dsDNA formation. Furthermore, the changes in the work function of the floating gate are believed to have contributed in the response of the sensor. As a control experiment, exposure of random DNA to the sensor was monitored as well, producing a negligible V_{T} shift. The sensitivity of the EG-OFET with the floating gate was studied by varying the concentrations of target ssDNA in a range from 10 nM to 10 μM. The results in Figure 2.13(b) show that the signal is saturated at high concentration (≥ 1 μM) but gradually decreases for lower concentrations. The data were fitted using a Langmuir isotherm with $K_{\mathrm{D}}=25$ nM by assuming that the ΔV decreases due to the smaller fraction of the surface converted to dsDNA. The observed limit of detection is 10 nM, assuming a background level equal to the signal from adding random DNA (<15 mV).

References

1. Kaltenbrunner M., Sekitani T., Reeder J., *et al.* An ultra-lightweight design for imperceptible plastic electronics. *Nature*. 2013;**499**(7459):458–463.
2. Torricelli F., Colalongo L., Raiteri D., Kovács-Vajna ZM., Cantatore E. Ultra-high gain diffusion-driven organic transistor. *Nature Communications*. 2016;**7**(10550):1–9.
3. Smits ECP., Mathijssen SGJ., van Hal PA., *et al.* Bottom-up organic integrated circuits. *Nature*. 2008;**455**(7215):956–959.
4. Liao C., Zhang M., Yao MY., Hua T., Li L., Yan F. Flexible organic electronics in biology: materials and devices. *Advanced Materials*. 2014;**27** (46):7493–7527.
5. Torricelli F., Ghittorelli M., Smits ECP., *et al.* Ambipolar organic tri-gate transistor for low-power complementary electronics. *Advanced Materials*. 2016;**28**(2):284–290.
6. Khodagholy D., Doublet T., Quilichini P., *et al.* In vivo recordings of brain activity using organic transistors. *Nature Communications*. 2013;**4**(1575):1–7.
7. Torsi L., Magliulo M., Manoli K., Palazzo G. Organic field-effect transistor sensors: a tutorial review. *Chemical Society Reviews*. 2013;**42**(22):8612–8628.
8. White SP., Dorfman KD., Frisbie CD. Label-free DNA sensing platform with low-voltage electrolyte-gated transistors. *Analytical Chemistry*. 2015;**87** (3):1861–1866.
9. Cramer T., Campana A., Leonardi F., *et al.* Water-gated organic field effect transistors – opportunities for biochemical sensing and extracellular signal transduction. *Journal of Materials Chemistry B*. 2013;**1**(31):3728–3741.
10. Spanu A., Lai S., Cosseddu P., Tedesco M., Martinoia S., Bonfiglio A. An organic transistor-based system for reference-less electrophysiological monitoring of excitable cells. *Scientific Reports*. 2015;**5**(8807):1–7.
11. Manoli K., Magliulo M., Mulla MY., *et al.* Printable bioelectronics to investigate functional biological interfaces. *Angewandte Chemie International Edition*. 2015;**54**(43):12562–12576.
12. Kergoat L., Piro B., Berggren M., Horowitz G., Pham, M-C. Advances in organic transistor-based biosensors: from organic electrochemical transistors to electrolyte-gated organic field-effect transistors. *Analytical and Bioanalytical Chemistry*. 2012;**402**(5):1813–1826.
13. Torsi L., Dodabalapur A. Organic thin-film transistors as plastic analytical sensors. *Analytical Chemistry*. 2005;**77**(19):380A–387A.
14. Li L., Gao P., Baumgarten M., *et al.* High performance field-effect ammonia sensors based on a structured ultrathin organic semiconductor film. *Advanced Materials*. 2013;**25**(25):3419–3425.

15. Angione MD., Cotrone S., Magliulo M., *et al.* Interfacial electronic effects in functional biolayers integrated into organic field-effect transistors. *Proceedings of the National Academy of Sciences*. 2012;**109**(17):6429–6434.
16. White HS., Kittlesen GP., Wrighton MS. Chemical derivatization of an array of three gold microelectrodes with polypyrrole: fabrication of a molecule-based transistor. *Journal of American Chemical Society*. 1984;**106**(25): 5375–5377.
17. Kim SH., Hong K., Xie W., *et al.* Electrolyte-gated transistors for organic and printed electronics. *Advanced Materials*. 2013;**25**(13):1822–1846.
18. Dhoot AS., Israel C., Moya X., Mathur ND., Friend RH. Large electric field effect in electrolyte-gated manganites. *Physical Review Letters*. 2009;**102** (13):136402-1–4.
19. Ye JT., Inoue S., Kobayashi K., *et al.* Liquid-gated interface superconductivity on an atomically flat film. *Nature Materials*. 2010;**9**(2): 125–128.
20. Lee J., Kaake LG., Cho JH., Zhu X-Y., Lodge TP., Frisbie CD. Ion gel-gated polymer thin-film transistors: operating mechanism and characterization of gate dielectric capacitance, switching speed, and stability. *Journal of Physical Chemistry C*. 2009;**113**(20):8972–8981.
21. White HS., Kittlesen GP., Wrighton MS. Chemical derivatization of an array of three gold microelectrodes with polypyrrole: fabrication of a molecule-based transistor. *Journal of the American Chemical Society*. 1984;**106**(18): 5375–5377.
22. Khodagholy D., Rivnay J., Sessolo M., *et al.* High transconductance organic electrochemical transistors. *Nature Communications*. 2013;**4**(2133):1–6.
23. Liao C., Yan F. Organic semiconductors in organic thin-film transistor-based chemical and biological sensors. *Polymer Reviews*. 2013;**53**(3):352–406.
24. Lin P., Yan F. Organic thin-film transistors for chemical and biological sensing. *Advanced Materials*. 2011;**24**(1):34–51.
25. Hammock ML., Knopfmacher O., Naab BD., Tok JB., Bao Z. Investigation of protein detection parameters using nanofunctionalized organic field-effect transistors. *ACS Nano*. 2013;**7**(5):3970–3980.
26. Torricelli F., Kovács-Vajna ZsM., Colalongo L. A charge control analytical model for organic thin film transistors. *Applied Physics Letters*. 2008;**92** (11):113306.
27. Torricelli F., Colalongo L. Unified mobility model for disordered organic semiconductors. *IEEE Electron Device Letters*. 2009;**30**(10):1048–1050.
28. Melzer K., Munzer AM., Jaworska E., Maksymiuk K., Michalska A., Scarpa, G. Polymeric ion-selective membrane functionalized gate-electrodes: ion-selective response of electrolyte-gated poly (3-hexylthiophene) field-effect transistors. *Organic Electronics*. 2014;**15**(2):595–601.
29. Magliulo M., Mallardi A., Mulla MY., *et al.* Electrolyte-gated organic field-effect transistor sensors based on supported biotinylated phospholipid bilayer. *Advanced Materials*. 2013;**25**(14):2090–2094.

30. Mulla MY., Tuccori E., Magliulo M., *et al.* Capacitance-modulated transistor detects odorant binding protein chiral interactions. *Nature Communications*. 2015;**6**(6010):1–9.
31. Palazzo G., De Tullio D., Magliulo M., *et al.* Detection beyond Debye's length with an electrolyte-gated organic field-effect transistor. *Advanced Materials*. 2015;**27**(5):911–916.
32. Hiroyuki O. Theory of electrostatics and electrokinetics of soft particles. *Science and Technology of Advanced Materials*. 2009;**10**(6):063001/1–13.
33. Borkovec M., Jönsson B., Kope GJM. Ionization processes and proton binding in polyprotic systems: small molecules, proteins, interfaces, and polyelectrolytes. In: *Surface and Colloid Science*, Ed.: E. Matijević, Springer US, Vol. 16, 2001, pp. 99–339.
34. Gailński M., Lewandowski A., Stępniak I. Ionic liquids as electrolytes. *Electrochimica Acta*. 2006;**51**(26):5567–5580.
35. Kergoat L., Herlogsson L., Braga D., *et al.* A water-gate organic field-effect transistor. *Advanced Material.* 2010;**22**(23):2565–2569.
36. Casalini S., Leonardi F., Cramer T., Biscarini F. Organic field-effect transistor for label-free dopamine sensing. *Organic Electronics*. 2013;**14**(1):156–163.
37. Chambers JP., Arulanandam BP., Matta LL., Weis A., Valdes JJ. Biosensor recognition elements. *Current Issues in Molecular Biology*. 2008;**10**(1):1–12.
38. Clark LC., Lyons C. Electrode systems for continuous monitoring in cardiovascular surgery. *Annals of the New York Academy of Sciences*. 2006;**102**(1):29–45.
39. Das P., Das M., Chinnadayyala SR., Singha IM., Goswami P. Recent advances on developing 3rd generation enzyme electrode for biosensor applications. *Biosensors Bioelectronics*. 2015;**79**:386–397.
40. Wang J. Electrochemical glucose biosensors. *Chemical Review*. 2008;**108**(2):814–825.
41. Bernards DA., Macaya DJ., Nikolou M., DeFranco JA., Takamatsu S., Malliaras G. Enzymatic sensing with organic electrochemical transistors. *Journal of Material Chemistry*. 2008;**18**(1):116–120.
42. Macaya DJ., Nikolou M., Takamatsu S., Mabeck JT., Owens RM., Malliaras GG. Simple glucose sensors with micromolar sensitivity based on organic electrochemical transistors. *Sensors Actuators B Chemistry*. 2007;**123**(1):374–378.
43. Shim NY., Bernards DA., Macaya DJ., *et al.* All-plastic electrochemical transistor for glucose sensing using a ferrocene mediator. *Sensors* (*Basel*). 2009;**9**(12):9896–9902.
44. Khodagholy D, Curto VF, Fraser KJ, *et al.* Organic electrochemical transistor incorporating an ionogel as a solid state electrolyte for lactate sensing. *Journal of Material Chemistry*. 2012; **14**;22(10): 4440.
45. Buth F., Donner A., Sachsenhauser M., Stutzmann M., Garrido JA. Biofunctional electrolyte-gated organic field-effect transistors. *Advanced Materials*. 2012;**24**(33):4511–4517.

46. Kergoat L., Piro B., Simon DT., Pham M-C., Noël V., Berggren M. Detection of glutamate and acetylcholine with organic electrochemical transistors based on conducting polymer/platinum nanoparticle composites. *Advanced Materials*. 2014;**26**(32):5658–5664.
47. Yalow RS., Berson SA. Assay of plasma insulin in human subjects by immunological methods. *Nature*. 1959;**184**(4699):1648–1649.
48. Holford TRJ, Davis F, Higson SPJ. Recent trends in antibody based sensors. *Biosensors Bioelectronics*. 2012;**34**(1):12–24.
49. Magliulo M., Mulla MY., Manoli K., *et al.* Ultrasensitive printable biosensors for point-of-care applications. *SPIE Newsroom* May 2015. doi: 10.1117/2.1201504.005961.
50. Casalini S., Dumitru AC., Leonardi F., *et al.* Multiscale sensing of antibody–antigen interactions by organic transistors and single-molecule force spectroscopy. *ACS Nano*. 2015;**9**(5):5051–5062.
51. Lin P., Luo X., Hsing I-M., Yan F. Organic electrochemical transistors integrated in flexible microfluidic systems and used for label-free DNA sensing. *Advanced Materials*. 2011;**23**(35):4035–40.
52. Kergoat L., Piro B., Berggren M., Pham M-C., Yassar A., Horowitz G. DNA detection with a water-gated organic field-effect transistor. *Organic Electronics*. 2012;**13**(1):1–6.
53. Song S., Wang L., Li J., Fan C., Zhao J. Aptamer-based biosensors. *Trends in Analytical Chemistry*. 2008;**27**(2):108–117.
54. Shi H., Yang F., Li W., *et al.* A review: fabrications, detections and applications of peptide nucleic acids (PNAs) microarray. *Biosensor Bioelectronics*. 2015;**66**:481–489.
55. Jonkheijm P., Weinrich D., Schröder H., Niemeyer CM., Waldmann H. Chemical strategies for generating protein biochips. *Angewandte Chemie – International Edition*. 2008;**47**(50):9618–9647.
56. Rusmini F., Zhong Z., Feijen J. Protein immobilization strategies for protein biochips. *Biomacromolecules*. 2007;**8**(6):1775–1789.
57. Goddard JM., Hotchkiss JH. Polymer surface modification for the attachment of bioactive compounds. *Progress in Polymer Science*. 2007;**32**(7):698–725.
58. Wong LS., Khan F., Micklefield J. Selective covalent protein immobilization: strategies and applications. *Chemical Review*. 2009;**109**(9):4025–4053.
59. Matharu Z., Bandodkar AJ., Gupta V., Malhotra BD. Fundamentals and application of ordered molecular assemblies to affinity biosensing. *Chemical Society Reviews*. 2012;**41**(3):1363.
60. Foulds NC., Lowe CR. Immobilization of glucose oxidase in ferrocene-modified pyrrole polymers. *Analytical Chemistry*. 1988;**60**(22):2473–2478.
61. Deore B., Chen Z., Nagaoka T. Potential-induced enantioselective uptake of amino acid into molecularly imprinted overoxidized polypyrrole. *Analytical Chemistry*. 2000;**72**(17):3989–3994.
62. Vikholm I., Viitala T., Albers WM., Peltonen J. Highly efficient immobilisation of antibody fragments to functionalised lipid monolayers. *Biochimica et Biophysica Acta*. 1999;**1421**(1):39–52.

63. Rabe M., Verdes D., Seeger S. Understanding protein adsorption phenomena at solid surfaces. *Advanced in Colloid and Interface Science*. 2011;**162**(1–2): 87–106.
64. Stoliar P., Bystrenova E., Quiroga SD., *et al.* DNA adsorption measured with ultra-thin film organic field effect transistors. *Biosensensor Bioelectronics.* 2009;**24**(9):2935–2938.
65. Iost RM., Crespilho FN. Layer-by-layer self-assembly and electrochemistry: applications in biosensing and bioelectronics. *Biosensensor Bioelectronics.* 2012;**31**(1):1–10.
66. Magliulo M., Altamura D., Di Franco C., *et al.* Structural and morphological study of a poly(3-hexylthiophene)/streptavidin multilayer structure serving as active layer in ultra-sensitive OFET biosensors. *Journal of Physical Chemistry C.* 2014;**118**(29):15853–15862.
67. Liao C., Mak C., Zhang M., Chan HLW., Yan F. Flexible organic electrochemical transistors for highly selective enzyme biosensors and used for saliva testing. *Advanced Materials.* 2015;**27**(4):676–681.
68. Chaki NK., Vijayamohanan K. Self-assembled monolayers as a tunable platform for biosensor applications. *Biosensensor Bioelectronics.* 2002; **17**(1–2):1–12.
69. Magliulo M., Pistillo BR., Mulla MY., *et al.* PE-CVD of hydrophilic-COOH functionalized coatings on electrolyte gated field-effect transistor electronic layers. *Plasma Processes and Polymers*. 2013;**10**(2):102–109.
70. Mulla M., Seshadri P., Torsi L., *et al.* UV crosslinked poly (acrylic acid): a simple method to bio-functionalize electrolyte-gated OFET biosensors. *Journal of Materials Chemistry B*. 2015;**3**:5049–5057.
71. Suspène C., Piro B., Reisberg S., *et al.* Copolythiophene-based water-gated organic field-effect transistors for biosensing. *Journal of Material Chemistry B.* 2013;**1**(15):2090.
72. Vikholm-Lundin I., Auer S., Paakkunainen M., *et al.* Cysteine-tagged chimeric avidin forms high binding capacity layers directly on gold. *Sensors Actuators B.* 2012;**171–172**:440–448.
73. Albers WM., Pelto JM., Suspène C., *et al.* Structural and functional characteristics of chimeric avidins physically adsorbed onto functionalized polythiophene thin films. *ACS Applied Material and Interfaces*. 2012; **4**(8):4067–4077.
74. Makaraviciute A., Ramanaviciene A. Site-directed antibody immobilization techniques for immunosensors. *Biosensensors Bioelectronics.* 2013;**50**:460–471.
75. Magliulo M., Mulla M., Singh M., *et al.* Printable and flexible electronics: from TFTs to bioelectronic devices. *Journal of Material Chemistry C.* 2015; **3**:12347–12363.
76. White SP., Dorfman KD., Frisbie CD. Label-free DNA sensing platform with low-voltage electrolyte-gated transistors. *Analytical Chemistry.* 2015; **87**:1861–1866.

Chapter 3

Conductive organic materials for DNA biosensors

Peng Lin[1], Weiwei Wei[1], Wenyan Tao[1], Shanming Ke[1] and Feng Yan[2]

Abstract

Conductive organic materials have attracted great interest due to their excellent electronic and biochemical properties, which make them suitable for promising applications in chemical and biological sensing. With many advantages like label-free and easy fabrication, DNA biosensors based on conductive organic materials have shown significant benefit for the rapid and highly sensitive genetic detection. Organic DNA sensors can be mainly divided into two types, which are based on electrochemical detection and organic thin-film transistors, respectively. This chapter gives a comprehensive description to these two types of DNA sensors, while the principles of sensors, relative organic materials, detection methods, transduction mechanisms, and their pros and cons are well addressed in detail. Moreover, some new developments and future perspectives in this field are also discussed.

3.1 Introduction

Deoxyribonucleic acid (DNA) is a very important biomolecule, which stores the genetic information for the growth and replication of living organisms and many viruses. Due to its significant physical, chemical, and biological properties, the detection and analysis of DNA play important roles in human health such as diseases diagnosis [1], drug screening [2,3], food safety [4], and environmental monitoring [5]. Therefore, more and more attention has been drawn to DNA technology in different research areas, including the discovery of genetic code and analysis of gene sequences.

Biosensor is composed of a biological element and a transducer. The analytes being tested, such as enzymes, antibodies, nucleic acids, and cell receptors, are

[1]College of Materials Science and Engineering, Shenzhen University, Shenzhen, People's Republic of China
[2]Department of Applied Physics, The Hong Kong Polytechnic University, Hong Kong, People's Republic of China

interacted with the biological elements. The measurable biological signals are then converted into electrical signals by the transducers. The analysis of gene sequences is mainly conducted by the methods of direct sequencing or DNA hybridization. DNA biosensors are the most important biosensors that have been used for the rapid and highly sensitive genetic detection. There are three key elements in a typical DNA biosensor: (i) immobilization of single-stranded DNA (ssDNA) probes onto a substrate; (ii) hybridization of complementary target DNA with probes to form double-stranded DNA (dsDNA); and (iii) signal readout. The hybridization between probe DNA and target DNA occurs through strong hydrogen bond formation between complementary bases, including adenine and thymine (A–T), guanine and cytosine (G–C). The complementary strands are two ssDNA strands with complete complementary base sequences.

Due to many advantages such as high accuracy and high throughput, fluorescence-based DNA sequencing has been widely investigated and commonly used as a commercial technique [6–8]. However, this method relies on expensive optical detection system and labeled DNA molecules, which is more complex and time consuming. Moreover, the low abundance of DNA molecules is difficult to be directly analyzed, which requires the assist of the polymerase chain reaction technique. Therefore, the advanced DNA biosensors with label-free, low cost, high sensitivity, and high speed should be developed to meet the ever-increasing market demand.

Conductive organic materials, which can be mainly grouped into two classes: conductive polymers and conductive small molecules, may have metallic conductivity or can be semiconductors. Such compounds have some significant characteristics, such as low cost, easy fabrication, and excellent biocompatibility. Due to their unique electronic and biochemical properties, conductive organic materials are the promising materials that can be integrated in biosensors, especially for DNA sensors. Organic DNA sensors have many advantages such as label-free, high sensitivity, high selectivity, high speed, and simple process. Therefore, this kind of sensor has been intensively studied with an amazing prospect.

There are two main types of organic DNA sensors, one is based on electrochemical detection and the other is based on organic thin-film transistors (OTFTs). In this chapter, we will focus on the development of these two types of DNA sensors. The principles of sensors, relative organic materials, transduction mechanisms, detection methods, and their pros and cons will be introduced and well discussed in different sections.

3.2 Electrochemical DNA sensors based on organic conducting polymers

3.2.1 Brief description

Organic conducting polymers (OCPs) are mainly organic compounds that possess unusual electronic properties, such as high electrical conductivity, low ionization potential, and high electron affinity, achieved by an extended π-orbital system, through which electrons can move from one end of a polymer to the other [9]. OCPs are also known to have considerable flexibility in their chemical structures. Through chemical modeling and synthesis, it is possible to modulate the required electronic and mechanical properties of conducting polymers. The most widely studied OCPs include polyaniline (PANI), polypyrrole (PPy), polythiophene (PTP), and their derivates, as shown in Figure 3.1. Among these OCPs, PTP is an important class of conducting polymers, because of its excellent thermal and environmental stability, solubility, processability, and high conductivity [10,11].

OCPs offer extraordinary possibilities for the development of highly sensitive and selective DNA sensors, because they can convert the behaviors of DNA hybridization into electrochemical output signals. Electrochemical sensing assays, based on the conformational effects of OCPs, have been extensively studied. Electrochemical properties of OCPs are very sensitive to small changes that may affect the conformation and interface properties of OCPs, including the charge density, pH value, ionic strength, and temperature of the media [12,13]. So, the hybridization of the complementary DNA can be clearly detected, without the need for an indicator, by a decrease in the impedance. The schematic of electrochemical DNA sensors based on OCPs is shown in Figure 3.2. The ssDNA probes are

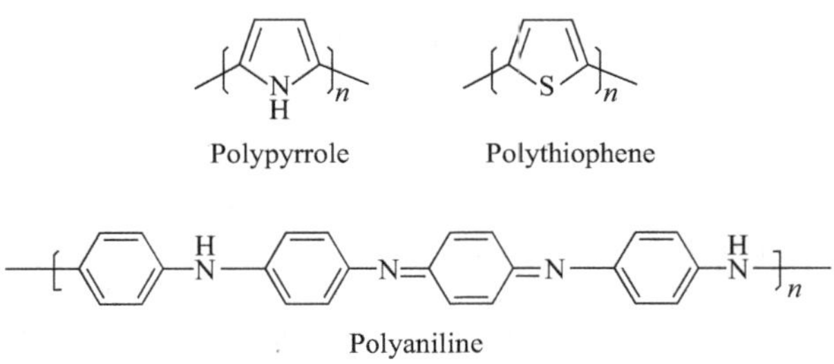

Figure 3.1 Structures of some common organic conducting polymers

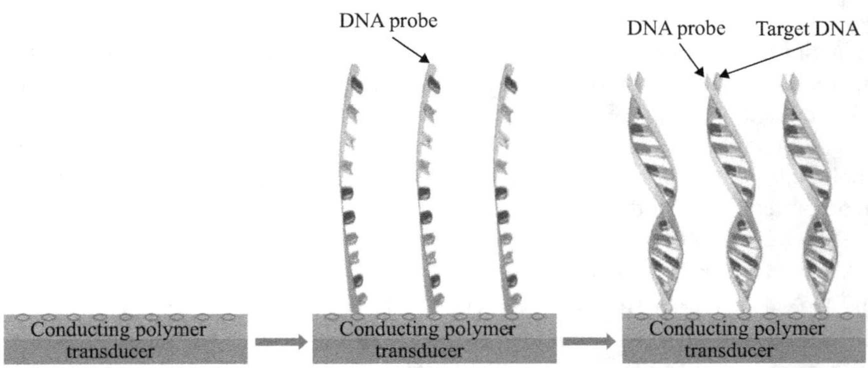

Figure 3.2 Schematic of electrochemical DNA sensors based on OCPs

immobilized on or within an OCP layer, while the target DNA is then captured by the ssDNA probe through base pairing. The recognition signal is generated and recorded by the transducer behind the OCP layer, which is generally integrated with a metal or glassy carbon electrode. Therefore, the quality of DNA immobilization with the OCP, as well as the properties of the OCP, plays an important role in the DNA sensor performance.

Especially, the association of a redox center with the OCP is also very interesting for DNA sensing, which allows a structure to be obtained with both ionic conductivity and electronic conductivity. It provides a material that possesses simultaneously mobile charges of electronic and ionic nature. The redox indicator can be methylene blue [14], methylene green [15], and ferrocene [16]. Among them, ferrocene associated with various conducting polymers, such as PPy [17,18], PTP [19], PANI [20], and poly(*p*-phenylene) [21], has been widely developed. The significant advantages of this method are its simplicity, fast electrochemical response time, and high sensitivity [22].

Here, we introduce the new progress in the applications of OCPs, specifically in DNA sensing, including water-soluble and non-soluble conducting polymers. Various cationic-type conjugated polymers, such as PANI, PPy, PTP, and poly (allylamine hydrochloride) derivatives with flexible molecular structures, have been well studied. The DNA immobilization methods, transduction mechanisms, and DNA sensing will be well discussed in different sections.

3.2.2 DNA immobilization methods

3.2.2.1 Electro-polymerization entrapment method

The electro-polymerization entrapment method, which is derived from the pioneering work on enzyme sensors [23], involves the electrochemical oxidation of a suitable monomer to the corresponding conducting polymer from a solution containing DNA probes. Riccardi *et al.* [24] immobilized the synthetic single-stranded 27-mer DNA (probe) within the 2,5-bis(2-thienyl)-*N*-(3-phosphorylpropyl)pyrrole film through electro-polymerization of its monomer solution, which contained a DNA probe on the PPy-modified microelectrode. Wang *et al.* demonstrated that a DNA probe could act as the sole dopant during the growth of PPy films, while maintaining its hybridization activity [25,26]. The advantage of this immobilization method is its simplicity, in which DNA probes can be immobilized in one-step polymerization. There are also some obvious drawbacks to this method, such as the increased possibility of damaging DNA probes, because of the high potentials employed during polymerization, and lower target DNA accessibility to the incorporated probe in the resulting film.

3.2.2.2 Layer-by-layer method

The layer-by-layer (LBL) method, which was firstly used in the polyelectrolyte self-assembling system, could overcome the disadvantages of the electro-polymerization entrapment method and improve probe accessibility by the target DNA. The main force in the LBL method is the static force between the positive charge on the polymer surface and the negative charge on the DNA chain surface. Liu *et al.* [27] reported that positively charged chitosan (CS) and negatively charged DNA were alternately adsorbed on the surface of pyrolytic graphite (PG) electrodes, forming $(CS/DNA)_n$ LBL films. The force was mainly by Coulombic attraction between them. The $(CS/DNA)_n$/PG electrode was used in the recognition of ssDNA and dsDNA, and also successfully applied to electrochemical detection of the damage of natural DNA caused by the Fenton reaction. The DNA molecules were easily implemented in the polyelectrolyte complexes due to high density of the negative charge of the phosphate residues. Zhang *et al.* [28] described a multilayered DNA sensor with self-doped PANI and Fe_2O_3 microspheres deposited on the carbon paste electrode by the LBL method. The specific target DNA sequence was electro-statically deposited on the electrode surface, which increased the charge-transfer resistance recorded in the presence of ferricyanide ions. The DNA sensor made it possible to determine DNA sequence from 1.0×10^{-13} to 1.0×10^{-7} M, with a detection limit of 2×10^{-14} M.

3.2.2.3 Cross-linker and affinity interaction method

When functionalized with –COOH, $-NH_2$, etc., DNA probes can be covalently attached to either a functionalized monomer or directly to a polymer. For example, Livache *et al.* [29] copolymerized the pyrrole monomer and the ssDNA probe on a surface of the electrode, which allowed the immobilization of multiple probes on electrode arrays. Another approach is to form an electro-polymerized conducting polymer film from a solution containing monomers on the bare substrate electrode, and then covalently attach the DNA probe onto the polymer functional groups. Therefore, the conducting polymer films can be prepared under hard conditions, such as with organic solvents and high potential, which may destroy the bioactivity of DNA. The following attachment of DNA could be performed under mild conditions, ensuring the integrity and bioactivity of DNA chains. Sui *et al.* [12] constructed a label-free DNA sensor using poly(methyl vinyl etheralt-maleic acid) (PMVEA)-doped poly(*o*-methoxyaniline) nanospheres. PMVEA provides a high density of carboxylic acid groups for covalent grafting. The DNA probe was covalent grafted on PMVEA through its –COOH groups. Similarly, the amino-substituted DNA probes were grafted onto the copolymer with pyrrole, poly(Py-co-PAA) film through the formation of covalent bonds between the carboxyl and amino groups catalyzed by EDC [30]. Thompson *et al.* [31] proposed a totally different approach, in which the ssDNA probe was linked to the OCPs by forming a bidentate complex between Mg^{2+} and an alkyl phosphoric acid group on the polymer and the phosphate group of the ssDNA.

3.2.3 Transduction mechanisms

To perform DNA hybridization sensing, including DNA damage, single-base mismatch, or two-base mismatch, the hybridization activity must be converted into an electrical signal. Methods for accomplishing these conversions are described in this section.

3.2.3.1 Amperometric and voltammetric detection

Amperometry is the most common approach used in DNA biosensors, because of its fast response and simplicity. In this approach, the signal is governed by the electron-transfer rate between the biomolecules and the underlying electrode surface involving the OCPs layer. The efficacy of modulating the electrical conductivity and the redox properties of the OCP by hybridization plays a key role in the case of DNA sensors. Kavanagh *et al.* [32] adapted an enzyme-amplified amperometric detection method for DNA hybridization on redox polymer and probe DNA tethered to the gold electrode. The hybridization between the immobilized probe DNA of the recognition surface and a biotin-conjugated target DNA sequence, which was followed by addition of an enzyme–avidin conjugate, resulted in electrical contact between the enzyme (glucose oxidase) and the mediating redox polymer. In the presence of glucose, the current generated due to the catalytic oxidation of glucose to gluconolactone was measured. The response current obtained was binding dependent. The sensor showed a linear range from 1×10^{-9} to 2×10^{-6} M for the target DNA, while the detection limit was approximately 1.4 fM.

Voltammetry is widely used in DNA sensors based on OCPs, because the doping and de-doping processes monitored during potential scanning can be efficiently modulated by the hybridization activity. Techniques such as square-wave voltammetry, differential pulse voltammetry, and cyclic voltammetry are adapted in OCP-based DNA sensors. It is found that better sensitivity can be achieved by using square-wave voltammetry [16,33]. Zhang *et al.* [16] described the development of a sequence-specific electrochemical DNA sensor using square-wave voltammetry. The ferrocene label provided the electrochemical signal, and the conjugated PTP helped transfer electrons from ferrocene to the electrode, as shown in Figure 3.3.

3.2.3.2 Impedimetric detection

Electrochemical impendence spectroscopy (EIS) can provide detailed information for the detection of DNA hybridization on the OCPs-modified electrodes. As for Faradic impedance in the presence of redox couples, such as $Fe(CN)_6^{3-/4-}$, the impendence spectra include a semicircle portion and a linear portion. The semicircle portion at higher frequencies corresponds to the electron-transfer-limited process (R_{ct}), and the linear portion at lower frequencies represents the diffusion-limited process, which is called the Warburg impedance. The immobilization of DNA probes and the hybridization event not only induce changes in the intrinsic properties of the OCPs film, but also bring changes in the various interfacial film properties, such as the capacitance, interfacial electron-transfer resistance, and the Warburg impedance. Compared with direct current (DC) techniques, EIS provides the following advantages: (i) a label-free redox indicator, such as MB or ferrocene, is not necessary to functionalize with OCPs or the ssDNA probe; (ii) a very small excitation amplitude (5–10 mV) is applied on the OCPs–DNA interface, which causes only minimal perturbation, reducing background current introduced by the measurement technique; (iii) EIS provides valuable mechanistic information and the data can be fitted by the suitable equivalent circuits, which is useful in DNA–DNA

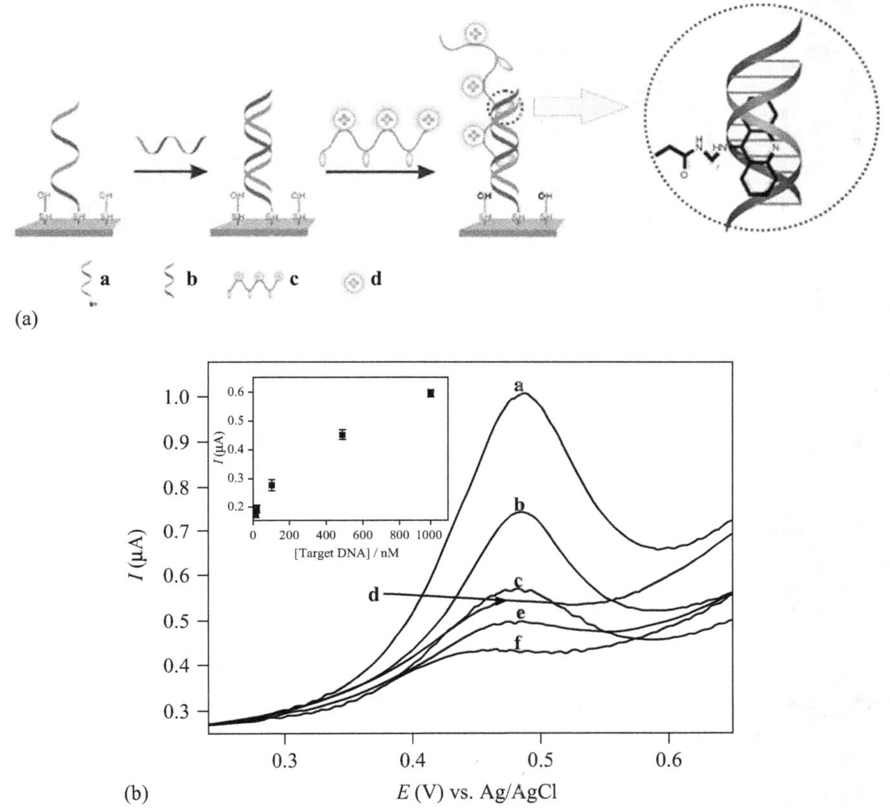

Figure 3.3 (a) Voltammetric detection of DNA through the intercalation of acridine derivatives into the DNA duplex and polythiophene as the "molecular wire" to transduce the hybridization signals. a: capture probe DNA, b: target DNA, c: multi-functional polythiophene, d: ferrocene unit in the multi-functional polymer. a=5′-SH-$(CH_2)_6$-ATGATGTTCGTTGTGTAGGATTTGC-3′; b=5′GCAAATCCTACACA ACGAACATAT-3′. (b) Square-wave voltammograms corresponding to the analysis of different concentrations of target DNA: a) 1 μM, b) 0.5 μM, c) 0.1 μM, d) 10 nM, e) 1 nM, and f) 0 M. Reproduced with permission from [16]

hybridization process studies. A label-free DNA sensor based on a new copolymer, poly(Py-co-PAA) film was reported [30]. PAA refers to 3-pyrrolylacrylic acid. The results showed a significant increase in the charge-transfer resistance upon the addition of a complementary target. It is especially important to mention that the non-Faradic electrochemical impedance spectroscopy without any redox species was also used in the detection of DNA hybridization [34]. The DNA probes were covalently attached to the precursor copolymer (3-acetic acidpyrrole,

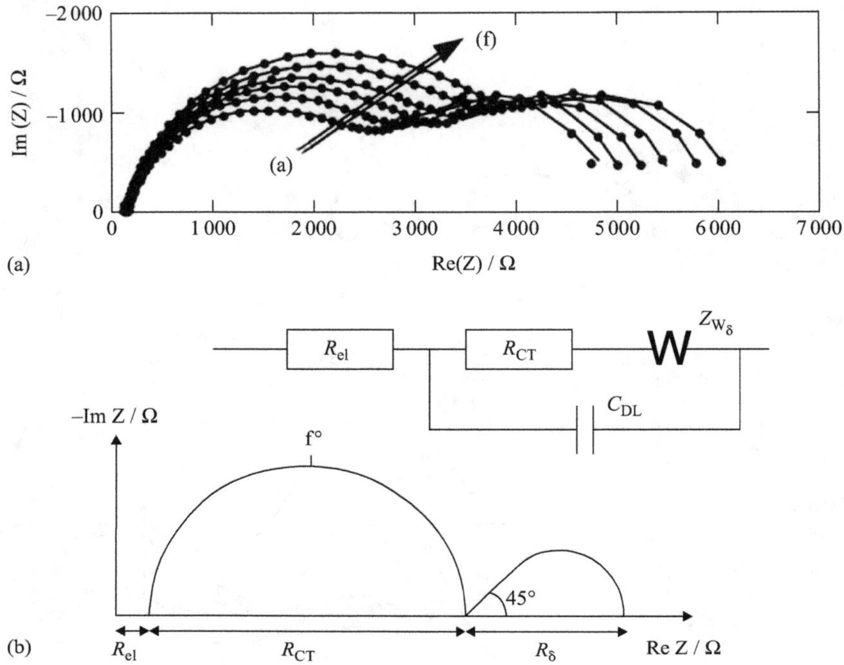

Figure 3.4 *(a) Nyquist plot of impedance spectra obtained for the DNA sensor at different target DNA concentrations: a) 0, b) 10^{-17}, c) 10^{-14}, d) 10^{-11}, e) 10^{-8}, f) 3.1×10^{-6} mol/L. Experimental (dots) and fitted data (lines) are presented. (b) Equivalent circuit: R_{el}, electrolyte resistance; R_{CT}, charge transfer resistance; C_{DL}, double-layer capacitance; $Z_{W\delta}$, finite length Warburg impedance. Reproduced with permission from [35]*

3-*N*-hydroxyphthalimide pyrrole). The measurement potential was set at a DC potential of -1.4 V, which minimized the Warburg impedance and emphasized the PPy–DNA interface impedance. Because PPy was in non-doped and semiconducting states at such low potential, no parasitic electrochemical reaction occurred. The suitable equivalent circuit was only composed of a solution resistance, charge-transfer resistance, and constant phase element (CPE). The charge-transfer resistance decreased upon the grafting of the DNA probes and increased upon hybridization with complementary DNA. A typical EIS response for the DNA sensor is shown in Figure 3.4 [35].

3.2.3.3 Conductometric and chronocoulometric detection

Conductometric detection is based on the change in the conformation of OCPs due to the formation of a DNA duplex, which results in the change of the OCPs conductivity. Ramanathan *et al.* [36] demonstrated a DNA biosensor based on direct incorporation of functional biological molecules (avidin- or streptavidin-conjugated ZnSe/CdSe) into the PPy nanowire. When DNA reacted with biotin-avidin, polypyrrole nanowires generated a rapid change in conductance. The DNA with a concentration of as low as 1 nM could be detected.

Chronocoulometry is a technique most commonly referred to as "bulk electrolysis." The working electrode is kept at a constant potential and the current that flows through the circuit is measured. This constant potential is applied long enough to fully reduce or oxidize all of the electroactive species in a given solution. A chronocoulometric DNA sensor [37] based on the screen-printed electrode was constructed with chitosan, PANI nanotubes (PANINTs), and ionic liquid (IL) $[C_{12}mim][PF_6]$. PANINTs and IL were doped into the commercial conductive ink, and chitosan and Au nanoparticles were covered on the electrode surface to improve the conductivity of the electrode. The signals were generated through chronocoulometric interrogation of $[Ru(NH_3)_6]^{3+}$ ions that bound to the DNA strands via electrostatic interactions. The target DNA could be quantitatively detected in the range from 1.0×10^{-16} to 1.0×10^{-14} M, and a detection limit of 8.0×10^{-17} M was obtained.

3.2.4 DNA sensing

3.2.4.1 DNA damage detection

Heavy metals and organic compounds, also classified as suspected carcinogens, may either cause DNA damage or hinder the DNA repair process [38]. The dsDNA and a photoactive probe assembled on the semiconductor electrodes as a photoelectrochemical DNA sensor were used to monitor the DNA damage reaction in styrene oxide and Fe^{2+}/H_2O_2 [39]. The photo-electrochemical DNA sensor can rapidly detect the two major DNA damages (DNA adducts and oxidative damage). Polyelectrolyte complexes have found applications in biosensors for the detection of DNA damage [40–42]. Polyelectrolyte complexes based on electro-polymerized phenothiazine dyes (methylene blue and methylene green), poly(allylamine hydrochloride), polystyrene sulfonate, and native DNA from salmon sperm were obtained by self-assembling them on the glassy carbon electrode. Changes in the redox status of the polyphenothiazines were followed by appropriate changes in the charge distribution and DNA damage. Du *et al.* [40] investigated the damage effect of Cd^{2+} on the DNA through the $(pLys/dsDNA)_n$-modified carbon paste electrode, as shown in Figure 3.5(a). The reduction peak current in differential pulse voltammetric (DPV) response decreased with the prolongation of the incubation time in $CdCl_2$ solution, indicating that Cd^{2+} could damage DNA, and the damaged DNA films could not reload methylene violet (a intercalation redox probe) to the same level as that of the intact DNA films. The DNA damage could be detected by DPV with this "loading/releasing/reloading" procedure. Wang *et al.* [43] studied DNA damage in the presence of catechol with the use of the $Co(phen)_3{}^{3+}$ electroactive probe on the polymer-modified glassy carbon electrode (GCE). A novel conjugated polymer, polydiphenylamine-4-sulfonic acid film, and dsDNA-modified GCE were used to construct the biosensor for the analysis of the effect of toxic catechol on DNA damage. Results indicated that the potentially toxic catechol and its metabolites were all responsible for DNA damage, as shown in Figure 3.5(b).

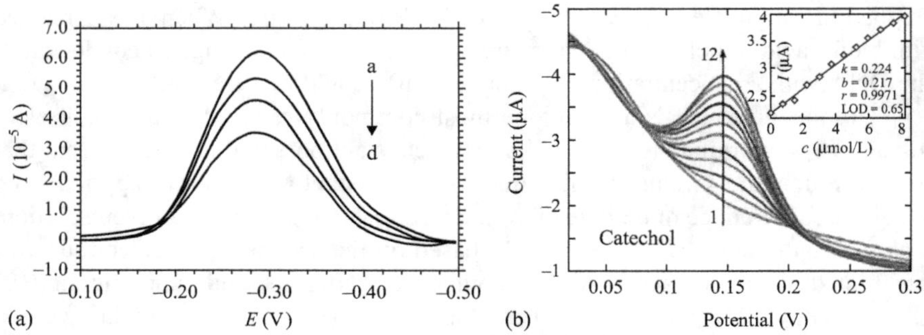

Figure 3.5 (a) DPVs in pH 7.0 B-R buffer at CPE/c-CNT/(pLys/dsDNA)3-MV films after incubation in 4.0×10^{-5}mol/L $CdCl_2$ solution for different times (min): a) 0, b) 10, c) 20, and d) 30; Reproduced with permission from [40]. (b) DPV responses of catechol at different concentrations (1–12): 0 to 8.25×10^{-6} M in steps of 0.75×10^{-6} M obtained at the novel DNA/PDPASA/GCE. Reproduced with permission from [43]

3.2.4.2 DNA mismatch detection

Target DNA detection is based on the principle of complementary hybridization, in which stability is dependent on the melting temperature. To ensure duplex formation, the hybridization process is commonly operated at a temperature of 5–10 °C below the melting temperature. However, many mismatches like point mutations exist and need to be discriminated. A conducting polymer, poly(JUG-co-JUGA) modified with a quinone group, which presented cation-exchange properties, was used to immobilize a probe strand DNA on a glassy carbon electrode [44]. The modified electrode could selectively detect a target DNA strand. This system was efficient for various probe and target lengths (10–30 bp) and can discriminate a single-mismatch. Another DNA biosensor, which was based on 21-mer DNA probes immobilizing on the PANI nanotube array, demonstrated good capabilities of differentiating the perfect matched target DNA from one-nucleotide mismatched DNA at a low concentration of 37.59 fM level [45]. Meanwhile, it was showed that TC mismatch was much less stable than the TG mismatch hybridization at the selected hybridization temperature, indicating that the DNA biosensor has the great potential to identify single-nucleotide polymorphism. Similarly, an electrochemical DNA sensor based on conjugated PTP was constructed, in which the acridine (a DNA intercalator) unit served as the basis for sequence discrimination, and the ferrocene label provided the electrochemical signal [16]. It was a good application for discriminating three single-base mismatch (T, A, G) mutants on the 25-mer DNA sequence at the concentration of 1 µM level by the voltammetric method. Furthermore, the effect of the varying position of single-base mismatch in the DNA was investigated and reported. As shown in Figure 3.6, a DNA sensor was

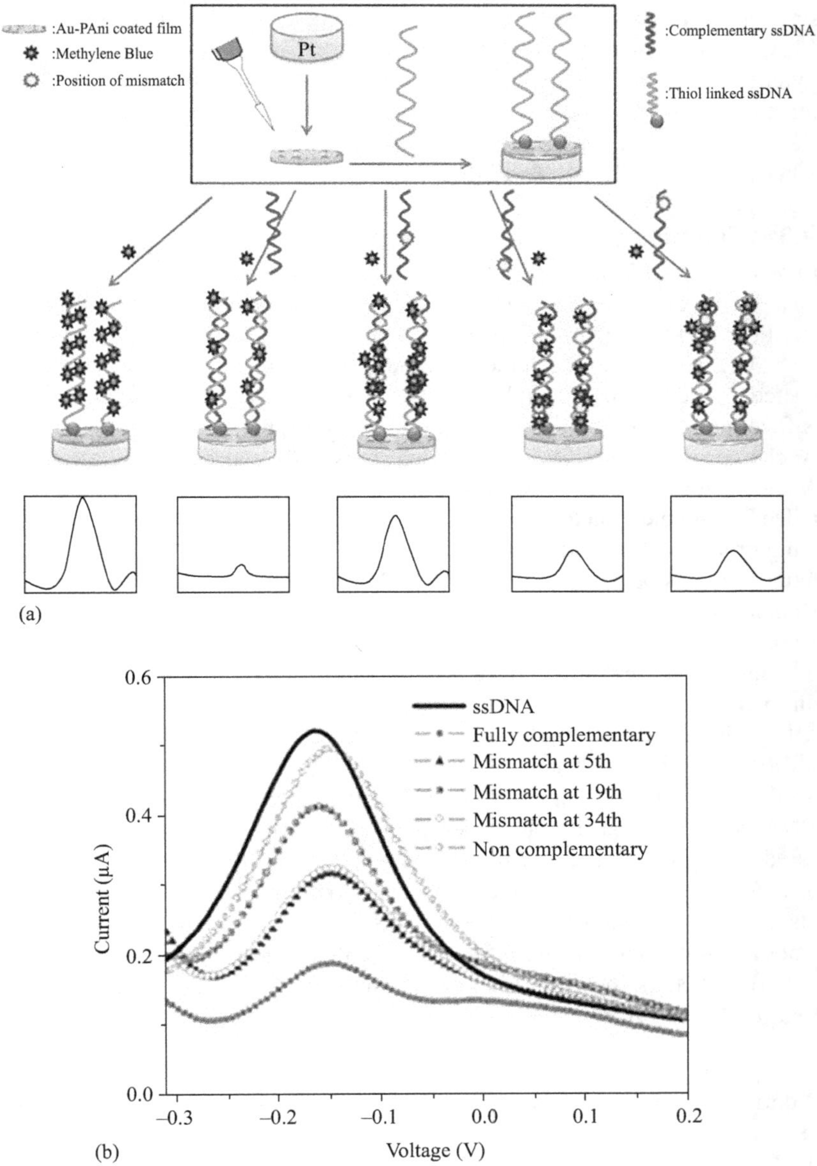

Figure 3.6 *(a) Schematic illustration of the fabrication steps of the electrochemical oligonucleotide biosensor for mismatch detection in different position. (b) DPV response of ssDNA strands having three positional mismatch (5th, 19th, 34th), with its fully complementary and fully non-complementary strands. Reproduced with permission from [46]*

constructed for the sensing of the positional change of DNA base mismatch on the gold-PANI nanocomposites modified Pt electrode [46]. It was achieved by using methylene blue as a redox indicator in the electrochemical DPV method. Position sensitive detection of a single-base mismatch in oligomeric duplexes was achieved, with a detection limit for concentration as low as 10^{-8} M for mixed oligomeric sequences and 10^{-6} M for dA–dT duplexes.

3.2.4.3 Complementary DNA detection

Peptide nucleic acid as probe for DNA detection

Compared with DNA, peptide nucleic acid (PNA) is an achiral and uncharged DNA mimic, in which the sugar–phosphate backbone has been replaced with a peptide like the *N*-(2-aminoethyl) glycine polyamide structure, and the nucleobases are connected by methylenecarbonyl linkages. PNA has gained much attention because of its favorable properties. Firstly, electrostatic repulsion between PNA and negatively charged DNA (or RNA) is avoided, thus PNA can be highly affined with DNA (or RNA). Secondly, PNA possesses high bio-stability. PNA oligomers are degraded by neither nucleases nor proteases. Finally, PNA is achiral, thereby avoiding problems of enantiomeric purity. These characteristics give PNA the great potential to act as biomolecular probes in DNA detection [47]. Leclerc *et al.* [33] developed a DNA sensor based on a ferrocene-functionalized, cationic PTP derivative and a PNA as a capture probe. They prepared gold electrodes with a monolayer of PNA capture probes and detected target DNA using square-wave voltammetry. The authors obtained a sensitivity of 5 fM with this method. Similarly, they developed an improved electrochemical DNA biosensor by modifying the electrode surface with gold nanoparticles [48]. The sensitivity was improved due to the use of gold nanoparticles, which significantly increased the quantity of the PNA capture probe. Prabhakar *et al.* [49] reported that 20-bases long NH_2-modified PNA probes, specific to a pathogen, were covalently immobilized onto a PANI/Au electrode to detect nucleic acid hybridization with complementary targets within 30 s using methylene blue as a redox probe. The PNA–PANI/Au electrode exhibited an improved specificity (1,000 times) and a detection limit of 0.125×10^{-18} M, as shown in Figure 3.7. These PNA–PANI/Au electrodes could be used for 13–15 times.

DNA as probe for DNA detection

Considering the high synthesis cost of PNA, DNA as probe is more economic. There are many biosensors reported that utilized DNA probe to detect the target DNA. Liu *et al.* [14] investigated a PTP integrated with a redox indicator (methylene blue) and its possibility as an indicator for electrochemical DNA hybridization detection. After hybridization with a complementary target DNA, the resulting negatively charged duplex interacted with the redox indicator-functionalized cationic PTP through an attractive electrostatic interaction. The hybridization always resulted in the increased current signal of PMT indicator, because of the electrostatic interaction between PMT-indicator and dsDNA.

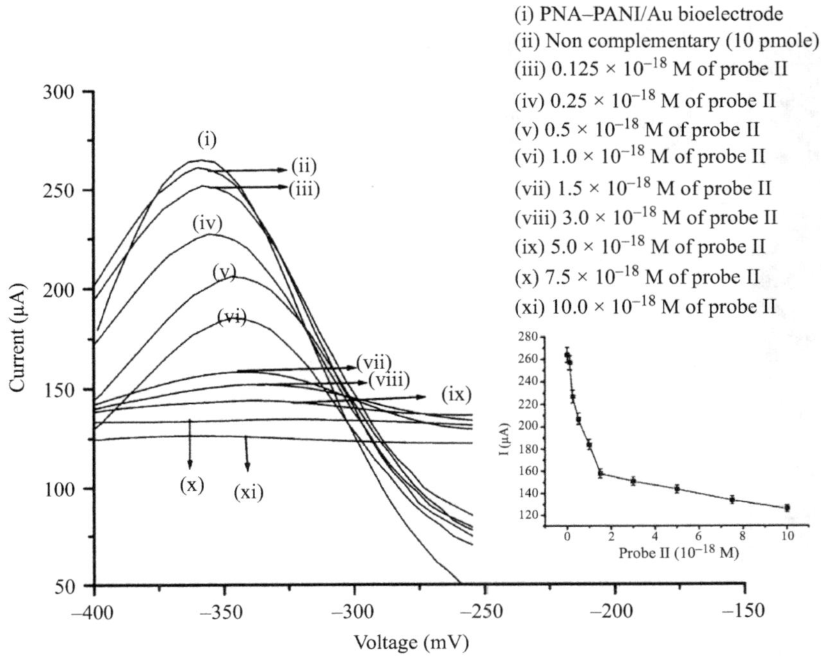

Figure 3.7 SWV of PNA–PANI/Au electrodes after hybridization with complementary target DNA strands in different concentrations. Inset shows plot of probe II concentration as a function of MB oxidation peak current. Phosphate buffer 0.05 M, pH 7.0 at pulse height: 60 mV, frequency: 150 Hz. Reproduced with permission from [49]

From the current change in DPV response, a linear range of concentration from 6.37 nM to 0.204 μM for DNA target was obtained under optimized condition.

Sui *et al.* [50] reported that PMVEA could be doped on poly(*o*-methoxyaniline) nanospheres and possessed larger specific area, thus more DNA probe molecules could be immobilized on the polymer. Therefore, a constructed label-free DNA sensor could detect complementary DNA with the concentration of as low as 0.10 nM. Tuan *et al.* [51] developed a highly sensitive DNA sensor based on PPy nanowire. The PPy nanowire was synthesized on the surface of the micro-sensor, using gelatin as the soft mold achieved through the potentio-static technique. N–H ends of the polypyrrole nanowire orientate upward from the surface, which would facilitate the DNA probe immobilization through the simple linkage with the phosphate groups of the probe DNA. Due to the high specific area of PPy nanowire, this sensor provided a new method for rapid and convenient detection of *Escherichia coli* DNA, with a detection limit of 0.1 nM. Zheng *et al.* [52] reported a novel DNA electrochemical biosensor based on elongated hexagonal-pyramid CdS

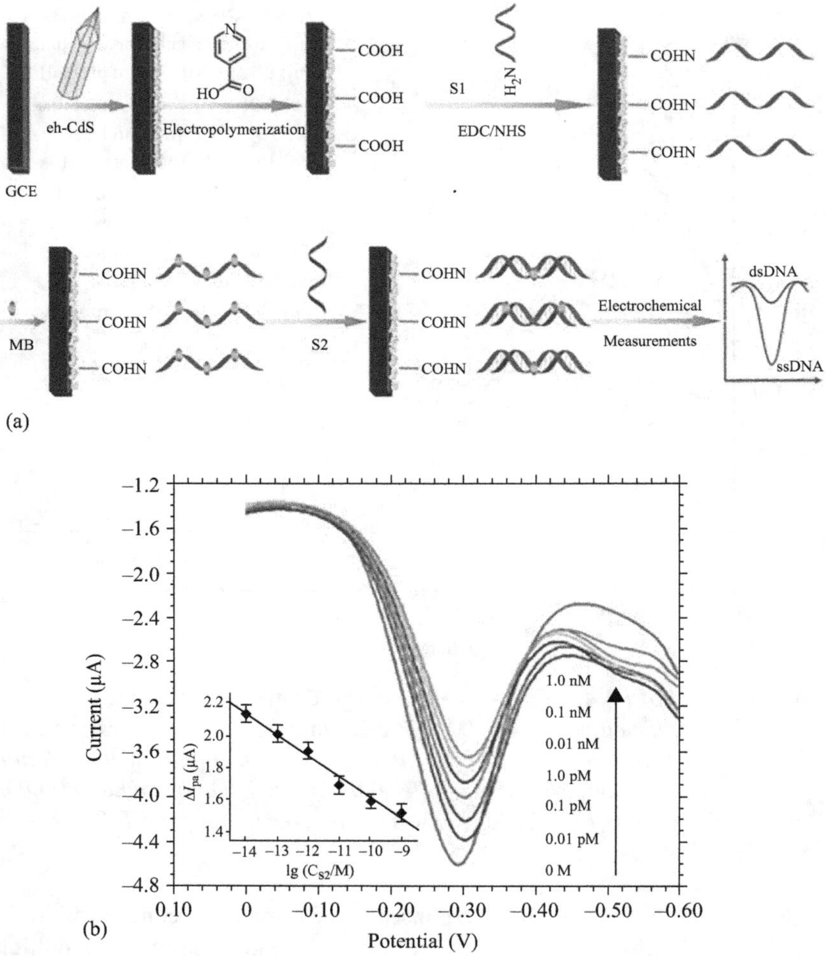

Figure 3.8 (a) The fabrication procedures of the eh-CdS/PIA composite film-based DNA biosensor. (b) DPVs of MB at probe DNA/PIA/eh-CdS/GCE upon hybridization with increasing amounts of target DNA. Inset shows the plot of ΔIpa versus the logarithm values of C_{S2} (lg C_{S2}). Reproduced with permission from [52]

and poly-isonicotinic acid composite film. Due to the high electrical conductivity of the elongated hexagonal pyramid CdS and the excellent stability of poly-isonicotinic acid, the sensor using methylene blue as a redox indicator could detect target DNA in a wide linear range from 10^{-14} to 10^{-9} M and with a low detection limit of 3.9 fM, as shown in Figure 3.8. Table 3.1 summarizes the characteristics of the electrochemical DNA sensors based on some common OCPs and their derivatives.

Table 3.1 Summary of electrochemical DNA sensors based on some common OCPs

Polymers	Immobilization method	Detection method	Detection aim	LOD	Ref.
Poly(pyrrole-nitrilotriacetic acid)	Electro-polymerization entrapment	AC impedance	Complementary DNA	~fM	[35]
Poly(pyrrole-co-3-pyrrolylacrylic acid)	Cross-linker and affinity	AC impedance	Complementary DNA	0.98 nM	[30]
PANI	Cross-linker and affinity	Chronocoulometry	Complementary DNA	8.0×10^{-17} M	[37]
PANI	Cross-linker and affinity	Square-wave voltammetry	Complementary DNA	0.125×10^{-18} M	[49]
PANI	Cross-linker and affinity	Differential pulse voltammetry	Complementary DNA and single-base mismatch	37.59 fM	[45]
Poly(*o*-methoxyaniline)	Cross-linker and affinity	AC impedance	Complementary DNA	0.1 nM	[50]
PTP	Cross-linker and affinity	Differential pulse voltammetry	Complementary DNA	6.37 nM	[14]
PTP	Cross-linker and affinity	Square-wave voltammetry	Complementary DNA and single-base mismatch	1 nM	[16]
PTP	Cross-linker and affinity	Square-wave voltammetry	Complementary DNA and single-base or two-base mismatch	0.5 nM	[33]
Poly(JUG-co-JUGA)	Cross-linker and affinity	Square-wave voltammetry	Complementary DNA and single-base mismatch	NA	[44]
PDDA/PSS	LBL method	AC impedance	DNA damage	NA	[15]
Poly-isonicotinic acid	Cross-linker and affinity	Differential pulse voltammetry	Complementary DNA and single-base mismatch	3.9×10^{-15} M	[52]

Abbreviations: PTP: polythiophene; PPy: polypyrrole; PANI: polyaniline; JUG: juglone; JUGA: 5-hydroxy-3-thioacetic-1,4-naphthoquinone; PDDA: poly(allylamine hydrochloride); PSS: polystyrene sulfonate.

3.3 DNA sensors based on OTFTs

3.3.1 Brief description

Since the discovery of the highly conductive polymer in the late 1970s [53], organic electronics has become revolutionary in the academic community and the industry, because of many merits of organic materials, such as low cost, flexibility, easy fabrication, and biocompatibility. Much progress has been made in the optimization of electrical, structural, chemical, mechanical, and optical properties of organic electronics, including organic light-emitting diodes [54], organic solar cells [55,56], and OTFTs [57,58].

It is notable that OTFTs have been intensively investigated for the applications in chemical and biological sensing [59–61]. Among the possible different transduction mechanisms, OTFTs allow a direct readout of an electrical signal. The signal-to-noise ratio can be improved by the inherent ability of transistors that can amplify the input signal, and it is important for sensors based on OTFTs to be able to detect low concentrations of specific analytes with high sensitivity and high selectivity. In addition, in the case of the analysis of electrically active biospecies in liquids, label-free detection of the analytes is possible, with a consequent reduction of costs and complexity in the sample preparation.

In an OTFT-based sensor, the active organic semiconductor in the channel is the critical element for the sensing performance. To date, several different organic semiconductors have been used in OTFTs sensing applications, such as pentacene [62], poly(3-hexylthiophene) (P3HT) [63], and poly(3,4-ethylenedioxythiophene): poly(styrene sulfonic acid) (PEDOT:PSS) [63], as shown in Figure 3.9.

Organic semiconductors, including conjugated polymers and small molecules, have normally exhibited superiority in biosensing. Polymer semiconductors normally contain a π-conjugated backbone and side chains that can increase the solubility of these materials in solvents. Therefore, OTFTs can be fabricated through the solution process, including spin-coating, screen printing [64], thermal evaporation, and inkjet printing [65], which can largely reduce the complexity and the costs of the manufacturing process. Because of the electronic properties (such as high electrical conductivity) that are similar to those of metals, some of these conjugated polymers are referred to as "synthetic metal" [66]. The intrinsically conductive polymers have been widely used in sensing applications by virtue of their inherent charge transport properties [67,68]. Some of the small molecule organic semiconductors have the carrier mobility which is comparable to or even higher than that of amorphous silicon [69,70]. Compared to the polymer counterparts, small molecule organic semiconductors exhibit some other advantages, such as high purity, well-defined molecular structures, and controllable molecular weight [71,72]. The precise organic synthesis can be used to fine-tune the OTFTs chemical and physical properties, which makes OTFTs have a higher sensitivity. Due to some significant advantages such as easy and cheap fabrication, flexibility, and biocompatibility, organic semiconductors obtain large achievements in OTFT devices, which make OTFTs excellent candidates for the application in disposable sensors.

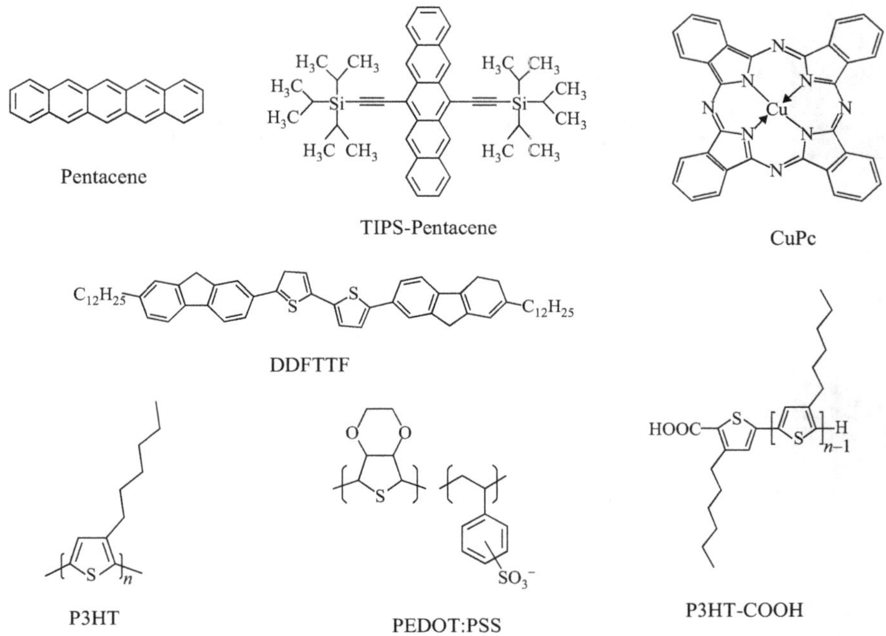

Figure 3.9 Chemical structures of the organic semiconductors commonly used in OTFT-based DNA sensors

As a kind of electric-based sensor, OTFTs open up opportunities for charged biomolecules' detection using an electrical test system. DNA is a polyanion with negative charges distribution along the chain, which can be sensed by OTFTs without labeling. OTFT-based DNA sensors have many significant advantages, such as easy fabrication, label-free detection, fast speed, high sensitivity, wide detection range, and low detection limit, which play important roles in the field of novel DNA sequencing and gene mapping technologies.

Research on OTFT devices for DNA sensors has emerged over the last decade, and various device structures have been studied. Among the various device structures, traditional OFET (Figure 3.10(a)) has been the most intensively used in the field of OTFT-based DNA sensing. This kind of sensor must be used in a dry environment; otherwise, its semiconductor layer could be corroded by buffer solution during the reaction, which could damage the performance of the sensor. In the last few years, some concept-new type OTFTs have emerged as a potential platform, such as charge-modulated organic field-effect transistor (CM-OFET) (Figure 3.10(b)), electrolyte-gated organic field-effect transistor (EG-OFET) (Figure 3.10(c)), and organic electrochemical transistor (OECT) (Figure 3.10(d)). These novel OTFTs can be operated in aqueous electrolytes, which are essential for real-time DNA sensing, and such devices are more suitable for in situ

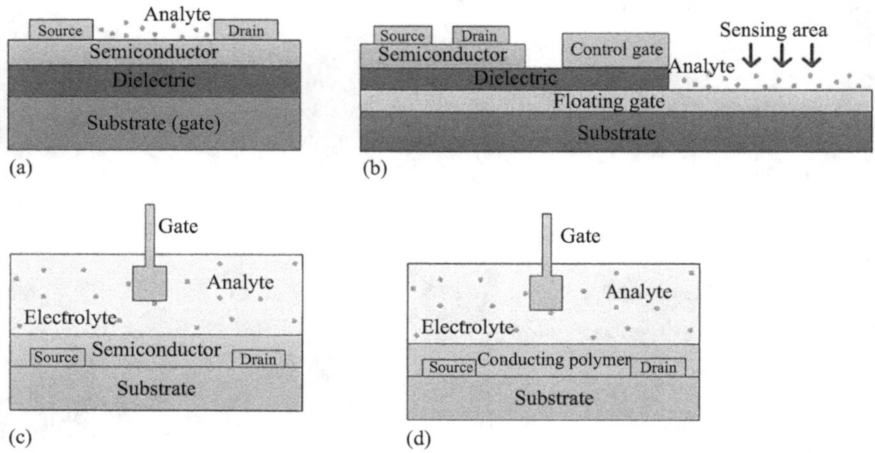

Figure 3.10 Different structures of OTFT DNA sensors: (a) traditional OFET structure, (b) CM-OFET structure, (c) EG-OFET structure, (d) OECT structure

measurements. Detailed discussions on each type of DNA sensor are carried out in the following sections.

Here, we introduce the research progress of label-free detection of DNA by OTFTs. We first describe the intrinsic detection principle of the OTFT-based DNA sensor. Sensors based on different types of OTFT, including traditional OFET and concept-new type OTFT, will be described separately in this section. All of the sensors will be classified and grouped by their applications and presented in sequence, and their pros and cons will be described in relative sections, respectively, to provide readers with a more intuitive and comprehensive understanding of the OTFT-based DNA sensor.

3.3.2 Mechanisms of OTFT sensors for DNA

OTFTs are three-terminal electronic devices including the source, drain, and gate electrodes. A thin film of organic semiconductor between the source and drain electrodes acts as the active channel [73]. The channel current (between source and drain electrodes) flows through the organic semiconductor and could be effectively modulated by the gate voltages [74]. In general, there are some differences in the operation principles and the device structures between traditional OFET and concept-new type OTFTs.

3.3.2.1 Traditional OFET

The basic components of a traditional OFET include an organic semiconducting thin film as the active channel, an insulating layer as the gate dielectric, and three electrodes (source, drain, and gate) (Figure 3.10(a)). In OFETs, the source–drain current is modulated by field-effect doping, in which the charge carrier density in the organic semiconductor is controlled by the gate electrode via an electric field applied across an insulating layer (gate dielectric). The operation of the OFET is based on the modulation of carrier density in the channel by the electric field from the gate electrode. Consequently, the channel current could be changed for several orders of magnitude by the gate voltage, indicating that the device is a type of amplifier or a transducer that can convert a voltage/potential signal into a current response [75]. OFET-based sensors are able to determine the target analytes, because of the physical or chemical changes in the semiconducting layers induced by the targets [76]. The OFET performance is significantly influenced by the properties of the grain boundaries and dielectric/insulator interface. Therefore, the properties and functionalities of OFET-based sensors can be tuned via surface engineering [59,77].

OFET normally operates in an accumulation region, in which the low off current is dominated by the inherent conductivity of the active layer. When a sufficient gate voltage is applied, the source–drain current flowing through the organic semiconductor film can be modulated by the field-effect doping process. The channel current of the OFET is given by [78]:

$$I_{DS} = \frac{W}{L}\mu C_i\left(V_g - V_t - \frac{V_{DS}}{2}\right)V_{DS}, \quad V_{DS} \ll V_g - V_t$$

$$I_{DS} = \frac{W}{2L}\mu C_i(V_g - V_t)^2, \quad V_{DS} > V_g - V_t$$

where I_{DS} is the channel current, V_{DS} is the source–drain voltage, V_g is the gate voltage, V_t is the threshold voltage of the device, W and L are the width and length of the channel, respectively, μ is the mobility of the carrier, and C_i is the capacitance of the gate insulator. The channel current of an OFET-based sensor relies on the threshold voltage V_t and carrier mobility μ, which both play key roles in sensing applications. When the detected analytes are interacted with the device, the threshold voltage V_t could be usually modulated by the doping effect of analytes [79,80]. Moreover, the carrier mobility μ could also be influenced by the analyte molecules which are diffused into the semiconductor grain boundaries [81,82].

3.3.2.2 Concept-new type OTFTs

Charge-modulated organic field-effect transistor

CM-OFET is a novel OFET. Compared with traditional OFET, CM-OFET has an elongated gate electrode, called floating gate, in which the uncovered part can be used as the biosensing area. Moreover, an additional fourth electrode (control gate) is introduced in the device, which forms a planar capacitor with the floating gate (Figure 3.10(b)). The control gate allows the biasing of the device, while the uncovered part of the floating-gate hosts the probe area of the sensor. During the sensing process, the device can be considered as two OFETs, one is operated as the "sensor," and the other is acted as the "reference" [83].

By setting the voltage drop on the insulating layer through the control gate voltage (V_c), the charges of analytes anchored on the floating gate can modulate the threshold voltage of the device. Therefore, the current flowing between the source and drain electrodes changes. Due to the charge conservation principle, the total charge Q_F trapped into the floating gate is given by [83]:

$$Q_F = C_{CF}(V_F - V_c) + C_{SF}V_F + C_{DF}(V_F - V_D) + Q_I$$

in which C_{CF}, C_{SF}, and C_{DF} are the capacitances between the floating gate and the other electrodes of control gate, source, and drain, respectively, V_F is the floating-gate voltage, V_D is the drain voltage, and Q_I is the charge induced by the molecules anchored on the probe area. The floating-gate voltage can be expressed by [83]:

$$V_F = \frac{C_{CF}}{C_{TOT}}V_C + \frac{C_{DF}}{C_{TOT}}V_D + \frac{Q_F - Q_I}{C_{TOT}}$$

where C_{TOT} is the sum of all the capacitance in the structure.

Electrolyte-gated organic field-effect transistor

EG-OFET is another novel concept-new type OFET. Different from the conventional OFET, EG-OFET has an organic semiconductor channel and a gate electrode that is separated by an electrolyte layer (Figure 3.10(c)) [84,85]. When the gate voltage is applied, anions and cations of the electrolyte are attracted to the electrolyte/gate interface and electrolyte/semiconductor interface, respectively, resulting in an electrical double layer (EDL) at the both interfaces [86,87]. Since the thickness of the EDL at the electrolyte/organic semiconductor interface in EG-OFET is much smaller than the distance between the gate and its insulator in traditional OFET, EG-OFET has much lower operation voltage ($<$1 V). The low-voltage operation and water-stable operation surrounding make EG-OFET a good candidate for in situ bio-detection.

One issue in this kind of device is that electrochemical doping and field-effect modulation often coexist. If the modulation of the channel current by the gate voltage in the device is mainly due to electrochemical doping instead of field-effect modulation in the active layer, it is better regarded as a type of OECT [88,89]. However, we still classify the EG-OFETs as a type of OFET in this section, according to the terminology.

Organic electrochemical transistor

OECT is a new kind of OTFT, which is also built-up by the combination of three electrodes (source, drain, and gate), similar to those of OFET. One difference is that an electrolyte medium rather than a solid-state insulating layer is incorporated between the gate electrode and the organic semiconductor channel in OECT, which is like EG-OFET. Therefore, the channel and gate electrodes are separated, which make the device be easily fabricated in an electronic circuit or integrated in a microfluidic system. The structures of OECT and EG-OFET are quite similar, only one difference is the material applied for the active layer. EG-OFET normally employs organic semiconductor, while OECT usually uses electrically conducting polymer that is not a typical semiconductor material in the traditional sense.

Compared with traditional OFET, OECT normally shows stable performance in aqueous environment at low voltage (<1 V). In OECT, the source–drain current is modulated by electrochemical doping or de-doping, in which the change in conductivity of the organic semiconductor is mediated by ions from an adjacent electrolyte. This process usually involves the migration of mobile ions into or out of the semiconducting layer. The channel current of the OECT is given by [60,90,91]:

$$I_{DS} = \frac{q\mu p_0 t W}{L V_p}\left(V_p - V_g^{eff} + \frac{V_{DS}}{2}\right)V_{DS}, \quad |V_{DS}| \ll \left|V_p - V_g^{eff}\right|$$

$$V_p = q p_0 t / C_i$$

$$V_g^{eff} = V_G + V_{offset}$$

where q is the electric charge, μ is the hole mobility, p_0 is the initial hole density in the active layer when gate voltage is zero, t is the thickness of the active layer, W and L are the channel width and length, respectively, C_i is the effective gate capacitance of the transistor, V_p is the pinch-off voltage, V_g^{eff} is the effective gate voltage, and V_{offset} is the offset voltage related to the potential drop at the gate/electrolyte interface and the electrolyte/channel interface [92].

In order to further understand the transient behavior of an OECT, the device can be divided into an electronic circuit and an ionic circuit [91]. The electronic transport of the device is related to the carrier density and mobility in the channel, while the ionic transport is determined by the ionic diffusion from the electrolyte into the organic semiconductor film. Moreover, the transient properties of the device can also be affected by the geometric size of the OECT [93,94]. In general, the smaller the device dimension is, the faster the response time can be obtained. A response time of several hundred microseconds has been achieved for the device with micro-dimensions [94], which enables fast detection and analysis of analytes in sensing applications.

3.3.2.3 The DNA sensing principles

So far, different possible mechanisms have been exploited to realize OTFT-based DNA sensors. Most of them are based on the interaction between DNA molecules and OTFT devices with the effect of modifying the electrical characteristics of the transistors. Firstly, an appropriate probe ssDNA should be immobilized on a particular area of the OTFT device, including organic semiconductor active layer, source–drain electrodes, gate electrode, and dielectric layer, via physical electrostatic attraction or chemical modification. This step is called bio-functionalization of the DNA sensor, which causes the highly selective detection of complementary target DNA through the hybridization process. When the DNA hybridizations occur on different areas of the OTFT device, different sensing principles for the recognization of DNA are carried out, such as the doping effect of organic semiconductor, the charge coupling in the dielectric layer, the contact resistance change on the source–drain electrodes, and the change of work function of the gate electrode. In general, the negative charges of DNA directly or indirectly affect the carrier density or mobility in the channel, which leads to the change of threshold voltage (V_t) of the device. Thus, the channel current I_D shift in transfer/output characteristic curves can be observed. Moreover, the hybridization of DNA induces more negative charges than the immobilization process, which causes more shift of I_D in transfer/output characteristic curves. Therefore, the immobilization and hybridization of DNA can be clearly distinguished by the change of I_D.

3.3.3 OTFT sensors for DNA

3.3.3.1 Traditional OFET-based DNA sensors

Pentacene-based OFET sensors for DNA

Pentacene has been proven to be the most widely used p-type organic semiconductor, because of its high hole mobility, excellent chemical stability, and adequate orbital overlapping of molecules. Moreover, it establishes a stable interface with the source and drain electrodes. As the highest occupied molecular orbital level of pentacene is close to the Femi level of some commonly used electrode metals, such as gold, copper, and aluminum, the holes can easily flow at the metal/semiconductor interface [95,96]. Zhang *et al.* [97] reported the first label-free DNA sensor based on OFET using pentacene as the active layer. The DNA molecules were physically immobilized on the surface of pentacene through a hydrophobic interaction between pentacene and DNA probes, which was a faster and easier method. The threshold voltage of the device, when tested under dry conditions, shifted to a more positive value upon the immobilization of DNA, as shown in Figure 3.11(a). It was found that the ssDNA and dsDNA could be successfully differentiated by recording the shift values of threshold voltage. The authors suggested that this phenomenon was induced by the electron-withdrawing nature of DNA. Moreover, the pulse-enhanced hybridization of DNA was also used to accelerate the detection process.

Although a traditional OFET device was able to roughly detect DNA immobilization and hybridization, the pentacene film morphology was an important property that greatly influenced the sensitivity of the device. The non-uniformity of the pentacene film made sensing performance vary from different transistors, different immobilization ability of DNA probes, and different reaction time of hybridization. Jagannathan *et al.* [81] statistically investigated the optimized conditions to improve the device sensitivity. They found that the thickness of pentacene film, substrate temperature, and input current influenced the morphology of pentacene film and played an important role in the sensitivity of the device. Enhanced performance was achieved by increasing the roughness of the pentacene film or reducing the film thickness. Later, Zhang *et al.* [98] investigated the sensing mechanism of the DNA sensors and found that the DNA molecules on the active layer could extract electrons and change the doping level in the pentacene film. They also implemented a microfluidic system to realize a self-supporting DNA sensor platform, enabling a disposable and low-cost sensor.

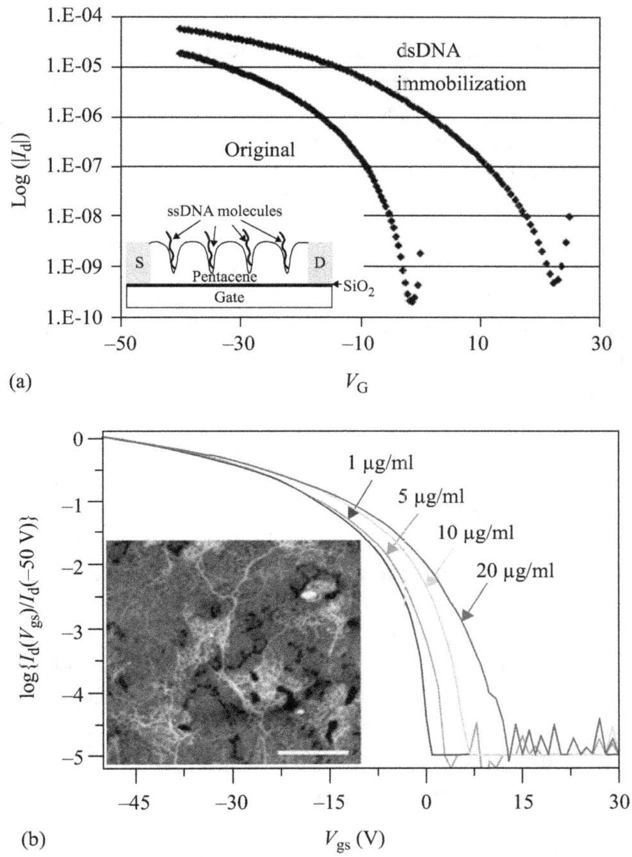

Figure 3.11 *(a) Transfer curve of an pentacene OFET before and after the immobilization of dsDNA; Reproduced with permission from [97]. (b) Transfer characteristics of a pentacene OFET in response to DNA immobilization with different concentrations. Inset: AFM image of pentacene film immobilized with DNA with a concentration of 10 μg/ml. The scale bar is 500 nm. Reproduced with permission from [82]*

Stoliar *et al.* [82] reported a label-free DNA sensor based on an ultrathin OFET using a two-monolayer-thick pentacene film. It was observed that the DNA molecules could be electrostatic adsorbed on the surface of pentacene film without any external binding agents or immobilization. A pinch-off voltage of the device shifted during the detection of different concentrations of DNA and the sensitivity reached to 1 μg/mL, as shown in Figure 3.11(b). This response could be ascribed to the increased positive charges within the pentacene film induced by DNA adsorption.

In order to improve the immobilization efficiency of DNA molecules on the surface of pentacene film, the electric field had been used to enhance DNA hybridization. Liu *et al.* [99] introduced different electric biases during the DNA molecules immobilization. An optimized immobilization with the time of 30 min at a bias of 50 V exhibited the highest immobilization efficiency. However, this technique has not proven to be useful in DNA sensors, since the high bias voltage may change the properties of the pentacene film. In terms of this technique, Chand *et al.* [100] studied electrical characteristics between the device and the adsorbed DNA as a function of immobilization voltage and time. They validated that higher and longer positive electric fields applied by gate voltage were more conducive to adsorption of DNA. A low electric field (5 V) to immobilize DNA on the pentacene surface led to a drastic reduction of immobilization time (2 min), compared to 60 min as reported earlier [101,102].

Other organic materials–based OFET sensors for DNA

Some alternative organic semiconductors with high stability in air and moisture have been used in OFET-based DNA sensors. Since pentacene is sensitive or vulnerable to external stimuli, such as oxygen and humidity, the direct immobilization of DNA molecules on the pentacene film may decrease the stability and electrical properties of the active layer. Khan *et al.* [103] proposed an in situ label-free DNA sensor based on OFET using 5,5′-bis-(7-dodecyl-9*H*-fluoren-2-yl)-2,2′-bithiophene (DDFTTF) as the semiconductor layer, which is aerotolerant and waterproof. An ultrathin poly-maleic anhydride film was deposited on the surface of the device for the attachment of PNA, which was then used for the in situ detection of target DNA in solution. The hybridization of complementary target DNA with PNA probes increased the surface charge density on the channel, which led to the change of channel current. Differentiating between the target DNA sequence with only one or two base mismatches was successfully achieved using this method, as shown in Figure 3.12(a). The device was able to detect the hybridization of target DNA down to 1 nM. The authors [104] also described a bilayer organic semiconductor OFET structure, which consisted of a pentacene active layer with copper phthalacyanine (CuPc) as the top surface passivation layer. In this case, CuPc acted as a surface passivation layer with efficient charge injection into the pentacene layer, thus providing a versatile method for the incorporation of many semiconductor materials as the sensing element. OTFTs fabricated in this kind of structure could be used for DNA sensing with stable performance in aqueous media. Gui *et al.* [105] applied this method to DNA sensing, and there was no significant initial performance drop after exposure to the ambient for 30 days in the OFET sensor.

Yan *et al.* [106] demonstrated a novel DNA sensor based on OFET with semiconducting polymer P3HT. Both single and double DNA molecules were immobilized on the surface of gold source–drain electrodes before coating the P3HT active layer, thus avoiding organic semiconductor corrosion by DNA buffer. A dramatic decrease of channel current was observed after the hybridization of DNA molecules, thus ssDNA and dsDNA were successfully differentiated in the device performance (Figure 3.12(b)). The sensing mechanism was attributed to the

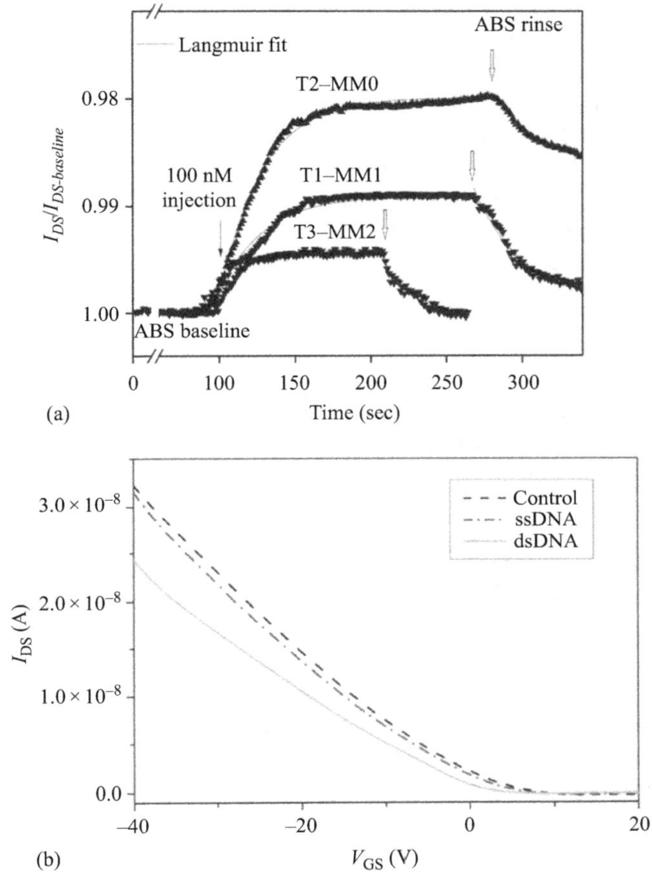

Figure 3.12 *(a) Channel current response vs. time when an OFET was exposed to DNA solutions with the sequences of T2-MM0, T1-MM1, and T3-MM2; Reproduced with permission from [103]. (b) Transfer curves of three P3HT OFETs with ssDNA, dsDNA or without DNA on the source/drain electrodes fabricated at the same condition. Reproduced with permission from [106]*

increase of the contact resistance at the source–drain electrodes, which was caused by the decrease of the work function of gold electrodes upon the introduction of DNA.

3.3.3.2 Concept-new type OTFT-based DNA sensors

CM-OFET-based sensors for DNA

CM-OFET DNA sensor is an organic implementation of the charge modulated field-effect transistor, whose sensitive area is completely separated from the OTFT [107].

Its working principle is independent of the semiconductor properties, as compared to traditional OFETs. As a DNA sensor, negatively charged DNA molecules anchored on the floating gate are detected because of the polarization induced in the gate dielectric. Moreover, the organic semiconductor does not need to be in contact with the solution that contains DNA strands, and, thus, can be protected from the external environment. In this way, it is possible to reduce the organic semiconductor degradation, which is due to its exposure to humidity, and, therefore, to extend the device's life span.

Demelas's group [83,108] proposed a CM-OFET DNA sensor deposited on a flexible substrate with floating-gate Au electrode, as shown in Figure 3.13(a) and (b). The DNA molecular was modified with thiol (HS) group as a thiol bound to connect with gold electrode. Negatively charged DNA anchored on the floating gate was detected based on the polarization induced in the floating gate, capacitively coupled to the organic semiconductor channel. This structure of device allowed protection of active layer from the external environment, thus preventing the device from degradation.

Lai *et al.* [109] reported a CM-OFET-based DNA hybridization sensor that could operate at ultralow voltage, the general overview of the device is described in Figure 3.13(c) and (d). Low operating voltage is considered crucial for the stability of the FET-based DNA sensor measured in aqueous medium. The authors had made further improvements to the previous device [83,108] for achieving low operating voltage. A double-layer dielectric consisting of ultrathin Al_2O_3 and Parylene C is adopted rather than the original one-layer dielectric. It was demonstrated that the sensor was able to distinguish fully not-complementary, single-mismatched, or complementary DNA targets between different target concentrations, with an unprecedented concentration as low as 0.1 nM.

EG-OFET-based sensors for DNA

EG-OFETs have attracted more attention lately, because of their low-voltage operation, as compared to conventional OFETs. After the success of the first EG-OFET [84], a series of EG-OFET biosensor devices came out in the last few years for the detection of DNA [110], enzymes [85], proteins [111], and others [112,113]. Kergoat *et al.* [110] reported an EG-OFET-based DNA sensor, in which the dielectric was constituted by a simple droplet of aqueous phosphate buffered saline (PBS) solution, or pure water, as described in Figure 3.14. In experiments, DNA probes were covalently anchored on the P3HT active layer via *N*-hydroxysuccinimide/1-ethyl-3-(3-dimethylaminopropyl)carbodiimide (NHS/EDC) chemistry. Clear changes in the electrical characteristics were observed upon DNA immobilization. Complementary DNA target strands and probe DNA could be easily distinguished by observing the shift of gate voltage. This behavior was attributed to the steric hindrance of DNA chains that prevent ion penetration into the bulk of the semiconductor [114]. The results showed the importance of the Debye length that can screen negative DNA charges. Pure water has a Debye length about two orders of magnitude larger than the general

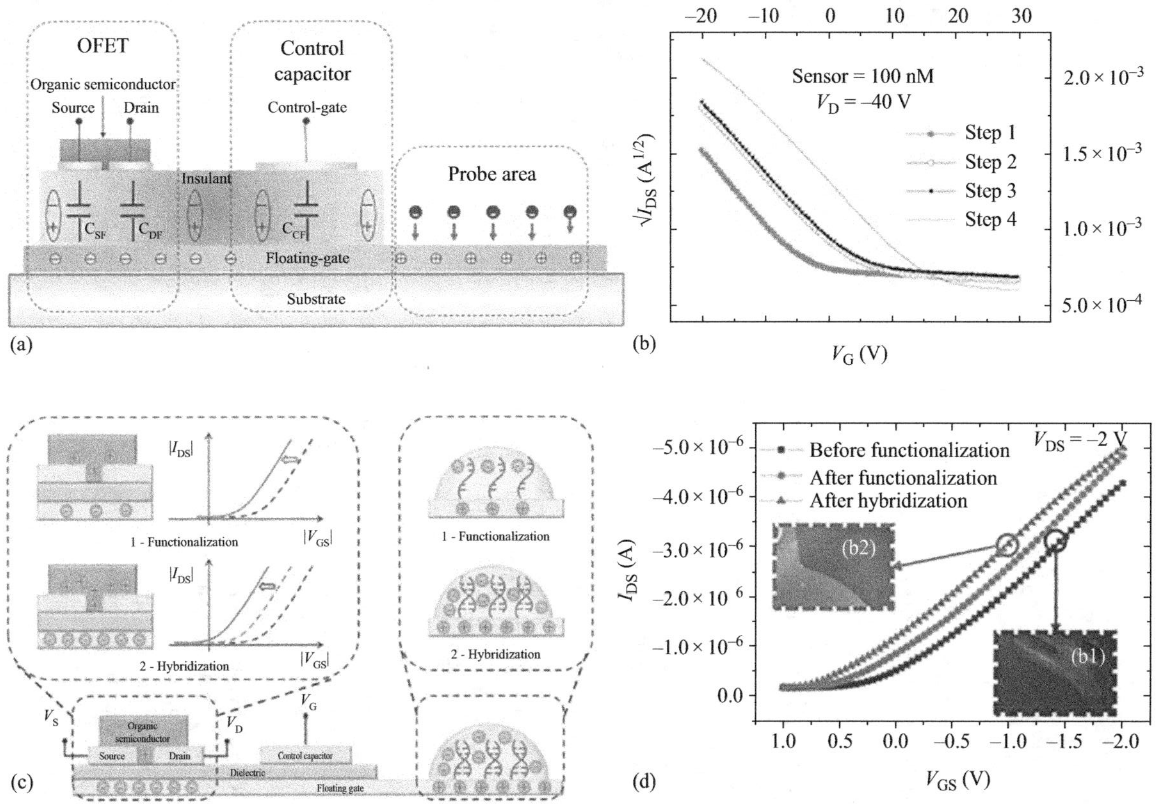

Figure 3.13 *(a) Vertical section of the CM-OFET sensor structure. (b) The trans-characteristics of a sensor taken before (Step 1) and after (Step 2) the immobilization of the HS–ssDNA, and before (Step 3) and after (Step 4) the deposition of the fully complementary strands of the HS–ssDNA; Reproduced with permission from [83]. (c) Structure and detection mechanism of the CM-OFET. (d) Transfer characteristics of the CM-OFET before and after DNA functionalization and hybridization. Reproduced with permission from [109]*

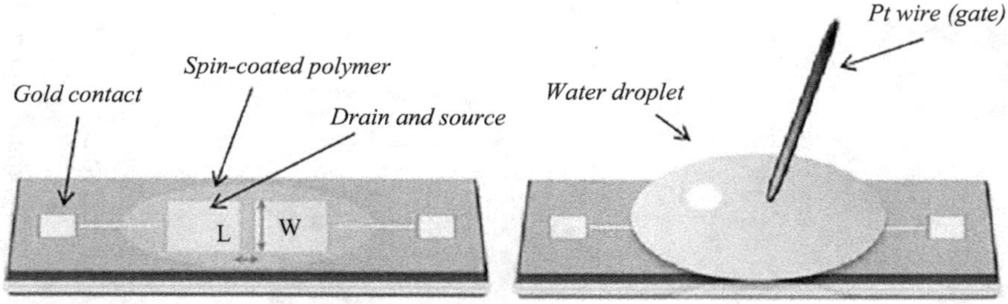

Figure 3.14 Schematic view of the WG-OFET device. Reproduced with permission from [110]

biological buffer. In this sense, the water-gated transistor appeared very pertinent for DNA detection.

One concern with OTFTs with electrolytes is slower switching time. To improve the device speed, Scott *et al.* [115] reported an EG-OFET-based DNA sensing platform, which used ion-gels as a gate dielectric by virtue of their extremely large capacitance (Figure 3.15). They also used a floating-gate electrode functionalized with ssDNA, whose potential was determined by both capacitive coupling with a primary, addressable gate electrode and the presence of adsorbed molecules. When DNA was hybridized at the floating gate, it neutralized the primary gate voltage felt by the P3HT semiconductor. The offset could be directly measured and quantitatively related to the density of probe DNA molecules. The detection limit for this configuration was 10 nM.

OECT-based sensors for DNA

The sensor based on an OECT normally operates with the principle of changing the potential drop across one of the two interfaces, i.e., gate/electrolyte and electrolyte/channel, under the effect of analytes. After the first OECT invented in 1984 [116], Krishnamoorthy *et al.* [117] reported the primary DNA sensing OECT based on conducting PEDOT as the active layer. The bio-functionalization was based on an electrostatic interaction between the ssDNA probe and the polymer active layer. The hybridization reaction between the probe DNA and the target DNA kept the polymer semiconductor in an oxidized state and simultaneously changed the conductivity of polymer, thus the complementary target DNA in PBS solution could be detected down to 8×10^{-8} g/mL.

Recently, Lin *et al.* [118] demonstrated a label-free DNA sensor by using PEDOT:PSS-based OECT (Figure 3.16). The devices could be fabricated on flexible polyethylene terephthalate substrates and integrated in a microfluidic system. The performance of the device showed little difference at different bending states, which was useful in sensing applications. The device could be used to detect complementary DNA targets down to 1 nM. The detection limit of the DNA sensor

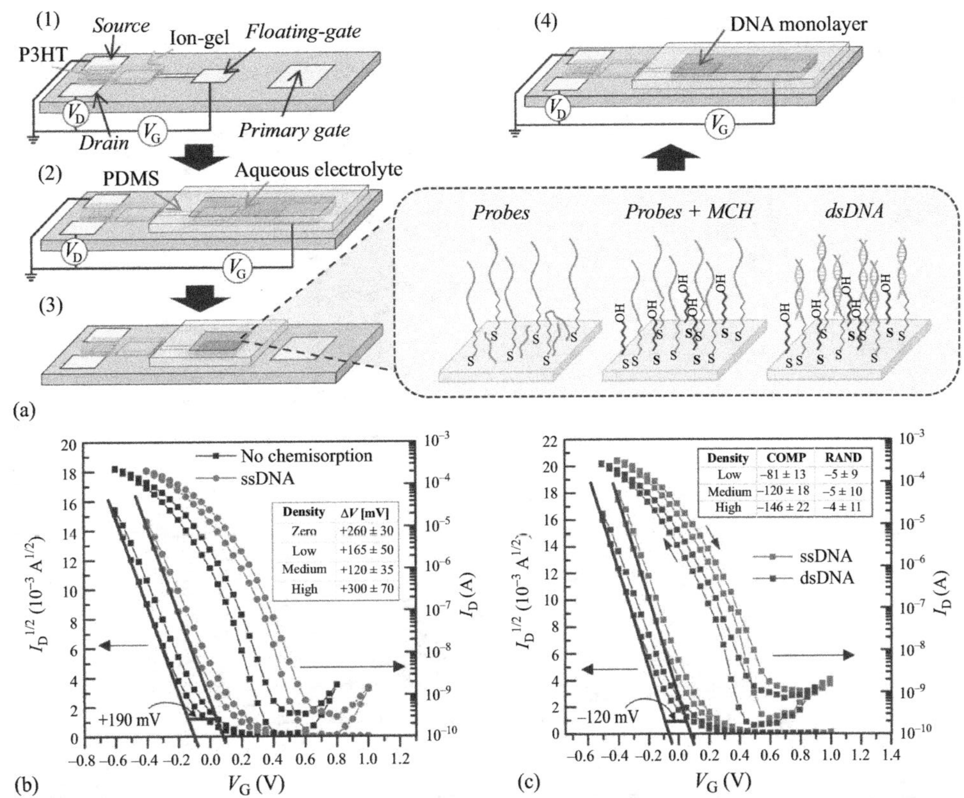

Figure 3.15 *(a) DNA detection scheme. (b) Sensor response of the device after chemisorption of ssDNA probes and complementary dsDNA. Reproduced with permission from [115]*

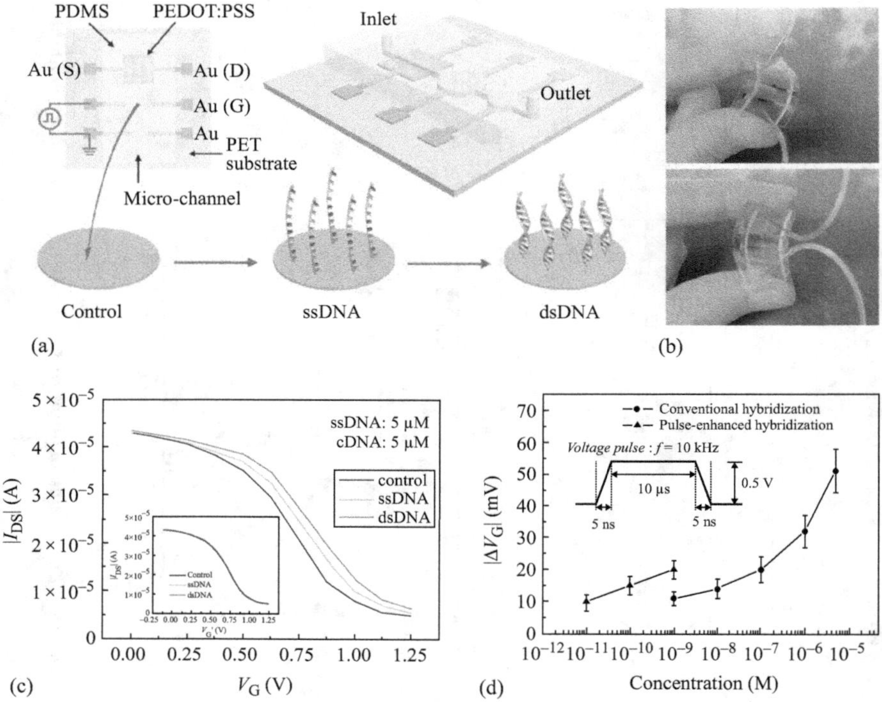

Figure 3.16 (a) Schematic of an OECT-based DNA sensor integrated in a PET flexible microfluidic system. (b) Photographs of the device bent to both sides. (c) Transfer curves of the device before and after the immobilization and the hybridization of DNA. Inset: the three transfer curves are shifted horizontally and merged into a universal curve. (d) $|\Delta V_G|$ as a function of the concentration of DNA targets in PBS solutions. Inset: the electric field pulse (frequency: 104 Hz, voltage: 0.5 V, pulse width: 10 μs, rise/fall time: 5 ns) applied during the hybridization of DNA. Reproduced with permission from [108]

was extended to 10 pM when the hybridization of DNA was enhanced by applying an electric pulse to the gate electrode. The sensing mechanism was normally induced by a surface dipole formed by the intrinsic negative charge of DNA molecules. It was expected that OECTs integrated in flexible microfluidic systems could have promising applications in various types of highly sensitive biosensors.

To allow readers to draw comparisons, examples of device structures, semiconductor materials, detection limits, and detection parameters of different types of OTFT-based DNA sensors are summarized in Table 3.2.

Table 3.2 Summary of OTFT-based DNA sensors

Sensor context	Semiconductor material	Detection limit	Solution condition	Ref.
Traditional OFET	Pentacene	Non	Sodium chloride, sodium citrate	[97]
	Pentacene	Non	Non	[81]
	Pentacene	Non	Sodium chloride, sodium citrate	[98]
	Pentacene	1 μg/mL	Tris(hydroxymethy) aminomethane	[82]
	Pentacene	Non	PBS (pH 7.4)	[99]
	Pentacene	Non	EtBr solution	[101]
	Pentacene	Non	EtBr solution	[102]
	Pentacene	Non	EtBr solution	[100]
	DDFTTF	1 μg/mL	Acetate buffer solution (pH 7.0)	[103]
	P3HT	Non	PBS (pH 7.4)	[106]
	Pentacene/CuPc	Non	PBS (pH 7.4)	[105]
CM-OFET	TIPS-pentacene	Non	PBS (pH 3.8)	[83,108]
	TIPS-pentacene	Non	KH_2PO_4	[109]
EG-OFET	P3PT-COOH	Non	DI water	[110]
	P3HT	Non	PBS (pH 8.0)	[115]
OECT	PEDOT	8×10^{-8} g/mL	PBS (pH 8.0)	[117]
	PEDOT:PSS	10 pg/mL	PBS (pH 7.0)	[118]

3.4 Conclusions

In this chapter, DNA biosensors based on conductive organic materials are well introduced and discussed, including electrochemical DNA sensors based on OCPs and DNA sensors based on OTFTs.

For electrochemical DNA sensors, we summarize the various devices based on different OCPs, mainly focusing on the common three conjugated polymers, including PANI, PPy, and PTP. Different immobilization methods for the DNA probes are introduced, while the frequently used method is the cross-linker and affinity interaction method, due to the enhancement of the bioactivity of the probe DNA. Meanwhile, different transduction mechanisms are well discussed. The effective electrochemical detection techniques are voltammetric and impedimetric methods. The sensing principle is that OCPs can convert the behavior of hybridization of a DNA probe with target DNA into electrochemical output signals, due to the change in the conformation and interface properties of OCPs.

Moreover, the sensitivity, selectivity, and bioactivity of probe DNA in OCPs were also comprehensively investigated. In can be found that less attention was given to the PNA probe combined with OCPs for application in the DNA sensors, perhaps due to the high cost of PNA, although it could be more sensitive for the detection of the target DNA. Another aspect of the OCPs-based DNA sensor that should be addressed is the stability that resulted from the weak physical adhesion between the polymer layer and an underlying electrode. Therefore, chemical modification of the working electrode with a monolayer would be beneficial for enhancing the stability of the OCPs-based DNA sensor. With the advancement of micro-electro-mechanical system (MEMS) technology in recent years, the geometrically well-designed and highly producible structure of OCPs can be achieved and the corresponding miniature DNA sensor devices or arrays would possess high sensitivity and multi-channel analysis abilities. This MEMS technology would also stimulate further research aimed at incorporating the OCPs based DNA sensor into a portable prototype for early gene disease diagnosis.

For OTFT-based DNA sensors, various types of devices with different sensing mechanisms are well discussed, including traditional OFETs and concept-new type OTFTs, such as CM-OFET, EG-OFET, and OECT. As the critical element for the sensing performance, organic semiconductors have emerged as viable candidates for the OTFT-based sensors. Typically, pentacene, PEDOT:PSS, and P3HT are the most commonly used for OTFT sensing application. Meanwhile, novel organic semiconducting materials are being developed gradually to further improve the performance of OTFT sensors. Due to their advantages in synthetic tailorability, tunable electronic conductivity, and simplified solution processable fabrication methods, as well as excellent biocompatibility, organic semiconductors have made OTFTs excellent candidates for DNA sensing.

The detection and analysis of DNA is one important research subject for OTFT-based sensors. As extensive investigation and development for many years, OTFTs with promising performance have been proven to be an effective means for DNA sensing. They have realized a direct, label-free, real-time, and sensitive detection of DNA with a detection limit down to the level of fM. Still, there are some aspects to be improved.

OFET, as a traditional OTFT, increases the diversity of devices and introduces the possibility of improving sensing selectivity, environmental adaptability, and degree of accuracy. While the development of the OFET-based DNA sensor is still in the exploratory stage, limit factors including the detection mechanisms of many OFET-based DNA sensors have not been researched in-depth. Meanwhile, the alternative organic semiconductors with high stability in air and moisture should be developed. The current focus is on pursuing new organic materials and optimizing the geometric features of devices, which will help to design novel and superior OFET-based DNA sensors.

It is worth noting that concept-new type OTFT sensors with stable performance in aqueous solutions are suitable for in situ or even in vivo bio-detections. Some of the new types of OTFT-based DNA sensors, such as CM-OFET, EG-OFET, and OECT, exhibit much better sensitivity than typical electrochemical detections. The main concern with EG-OFET or OECT with electrolytes is the slower switching time, usually limited to 100 Hz, compared to traditional OFETs with solid dielectrics. To improve the device speed, it is proposed to reduce the channel size, or use ion gels, to form gelation of a block copolymer in an ionic liquid.

The next step in research of DNA sensors based on OTFTs is to further improve these devices for real applications. On one hand, the sensitivity, selectivity, and stability characteristics of existing OTFTs based DNA sensors should be improved. For this, new materials and fabrication techniques should be developed, the geometric features of the devices should be optimized, and the device reusability should be enhanced. On the other hand, novel types of OTFTs based DNA sensors should be developed, which are potentially low cost, highly sensitive, flexible, biocompatible, and easy to fabricate.

References

[1] Vattanaviboon P., Sangseekhiow K., Winichagoon P., Promptmas C. 'Detection and haplotype differentiation of Southeast Asian alpha-thalassemia using polymerase chain reaction and a piezoelectric biosensor immobilized with a single oligonucleotide probe'. Translational Research. 2008; 151: 246–254.

[2] Minunni M., Tombelli S., Mascini M., Bilia A., Bergonzi M. C., Vincieri, F. F. 'An optical DNA-based biosensor for the analysis of bioactive constituents with application in drug and herbal drug screening'. Talanta. 2005; 65: 578–585.

[3] Keusgen M. 'Biosensors: new approaches in drug discovery'. Naturwissenschaften. 2002; 89: 433–444.

[4] Logrieco A., Arrigan D. W., Brengel-Pesce K., Siciliano P., Tothill I. 'DNA arrays, electronic noses and tongues, biosensors and receptors for rapid detection of toxigenic fungi and mycotoxins: a review'. Food Additives and Contaminants. 2005; 22: 335–344.

[5] LaGier M. J., Fell J. W., Goodwin K. D. 'Electrochemical detection of harmful algae and other microbial contaminants in coastal waters using hand-held biosensors'. Marine Pollution Bulletin. 2007; 54: 757–770.

[6] Wang J. 'Survey and summary from DNA biosensors to gene chips'. Nucleic Acids Research. 2000; 28: 3011–3016.

[7] Robin L., Stears T. M., Mark S. 'Trends in microarray analysis'. Nature Medicine. 2003; 9: 140–145.

[8] Sahoo S. K., Sharma D., Bera R. K., Crisponi G., Callan J. F. 'Iron(III) selective molecular and supramolecular fluorescent probes'. Chemical Society Reviews. 2012; 41: 7195–7227.

[9] Menon R., Yoon C. O., Moses D., Heeger J. 'Metal–insulator transition in doped conducting polymer' in: Skotheim T. A., Elsenbaumer T. A., Reynold J. R. (Eds), 'Handbook of conducting polymers', 2nd Ed., New York: CRC Press Taylor & Francis Group; 1998. p. 27–34.

[10] Blanchard P., Leriche P., Frere P., Roncali J. 'Advanced functional polythiophene based on tailored precursors' in: Skotheim T. A, Reynolds J. R. (Eds), 'Handbook of conducting conjugated polymers', 3rd Ed., Florida: CRC Press; 2006. Chapter 13, p. 1–77.

[11] Jeffries-El M., Mccullough D. 'Regioregular polythiophene' in: Skotheim T. A., Reynolds J. R. (Eds), 'Handbook of conducting polymers, conjugated polymers', 3rd Ed., Florida: CRC Press; 2006. Chapter 9, p. 1–49.

[12] Sui J., Zhang L. J., Peng H. 'Label-free DNA sensor construction using self-assembled poly(o-methoxyaniline) hollow nanospheres'. European Polymer Journal. 2013; 49: 139–146.

[13] Lee Y. T., Shim Y. B. 'Direct DNA hybridization detection based on the oligonucleotide functionalized conductive polymer'. Analytical Chemistry. 2001; 73: 5629–5632.

[14] Liu M. Q., Luo C. H., Peng H. 'Electrochemical DNA sensor based on methylene blue functionalized polythiophene as a hybridization indicator'. Talanta. 2012; 88: 216–221.

[15] Evtugyn G. A., Stepanova V. B., Porfireva A. V., Zamaleeva A. I., Fakhrullin R. R. 'Electrochemical DNA sensors based on nanostructured organic dyes/DNA/polyelectrolyte complexes'. Journal of Nanoscience and Nanotechnology. 2014; 14: 6738–6747.

[16] Zhang L. Y., Sun H., Li D., Song S. P., Fan C. H., Wang S. 'A conjugated polymer-based electrochemical DNA sensor: design and application of a multi-functional and water-soluble conjugated polymer'. Macromolecular Rapid Communications. 2008; 29: 1489–1494.

[17] Saint-Aman E., Visan M. T., Moutet J. C. 'Investigation of electrochemical reversibility and redox-active polypyrrole film formation of amide ferrocene–pyrrole derivatives'. Electrochimica Acta. 1997; 42: 1829–1837.

[18] Korri-Youssoufi H., Makrouf B. 'Electrochemical biosensing of DNA hybridization by electroactive ferrocene functionalized polypyrrole'. Synthetic Metals. 2001; 119: 265–266.

[19] Floch F. L., Ho H. A., Harding-Lepage P., Leclerc M. 'Label-free electrochemical detection of protein based on a ferrocene-bearing cationic polythiophene and aptamer'. Analytical Chemistry. 2006; 78: 4727–4731.

[20] Wang C. L., Mulchandani A. 'Ferrocene-conjugated polyaniline-modified enzyme electrodes for determination of peroxides in organic media'. Analytical Chemistry. 1995; 67: 1109–1114.

[21] Ramanaviciene A., Ramanavicius A. 'Pulsed amperometric detection of DNA with an ssDNA/polypyrrole-modified electrode'. Analytical Bioanalytical Chemistry. 2004; 379: 287–293.

[22] Bizid S., Mlika R., Haj Said A., Chemli M., KorriYoussoufi H. 'Investigations of poly(p-phenylene) modified with ferrocene and their application in electrochemical DNA sensing'. Sensors and Actuators B: Chemical. 2016; 226: 370–380.

[23] Umana M., Waller J. 'Protein-modified electrodes. The glucose oxidase/polypyrrole system'. Analytical Chemistry. 1986; 58: 2979–2983.

[24] Riccardi C. S., Yamanaka H., Josowicz M., Kowalik J., Mizaikoff B., Kranz C. 'Label-free DNA detection based on modified conducting polypyrrole films at microelectrodes'. Analytical Chemistry. 2006; 78: 1139–1145.

[25] Wang J., Jiang M., Fortes A., Mukherjee B. 'New label-free DNA recognition based on doping nucleic-acid probes within conducting polymer films'. Analytical Chimica Acta. 1999; 402: 7–12.

[26] Wang J., Jiang M. 'Toward genoelectronics: nucleic acid doped conducting polymers'. Langmuir. 2000; 16: 2269–2274.

[27] Liu Y., Hu N. F. 'Loading/release behavior of (chitosan/DNA)$_n$ layer-by-layer films toward negatively charged anthraquinone and its application in electrochemical detection of natural DNA damage'. Biosensors and Bioelectronics. 2007; 23: 661–667.

[28] Zhang W., Yang T., Li X., Wang D., Jiao K. 'Conductive architecture of Fe_2O_3 microspheres/self-doped polyaniline nanofibers on carbon ionic liquid electrode for impedance sensing of DNA hybridization'. Biosensors and Bioelectronics. 2009; 25:428–434.

[29] Lassalle N., Roget A., Livache T., Mailley P., Vieil E. 'Electropolymerisable pyrrole oligonucleotide: synthesis and analysis of ODN hybridization by fluorescence and QCM'. Talanta. 2001; 55: 993–1004.

[30] Peng H., Soeller C., Vigar N. A., Caprio V., Travas-Sejdic J. 'Label-free detection of DNA hybridization based on a novel functionalized conducting polymer'. Biosensors and Bioelectronics. 2007; 22: 1868–1873.

[31] Thompson L. A., Kowalik J., Josowicz M., Janata J. 'Label-free DNA hybridization probe based on a conducting polymer'. Journal of American Chemical Society. 2003; 125: 324–325.

[32] Kavanagh P., Leech D. 'Redox polymer and probe DNA tethered to gold electrodes for enzyme-amplified amperometric detection of DNA hybridization'. Analytical Chemistry. 2006; 78: 2710–2716.

[33] Floch F. L., Ho H. A., Harding-Lepage P., Bedard M., Neagu-Plesu R., Leclerc M. 'Ferrocene-functionalized cationic polythiophene for the label-free electrochemical detection of DNA'. Advanced Materials. 2005; 17: 1251–1256.

[34] Tlili C., Korri-Youssoufi H., Ponsonnet L., Martelet C. Jaffrezic-Renault N. J. 'Electrochemical impedance probing of DNA hybridization on oligonucleotide-functionalized polypyrrole'. Talanta. 2005; 68:131–137.

[35] Baur J., Gondran C., Holzinger M., Defrancq E., Perrot H., Cosnier S. 'Label-free femtomolar detection of target DNA by impedimetric DNA sensor based on poly(pyrrole-nitrilotriacetic acid) film'. Analytical Chemistry. 2010; 82: 1066–1072.

[36] Ramanathan K., Bangar M. A., Yun M., Chen W., Myung N. V., Mulchandani A. 'Bioaffinity sensing using biologically functionalized conducting-polymer nanowire'. Journal of American Chemical Society. 2005; 127: 496–497.

[37] Ren R., Leng C. C., Zhang S. S. 'A chronocoulometric DNA sensor based on screen-printed electrode doped with ionic liquid and polyaniline nanotubes'. Biosensors and Bioelectronics. 2010; 25: 2089–2094.

[38] Shosuke K., Yusuke H., Mariko M., Shinji, O. 'The role of metals in site-specific DNA damage with reference to carcinogenesis'. Free Radical Biology and Medicine. 2002; 32: 822–832.

[39] Liang M. M., Guo L. H. 'Photoelectrochemical DNA sensor for the rapid detection of DNA damage induced by styrene oxide and the Fenton reaction'. Environmental Science & Technology 2007; 41: 658–664.

[40] Du M., Yang T., Jiao K. 'Carbon nanotubes/(pLys/dsDNA)n layer-by-layer multilayer films for electrochemical studies of DNA damage'. Journal of Solid State Electrochemistry. 2010; 14: 2261–2266.

[41] He P., Bayachou M. 'Layer-by-layer fabrication and characterization of DNA-wrapped single-walled carbon nanotube particles'. Langmuir. 2005; 21: 6086–6092.

[42] Ma H. Y., Zhang L. P., Pan Y., Zhang K. Y., Zhang Y. Z. 'A novel electrochemical DNA biosensor fabricated with layer-by-layer covalent attachment of multiwalled carbon nanotubes and gold nanoparticles'. Electroanalysis. 2008; 20: 1220–1226.

[43] Wang P. P., Ni Y. N., Kokot S. 'A novel dsDNA/polydiphenylamine-4-sulfonic acid electrochemical biosensor for selective detection of the toxic catechol and related DNA damage'. Analyst. 2013; 138: 1141–1148.

[44] Reisberg S., Piro B., Noel V., Pham M. C. 'DNA electrochemical sensor based on conducting polymer: dependence of the "signal-on" detection on the probe sequence localization'. Analytical Chemistry. 2005; 77: 3351–3356.

[45] Chang H. X., Yuan Y., Shi N. L., Guan Y. F. 'Electrochemical DNA biosensor based on conducting polyaniline nanotube array'. Analytical Chemistry. 2007; 79: 5111–5115.

[46] Chowdhury A. D., AmitabhaDe N. A., Sarkar M. 'Detection of positional mismatch in oligonucleotide by electrochemical method'. Sensors and Actuators B: Chemical. 2014; 202: 917–923.

[47] Shakeel S., Karim S., Ali A. 'Peptide nucleic acid (PNA) – a review'. Journal of Chemical Technology and Biotechnology. 2006; 81:892–899.

[48] Fang B., Jiao S., Li M., Qu Y. X. 'Label-free electrochemical detection of DNA using ferrocene-containing cationic polythiophene and PNA probes on nanogold modified electrodes'. Biosensors and Bioelectronics. 2008; 23: 1175–1179.

[49] Prabhakar N., Arora K., Singh H., Malhotra B. D. 'Polyaniline based nucleic acid sensor'. Journal of Physical Chemistry B. 2008; 112: 4808–4816.

[50] Sui J., Zhang L. J., Peng H. 'Label-free DNA sensor construction using self-assembled poly(*o*-methoxyaniline) hollow nanospheres'. European Polymer Journal. 2013; 49: 139–146.

[51] Tuan M. A., Pham D. T., Chu T. X., Hieu N. M., Hai N. H. 'Highly sensitive DNA sensor based on polypyrrole nanowire'. Applied Surface Science. 2014; 309: 285–289.

[52] Zheng D. L., Wang Q. X., Gao F., Wang Q. H., Qiu W. W. 'Development of a novel electrochemical DNA biosensor based on elongated hexagonal-pyramid CdS and poly-isonicotinicacid composite film'. Biosensors and Bioelectronics. 2014; 60: 167–174.

[53] Shirakawa H., Louis E. J., Macdiarmid A. G., Chiang C. K., Heeger A. J. 'Synthesis of electrically conducting organic polymers: halogen derivatives of polyacetylene, (CH)x'. Journal of the Chemical Society – Chemical Communications. 1977; 578–580.

[54] Friend R. H., Gymer R. W., Holmes A. B., *et al.* 'Electroluminescence in conjugated polymers'. Nature. 1999; 397: 121–128.

[55] Li G., Shrotriya V., Huang J. S., *et al.* 'High-efficiency solution processable polymer photovoltaic cells by self-organization of polymer blends'. Nature Materials. 2005; 4: 864–868.

[56] Kim J. Y., Lee K., Coates N. E., *et al.* 'Efficient tandem polymer solar cells fabricated by all-solution processing'. Science. 2007; 317: 222–225.

[57] Horowitz G. 'Organic thin film transistors: From theory to real devices'. Journal of Materials Research. 2004; 19: 1946–1962.

[58] Facchetti A., Yoon M. H., Marks T. J. 'Gate dielectrics for organic field-effect transistors: new opportunities for organic electronics'. Advanced Material. 2005; 17: 1705–1725.

[59] Angione M. D., Cotrone S., Magliulo M., *et al.* 'Interfacial electronic effects in functional biolayers integrated into organic field-effect transistors'. Proceedings of the National Academy of Sciences of the United States of America. 2012; 109: 6429–6434.

[60] Bernards D. A., Macaya D. J., Nikolou M., De Franco J. A., Takamatsu S., Malliaras G. G. 'Enzymatic sensing with organic electrochemical transistors'. Journal of Materials Chemistry. 2008; 18: 116–120.

[61] Berggren M., Dahlfors A. R. 'Organic bioelectronics'. Advanced Materials. 2007; 19: 3201–3213.

[62] Zhu Z. T., Mason J. T., Dieckmann R., Malliaras G. G. 'Humidity sensors based on pentacene thin-film transistors'. Applied Physics Letters. 2002; 81: 4643–4645.

[63] Roberts, M. E., Sokolov, A. N., Bao, Z. 'Material and device considerations for organic thin-film transistor sensors'. Journal of Materials Chemistry. 2009; 19: 3351–3363.

[64] Rogers J. A., Bao Z., Baldwin K., *et al.* 'Paper-like electronic displays: large-area rubber-stamped plastic sheets of electronics and micro-encapsulated electrophoretic inks'. Proceedings of the National Academy of Sciences of the United States of America. 2001; 98: 4835–4840.

[65] Sirringhaus H., Kawase T., Friend R. H., *et al.* 'High-resolution inkjet printing of all-polymer transistor circuits'. Science. 2000; 29: 2123–2126.

[66] Boer B. de, Facchetti A. 'Semiconducting polymeric materials'. Polymer Reviews. 2008; 48: 423–431.

[67] Janata J., Josowicz M. 'Conducting polymers in electronic chemical sensors'. Nature Materials. 2003; 2:19–24.

[68] Adhikari B., Majumdar S. 'Polymers in sensor applications'. Progress in Polymer Science. 2004; 29: 699–766.

[69] Lucas B., Trigaud T., Videlot-Ackermann C. 'Organic transistors and phototransistors based on small molecules'. Polymer International. 2012; 61: 374–389.

[70] Lin Y., Li Y., Zhan X. 'Small molecule semiconductors for high-efficiency organic photovoltaics'. Chemical Society Reviews. 2012; 41: 4245–4272.

[71] Wu W., Liu Y., Zhu D. 'π-Conjugated molecules with fused rings for organic field-effect transistors: Design, synthesis and applications'. Chemical Society Reviews. 2010; 39: 1489–1502.

[72] Mishra A., Bäuerle P. 'Small molecule organic semiconductors on the move: Promises for future solar energy technology'. Angewandte Chemie International Edition. 2012; 51: 2020–2067.

[73] Dimitrakopoulos C. D., Mascaro D. J. 'Organic thin-film transistors: a review of recent advances'. IBM Journal of Research and Development. 2001; 45: 11–27.

[74] Lin P., Yan F. 'Organic thin-film transistors for chemical and biological sensing'. Advanced Materials. 2012; 24: 34–51.

[75] Li J., Liu D. Miao Q., Yan F. 'The application of a high-k polymer in flexible low-voltage organic thin-film transistors'. Journal of Materials Chemistry. 2012; 22: 15998–16004.

[76] Yan F., Li J. H., Mok S. M. 'Highly photosensitive thin film transistors based on a composite of poly(3-hexylthiophene) and Titania nanoparticles'. Journal of Applied Physics. 2009; 106(7): 074501.

[77] Lakshmi D., Bossi A., Whitcombe M. J., *et al.* 'Electrochemical sensor for catechol and dopamine based on a catalytic molecularly imprinted polymer-conducting polymer hybrid recognition element'. Analytical Chemistry. 2009; 81:3576–3584.

[78] Horowitz G. 'Organic field-effect transistors'. Advanced Materials. 1998; 10: 365–377.

[79] Scarpa G., Idzko A. L., Yadav A., Thalhammer, S. 'Organic ISFET based on poly (3-hexylthiophene)'. Sensors. 2010; 10: 2262–2273.

[80] Roberts M. E., Mannsfeld S. C. B., Queralto N., *et al.* 'Water-stable organic transistors and their application in chemical and biological sensors'. Proceedings of the National Academy of Sciences USA. 2008; 105: 12134–12139.

[81] Jagannathan L., Subramanian V. 'DNA detection using organic thin film transistors: Optimization of DNA immobilization and sensor sensitivity'. Biosensors and Bioelectronics. 2009; 25: 288–293.

[82] Stoliar P., Bystrenov E., Quirog S. D., *et al.* 'DNA adsorption measured with ultra-thin film organic field effect transistors'. Biosensors and Bioelectronics. 2009; 24: 2935–2938.

[83] Demelas M., Lai S., Casula G., Scavetta E., Barbaro M., Bonfiglio A. 'An organic, charge-modulated field effect transistor for DNA detection'. Sensors and Actuators B. 2012; 171: 198–203.

[84] Kergoat L., Herlogsson L., Braga D., *et al.* 'A water-gate organic field-effect transistor'. Advanced Materials. 2010; 22: 2565–2569.

[85] Buth F., Donner A., Sachsenhauser M., Stutzmann M., Garrido J. A. 'Bio-functional electrolyte-gated organic field-effect transistors'. Advanced Materials. 2012; 24: 4511–4517.

[86] Said E., Larsson O., Berggren M., Crispin X. 'Effects of the ionic currents in electrolyte-gated organic field-effect transistors'. Advanced Functional Materials. 2008; 18: 3529–3536.

[87] Larsson O., Said E., Berggren M., Crispin X. 'Insulator polarization mechanisms in polyelectrolyte-gated organic field-effect transistors'. Advanced Functional Materials. 2009; 19: 3334–3341.

[88] Yuan J. D., Dhoot A. S., Namdas E. B., *et al.* 'Electrochemical doping in electrolyte-gated polymer transistors'. Journal of the American Chemical Society. 2007; 129: 14367–14371.

[89] Lee J., Kaake L. G., Cho J. H., Zhu X. Y., Lodge T. P., Frisbie C. D. 'Ion gel-gated polymer thin-film transistors: operating mechanism and characterization of gate dielectric capacitance, switching speed, and stability'. The Journal of Physical Chemistry C. 2009; 113: 8972–8981.

[90] Nilsson D., Kugler T., Svensson P., Berggren M. 'An all-organic sensor–transistor based on a novel electrochemical transducer concept printed electrochemical sensors on paper'. Sensors and Actuators B. 2002; 86: 193–197.

[91] Bernards D. A., Malliaras G. G. 'Steady-state and transient behavior of organic electrochemical transistors'. Advanced Functional Materials. 2007; 17: 3538–3544.

[92] Cicoira F., Sessolo M., Yaghmazadeh O., DeFranco J. A., Yang S. Y., Malliaras G. G. 'Influence of device geometry on sensor characteristics of planar organic electrochemical transistors'. Advanced Materials. 2010; 22: 1012–1016.

[93] Zhang M., Lin P., Yang M., Yan F. 'Fabrication of organic electrochemical transistor arrays for biosensing'. Biochimica et Biophysica Acta. 2013; 1830 (9): 4402–4406.

[94] Khodagholy D., Gurfinkel M., Stavrinidou E., *et al.* 'High speed and high density organic electrochemical transistor arrays'. Applied Physics Letters. 2011; 99(16): 163304.

[95] Gundlach D. J., Zhou L., Nichols J. A., Jackson T. N. 'An experimental study of contact effects in organic thin film transistors'. Journal of Applied Physics. 2006; 100: 024509 .

[96] Cosseddu P., Bonfiglio A. 'A comparison between bottom contact and top contact all organic field effect transistors assembled by soft lithography'. Thin Solid Films, 2007; 515: 7551–7555.

[97] Zhang Q. T., Subramanian V. 'DNA hybridization detection with organic thin film transistors: Toward fast and disposable DNA microarray chips'. Biosensors and Bioelectronics. 2007; 22: 3182–3187.

[98] Zhang Q. T., Jagannathan L., Subramanian V. 'Label-free low-cost disposable DNA hybridization detection systems using organic TFTs'. Biosensors and Bioelectronics. 2010; 25: 972–977.

[99] Liu N., Hua Y., Zhang J., Cao J., Liu Y., Wang J. 'A label-free, organic transistor-based biosensor by introducing electric bias during DNA immobilization'. Organic Electronics. 2012; 13: 2781–2785.

[100] Chand R., Jeun J. H., Park M. H., Kim J. M., Shin I. S. 'Electro-immobilization of DNA for ultrafast detection on a microchannel integrated pentacene TFT'. Journal of Industrial and Engineering Chemistry. 2015; 21: 126–128.

[101] Kim J. M., Jha S. K., Chand R., Lee D. H., Kim Y. S. 'DNA hybridization sensor based on pentacene thin film transistor'. Biosensors and Bioelectronics. 2006; 26: 2264–2269.

[102] Kim J. M., Jha S. K., Lee D. H., Chand R., Jeun J. H., Kim Y. S. 'A flexible pentacene thin film transistors as disposable DNA hybridization sensor'. Journal of Industrial and Engineering Chemistry. 2012; 18: 1642–1646.

[103] Khan H. U., Roberts M. E., Johnson O., Forch R., Knoll W., Bao Z. N. 'In Situ, Label-Free DNA Detection Using Organic Transistor Sensors'. Advanced Materials. 2010; 22: 4452–4456.

[104] Khan H. U., Roberts M. E., Knoll W., Bao Z. 'Pentacene based organic thin film transistors as the transducer for biochemical sensing in aqueous media'. Chemistry of Materials. 2011; 23: 1946–1953.

[105] Gui H. Y., Wei B., Wang J. 'High sensitivity and air stability in an organic transistor-based biosensor by inserting a CuPc layer'. Physica Status Solid A. 2014, 11:2499–2502.

[106] Yan F., Mok S. M., Yu J. J., Chan H. L. W., Yang M. 'Label-free DNA sensor based on organic thin film transistors'. Biosensors and Bioelectronics. 2009; 24: 1241–1245.

[107] Barbaro M., Bonfiglio A., Raffo L. 'A charge-modulated FET for detection of biomolecular processes: conception, modeling, and simulation'. IEEE Transaction on Electron Devices. 2006; 53:158–166.

[108] Demelas M., Lai S., Barbaro M., Bonfiglio A. 'DNA hybridization detection based on an organic charge modulated field effect transistor'. 10th IEEE Conference on Sensors; Limerick, Ireland, Oct 2011. 1917–1920.

[109] Lai S., Demelas M., Casula G., Cosseddu P., Barbaro M., Bonfiglio A. 'Ultra-low voltage, OTFT-based sensor for label-free DNA detection'. Advanced Materials. 2013; 25:103–107.

[110] Kergoat L., Piro B., Berggren M., Pham M. C., Yassar A., Horowitz G. 'DNA detection with a water-gated organic field-effect transistor'. Organic Electronics. 2012; 13: 1–6.

[111] Magliulo M., Mallardi A., Mulla M. Y., *et al.* 'Electrolyte-gated organic field effect transistor sensors based on supported biotinylated phospholipid bilayer'. Advanced Materials. 2013; 25: 2090–2094.

[112] Schmoltner K., Kofler J., Klug A., List-Kratochvil E. J. W. 'Electrolyte-gated organic field effect transistor for selective reversible ion detection'. Advanced Materials. 2013; 25: 6895–6899.

[113] Buth F., Kumar D., Stutzmann M., Garrido J. A. 'Electrolyte-gated organic field-effect transistors for sensing applications'. Applied Physics Letters. 2011; 98: 153302.

[114] Reisberg S., Piro B., Noel V., Pham M. C. 'DNA electrochemical sensor based on conducting polymer: dependence of the "signal-on" detection on the probe sequence localization'. Analytical Chemistry. 2005; 77: 3351–3356.

[115] White S. P., Dorfman K. D., Frisbie C. D. 'Label-free DNA sensing platform with low-voltage electrolyte-gated transistors'. Analytical Chemistry. 2015; 87:1861–1866.

[116] White H. S., Kittlesen G. P., Wrighton M. S. 'Chemical derivatization of an array of three gold microelectrodes with polypyrrole: fabrication of a molecule-based transistor'. Journal of American Chemical Society. 1984; 106: 5375–5377.

[117] Krishnamoorthy K., Gokhale R. S., Contractor A. Q., Kumar A. 'Novel label-free DNA sensors based on poly (3,4-ethylenedioxythiophene)'. Chemical Communications. 2004; 7: 820–821.

[118] Lin P., Luo X., Hsing I., Yan F. 'Organic electrochemical transistors integrated in flexible microfluidic systems and used for label-free DNA sensing'. Advanced Materials. 2011; 23: 4035–4040.

Chapter 4

Organic transistor-based mechanical sensors for artificial skin applications

Piero Cosseddu[1], Stefano Lai[1], Alberto Loi[2] and Annalisa Bonfiglio[1]

Abstract

The human skin is endowed with a big amount of tactile receptors, which perceive mechanical and thermal stimuli and allow the arising of different feelings (smoothness/roughness, softness/hardness, hotness/coldness, pain, etc.), also detecting stimulus position, size and duration. The biggest challenges that come from the development of tactile systems concern not only the need of fabricating multimodal tactile transducers, but also the fact that these sensing capabilities must be distributed on large, complex areas and possibly at low fabrication costs, dealing with still open issues such as the reduction of the numbers of connecting cables, the realization of a dedicated network architecture, the reduction of power dissipation and last but not the least the reduction of fabrication costs. From this point of view, Organic Electronics represents a step forward as conjugated polymers join the mechanical characteristics of plastics, therefore highly flexible and easily transferrable on whatever kind of surface, with the possibility of being solution processable. This property is particularly important, because it allows processing and deposition of these materials on large areas with very easy and cost-efficient technologies suitable for mass production, such as inkjet printing. The focus of this chapter is in particular on the employment of electronic devices based on conjugated materials for the realization of strain/pressure sensors that may be used for the fabrication of artificial skin.

4.1 Introduction

Organic Electronics has attracted a huge attention along these past decades as it offers the chance to fabricate electronic systems with multiple functionalities, over large areas, at low temperatures, on flexible substrates, and, in particular, using cost-efficient technologies [1–4].

[1]Department of Electrical and Electronic Engineering, University of Cagliari, Italy
[2]Cetemmsa Technological Center, Mataró, Spain

After organic light-emitting diodes and displays based on organic devices, the era of another killer application for this technology, namely sensors based on organic devices, is now emerging. Due to the fact that such technology allows the employment of fast and cheap fabrication procedures on very large areas, Organic Electronics is paving the way towards their ubiquitous distribution in the environment.

Very efficient sensors may be fabricated starting from the concept of organic thin-film transistor (OTFT). In particular, it is worth mentioning that one of the main advantages of employing these devices for sensing applications is that they are active devices; many different electronic parameters (mobility, threshold voltage, contact resistance, etc.), not only one, as for instance in chemo/piezoresistive sensors, can be extracted from their electrical characterization. Therefore, OTFT-based sensors are multiparametric sensors, offering the possibility of using a combination of variables in order to characterize their response to the parameter to be sensed. Moreover, active sensors combine in the same device both switching and sensing functions and the electrical response of the device to a certain external stimulus can be locally amplified by the transistor itself allowing one to easily obtain sensing matrices of limited size and improved reliability.

One more important feature is that an OTFT is in fact a multilayered structure, dominated by the physical properties of several material interfaces, and each one of these can be employed for testing different parameters; as a consequence, it is virtually possible to use a single device for achieving a multimodal sensing ability.

For instance, capacitive changes in the gate dielectric or conduction changes in the active layer can be employed for directly detecting an external mechanical stimulus. In addition, these devices can be connected in series with a pyro/piezoresistive material in order to indirectly detect some sort of external stimulus. Moreover, the device electrodes may also be functionalized with specific self-assembled monolayers (SAM) in order to achieve chemical sensing functionalities.

A panoramic description of these different possibilities is given in Figure 4.1.

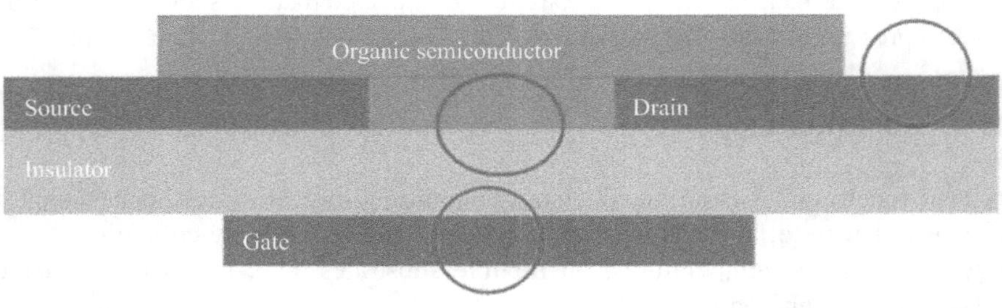

Figure 4.1 Schematic representation of a bottom gate, bottom contact OTFT, showing the different areas of the device that could be employed for sensing applications

4.2 Influence of mechanical deformation on the OTFTs electrical properties

It is a common thought that conjugated polymer-based devices can be employed for the fabrication of highly flexible electronics. However, despite the intrinsic mechanical flexibility of the most common semiconducting polymers (when deposited in thin films), the electrical stability of these materials during mechanical deformation is still under debate and still constituting a big issue for the realization of highly flexible electronic circuitry. In fact, there are several reports that demonstrate a pronounced variation of charge transport properties in OTFTs during mechanical stress. One of the first studies on this issue was realized by Sekitani *et al.* [5]. In their investigations, they demonstrated that by applying a tensile (compressive) stress in Pentacene-based OTFTs a pronounced decrease (increase) in the device output current was observed. They stated that this phenomenon can be attributed to a variation of the active layer morphological and/or structural properties of the polycrystalline semiconductor film induced by the applied mechanical stress. Indeed, at the micro-scale, a tensile stress could lead to an increase in the distance between grains in the channel of the devices, whereas, at the nano-scale, it could also be possible to induce an increase in the lateral spacing between the molecules, thus affecting the π-stacking in the film. In both cases, the immediate consequence should be a reduction of the hopping probability for charge carriers, thus leading to a lowering, as experimentally observed, of the channel mobility.

This should well explain the observed behavior but unfortunately, no direct evidence for this phenomenon was shown. More recently, two interesting studies have shed more light onto this behavior. In particular, Yang *et al.* [6] demonstrated that the observed changes in the field-effect mobility under strain are due to bending-stress-driven phase transitions between the bulk phase and the thin-film phase in the Pentacene film (Figures 4.2 and 4.3). For instance, it was found that the volume fraction of bulk phase in the Pentacene film increases from 10.7% to 27.7% under 1.1% of tensile strain but decreases to 3.5% under 1.0% of compressive strain.

We have also investigated this issue, and we have observed a clear, marked and reversible variation of the morphological features in Pentacene thin films during tensile deformation. In a few words, as reported by Scenev *et al.* [7], tensile stress induces a stretching of the Pentacene film morphology, thus increasing the distance between adjacent domains within the active channel (see Figure 4.4), that, again, could very well explain the reduction of the carriers mobility observed in the electromechanical characterization of the fabricated OTFTs.

Starting from these considerations, we went a little deeper into the influence of the morphological features on the electrical response of OTFTs to mechanical deformation [8]. In particular, since the sensitivity seems to be strongly correlated with the morphological variation in the active channel, we fabricated Pentacene-based OTFTs and intentionally modified their morphological features by changing the deposition parameters. We fabricated four different sets of OTFTs using always the same geometry, but depositing Pentacene at different deposition rates (Table 4.1). In this way, we were able to induce a controlled variation of the average domain dimensions, as shown in Figure 4.5.

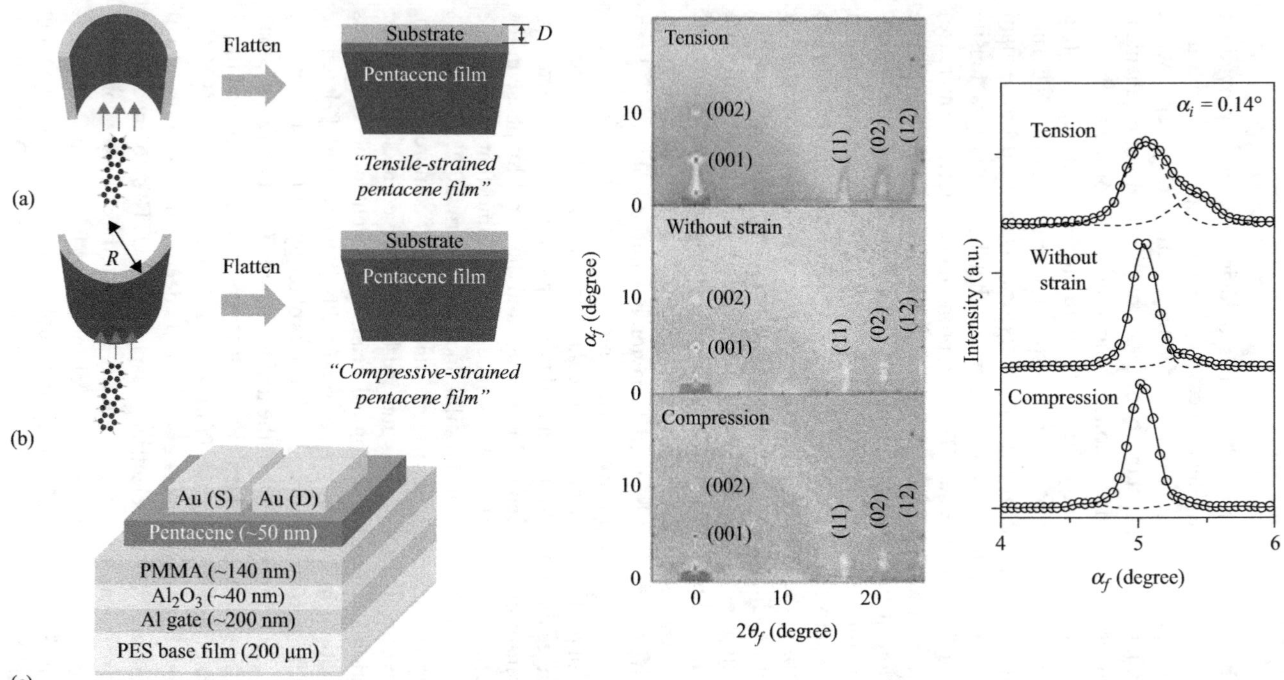

Figure 4.2 2 (Left) Schematic diagrams of the production of bending-strained Pentacene films. (Right) 2D GIXS patterns measured at $\alpha_i = 0.14°$ *for bending-strained Pentacene thin films (~50 nm thick) in a flat state under tensile strain* ($\varepsilon = -1.1\%$)*, without strain* ($\varepsilon = 0$)*, and compressive strain* ($\varepsilon = 1.0\%$)*. (b) Out-of-plane GIXS profiles extracted along the* α_f *direction at* $2\theta_f = 0°$*. Reprinted with permission from [6]. © 2008, AIP Publishing LLC*

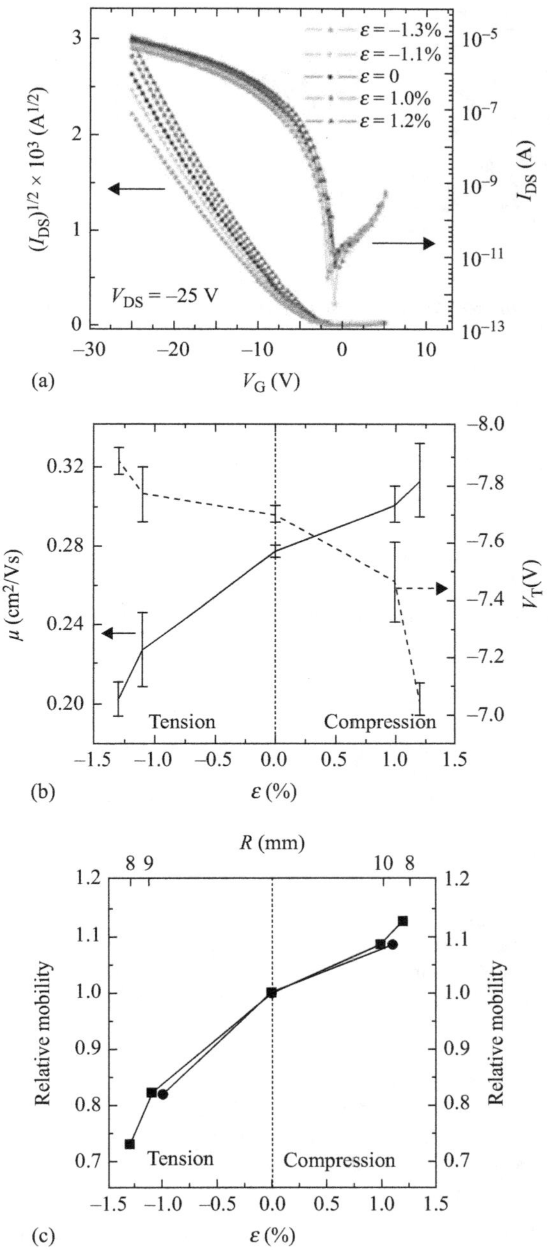

Figure 4.3 Transfer characteristics of Pentacene FETs and relative changes in the mobility a threshold voltage of the measured devices before and during mechanical deformation. Reprinted with permission from [6]. © 2008, AIP Publishing LLC

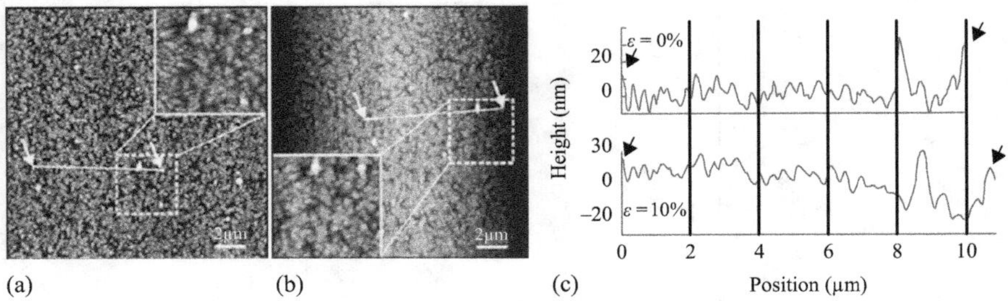

Figure 4.4 AFM micrographs of Pentacene a thin film before (a) and during (b) bending, and the relative cross-section showing the morphological changes induced by mechanical deformation. Reprinted from Organic Electronics, Vol. 14, V. Scenev, P. Cosseddu, A. Bonfiglio, et al., "Origin of mechanical strain sensitivity of Pentacene thin-film transistors", 1323–1329, Copyright (2013), with permission from Elsevier

Table 4.1 Average morphological parameters estimated in Pentacene thin films deposited using different deposition rates

Deposition rate	**11 A/s**	**5 A/s**	**0.5 A/s**	**0.08 A/s**
Length (nm)	91 ± 53	115 ± 68	197 ± 118	236 ± 143
Area (nm^2)	4,833 ± 3,000	7,732 ± 4,805	21,180 ± 12,173	40,369 ± 25,131
Perimeter (nm)	271 ± 159	383 ± 229	727 ± 440	893 ± 546

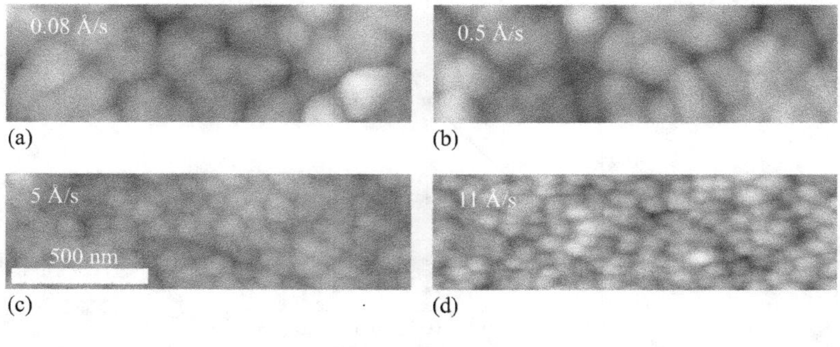

Figure 4.5 AFM micrographs of different Pentacene films, deposited at different deposition rates. Reprinted from Organic Electronics, Vol. 14, P. Cosseddu, G. Tiddia, S. Milita, A. Bonfiglio, "Continuous tuning of the mechanical sensitivity of Pentacene OTFTs on flexible substrates: from strain sensors to deformable transistors", 206–211, Copyright (2013), with permission from Elsevier

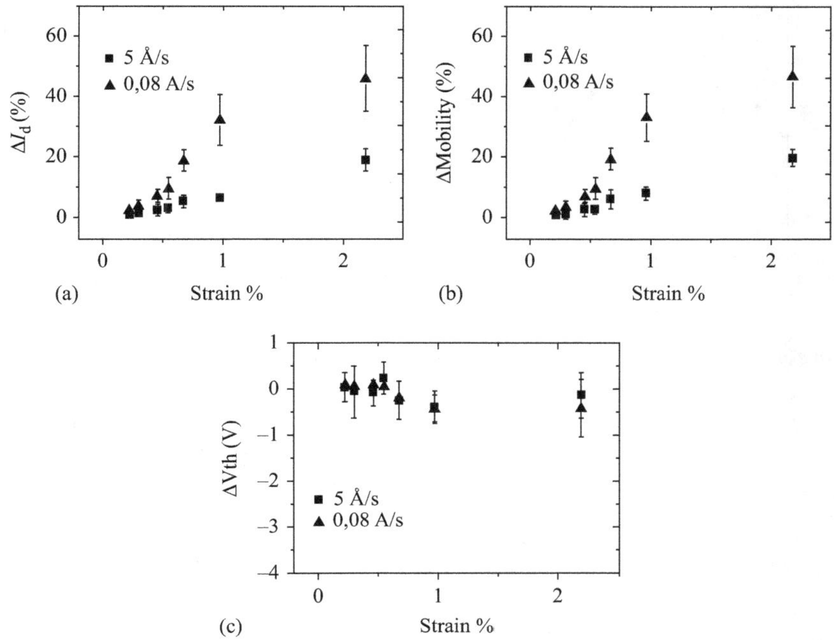

Figure 4.6 Average current (a), mobility (b) and threshold voltage (c) variation measured in two sets of transistors fabricated using different Pentacene deposition rates

The electromechanical characterization of the different sets of devices gave rise to very interesting results. First of all, a clear correlation between the output current variation induced by tensile stress and the carriers mobility variation was found, whereas, on the contrary, no evidence of a correlation with the threshold voltage changes was visible (Figure 4.6). Most importantly, it was also noticed that the larger are the average grain dimensions, the higher is the observed sensitivity to surface strain.

In particular, as summarized in Figure 4.7, it is possible to finely tune the final sensitivity of the OTFTs by properly controlling the morphology of the Pentacene films going from devices with a very poor sensitivity, characterized by very small grain dimensions, that could be employed for the fabrication of flexible transistors, immune to strain effects, to devices with a very high sensitivity, characterized by larger grain dimensions, that could be used for mechanical sensing applications.

In another work, we compared the electrical response of two very different systems, a highly disordered, almost amorphous, thin film, based on spin-coated P3HT, and a more ordered, polycrystalline film made of thermally evaporated Pentacene [9]. In this case, a pronounced decrease of the sensitivity to mechanical deformation can be observed when the devices are made by a highly disordered P3HT active layer, characterized by very small grain dimensions, which seems to be in full agreement with what already observed in [8] (Figure 4.8).

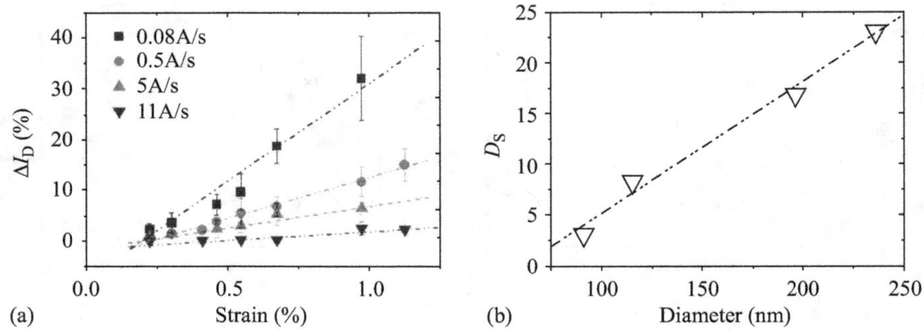

Figure 4.7 Sensitivity variation vs. strain (a) in the four different sets of devices and correlation between sensitivity and grain dimensions (b). Reprinted from Organic Electronics, Vol. 14, P. Cosseddu, G. Tiddia, S. Milita, A. Bonfiglio, "Continuous tuning of the mechanical sensitivity of Pentacene OTFTs on flexible substrates: from strain sensors to deformable transistors", 206–211, Copyright (2013), with permission from Elsevier

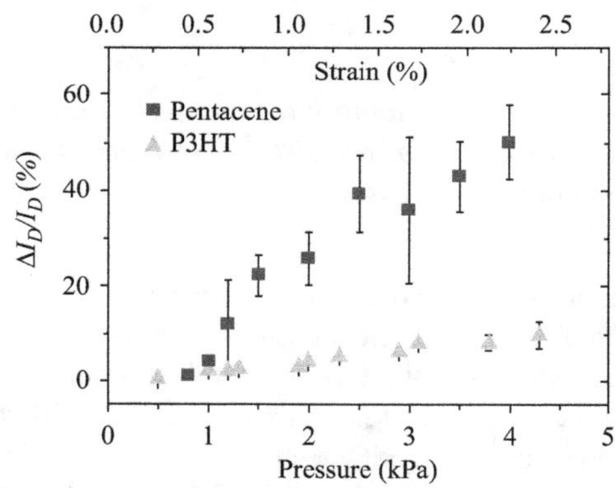

Figure 4.8 Electrical response to mechanical deformation of the two different sets of samples

In recent years, other groups have also investigated this phenomenon demonstrating that morphological/structural changes in the active channel are not the only issues to be taken into account. For instance, Jedaa *et al.* [10] demonstrated that the presence of alkyl chains in organic semiconducting small molecules could have a dramatic impact in the sensitivity to mechanical deformation. In particular, they fabricated OTFTs using different small molecules such as Pentacene and two different alkyl-terminated sexithiophene molecules, which differ for the length of alkyl chains, namely ET-6T and DEC-6T. A detailed electromechanical characterization was performed in different sets of devices and, very interestingly, they observed that the mechanical sensitivity was dramatically reduced in the alkyl-terminated materials (Figure 4.9). They speculated that this particular behavior

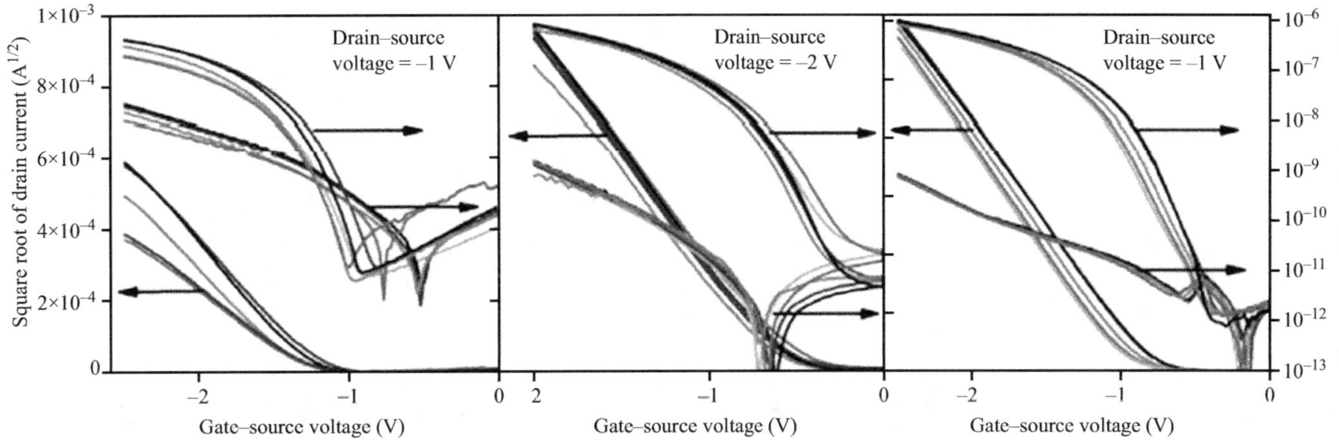

Figure 4.9 *(Color online) Drain current, square root of drain current, and leakage current of Pentacene (left), Et-6T-Et (middle), and Dec-6T-Dec TFTs (right) with AlO_x/C10-PA as gate dielectrics as a function of strain in concave bending from flat pristine to flat after bending, considering strain values of 1.1%, 1.8% and 2.5%. Reprinted with permission from [10]. © 2009, AIP Publishing LLC*

could be due to the fact that the largest part of the induced surface tension is accommodated by the flexible alkyl chains located at the end of the molecule and directly interfaced with the gate dielectric.

A very interesting study reported by Sokolov *et al.* [11] demonstrated that the interfacial properties between the gate dielectric and the organic semiconductor can also dramatically affect the sensitivity to surface deformation. Indeed, the induction of strain into polymeric systems can lead to a rearrangement of the polymer structure giving rise to a change in surface heterogeneity and a change in surface energy. Since it is well known that the surface energy of the dielectric, as well as the surface dipole, plays a large role in controlling charge-carrier density within organic semiconductors, it is expected that the field-effect mobility will change in response to polymer rearrangement at the surface, manifested as a change in the surface energy or surface dipole reorientation induced by mechanical stress. Moreover, as previously observed by Jedaa *et al.*, they also observed that the introduction of an SAM of a linear alkyl chain onto the surface of the dielectric material may serve as a general method to drastically decrease the effects of strain on the semiconductor mobility.

4.3 OTFT-based mechanical sensors

As introduced in the previous section, charge transport properties in organic semiconductors can be severely affected by mechanical deformation. Fortunately, several publications have demonstrated that this behavior is very reproducible and recoverable, at least within a certain range of mechanical deformation, thus allowing one to employ these structures for the fabrication of mechanical sensors.

The most employed device structure, for this kind of application, is certainly the thin film transistor configuration. Indeed, employing a transistor, as stated in the Introduction of this chapter, has several advantages compared to using two terminal devices such as purely capacitive sensors, piezoelectric sensors or piezoresistive sensors. In the following, an overview of the main examples of OTFT-based mechanical sensors reported in the literature is given.

One of the first examples was reported by Rang *et al.* [12]. The authors have investigated the hydrostatic-pressure dependence of *I–V* curves in organic transistors and found a large and reversible dependence of drain current and hole mobility on hydrostatic pressure, suggesting that this kind of device could be suitable for sensor applications. A different approach was used by Darlinsky *et al.* [13]. They fabricated OTFTs on glass substrates and characterized them as force sensors by applying a controlled force on the device area using a micro-needle whose movement was controlled by a step motor. In the study, they employed two device configurations with different channel lengths. In this way, they managed to apply the force directly on the channel area (for large channel devices) or simultaneously on the active layer and on the source and drain electrodes. Interestingly, they observed a marked difference in the two different cases (Figure 4.10).

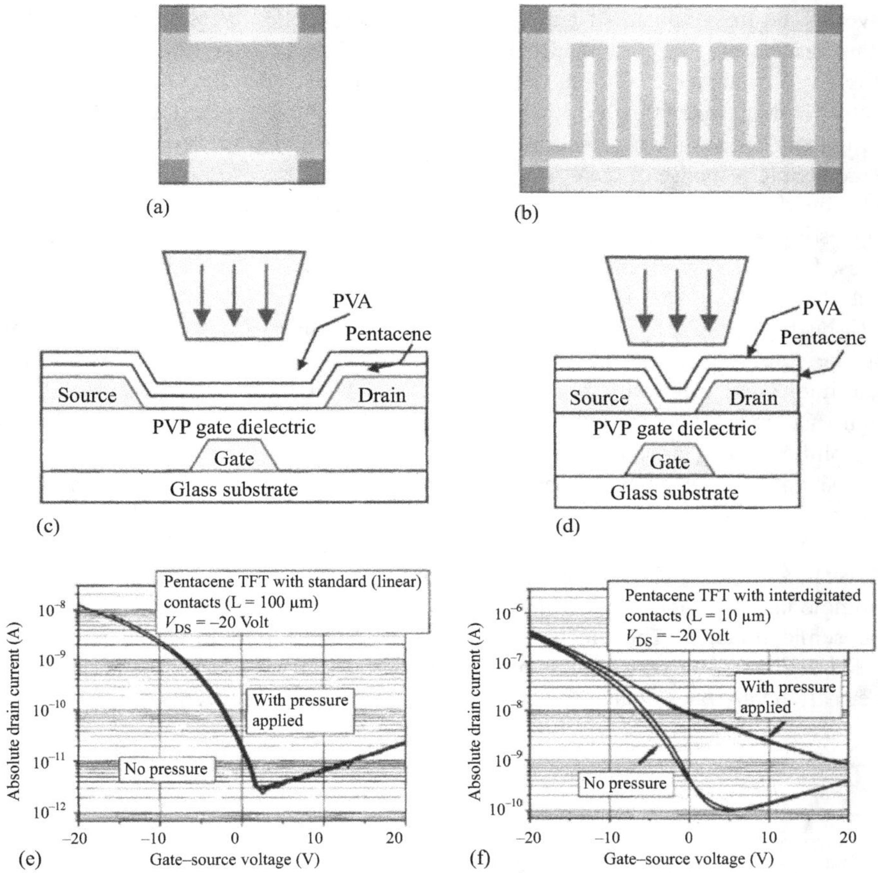

Figure 4.10 Optical images (a, b) and cross-sections (c, d) of the different devices fabricated by Darlinki et al., and the response to the applied force in the two cases [13]. Reprinted with permission from [13].

In particular, no significant change in the electrical characteristics was observed when the force was exerted only in the channel area, whereas a clear difference was observed in the second case. This behavior was attributed to a possible activation of trap states at the dielectric/semiconductor interface induced by the applied mechanical stimulus.

More recently, strain sensors using an organic semiconductor as the sensor (resistive) element of a strain gauge have been also reported by Jung *et al.* [14,15]. In this case, the strain sensor consisted of a Wheatstone bridge structure where the Pentacene film acts as a sensing layer of a strain gauge. Our group has also investigated the response of flexible OTFTs to the applied pressure. In particular, Manunza *et al.* [16,17] fabricated highly flexible OTFTs on a 1.5-μm thick polyethyleneterephtalate (PET) foil, called Mylar®, which acted at the same time as gate dielectric, but also as flexible mechanical support for the entire device structure, and using Pentacene as the organic semiconductor. Interestingly, we found a marked and reproducible response of the device to the applied pressure. Moreover, this behavior was found to be linear within a certain range of applied pressure (0–10 kPa), thus suggesting a possible employment of this kind of device for practical applications. These findings were recently further confirmed by Cosseddu *et al.* who demonstrated that also by using solution processable organic semiconductors, such as P3HT or 6,13-bis(triisopropylsilylethynyl)-Pentacene (TIPS-Pentacene), very similar results can be achieved [9]. In this case, we have fabricated OTFTs on a flexible PET substrate. Gold electrodes have been deposited by thermal evaporation under a high-vacuum condition and patterned using a shadow mask (for the gate) and standard photolithography (for source and drain).

The gate dielectric layer, namely polypara-xylylene (Parylene C, purchased by Specialty Coating Systems), was deposited from the vapor phase, and finally, few drops of a TIPS-Pentacene solution at 0.5 wt% in toluene were deposited on the transistors channel while keeping the substrate on a hotplate at 90 °C, in order to promote fast solvent evaporation and subsequently to obtain good crystallization of the semiconductor. The electromechanical characterization of TIPS-Pentacene-based devices gave very similar results compared to what previously observed in thermally evaporated Pentacene devices, as reported in Figure 4.11 [18].

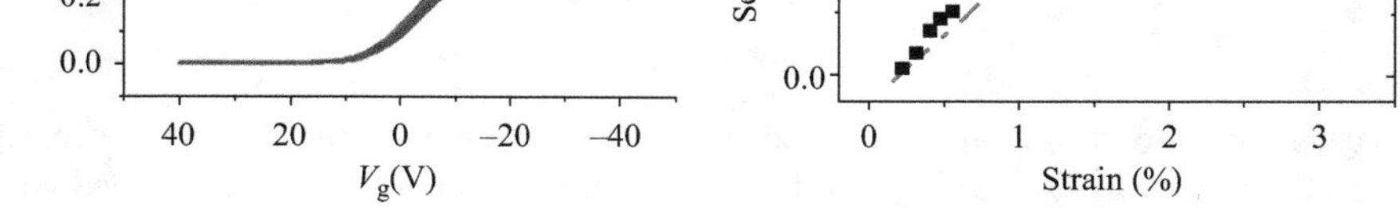

Figure 4.11 Normalized transfer characteristics (with respect to the maximum current recorded in the unbent device) variations and sensitivity vs. strain measured in TIPS-Pentacene devices. A. Loi, L. Basiricò, P. Cosseddu, et al. "Matrices of inkjet printed OFETs for the realization of artificial robotic skin", Materials Research Society Symposia Proceedings, MRS Proceedings, mrsf111401t0402, reproduced with permission

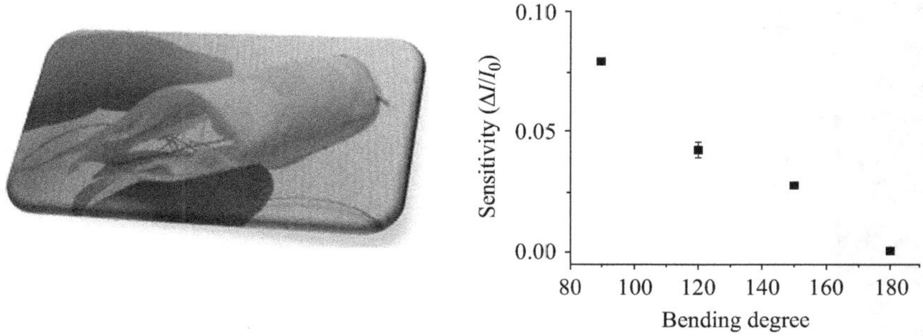

Figure 4.12 Example of an OTFT sensor transferred on a ribbon and sewn on an elastic bend (a); electrical response of the sensor to different bending angles (b)

Thanks to the high flexibility of the employed structure, these devices can be easily transferred onto garments or elastic bends for monitoring bio-mechanical parameters such as breath rhythm or joints motion. In one of the possible applications, as shown in Figure 4.12, the OFET-based sensor has been transferred on a ribbon, and the final system has been sewn on an elastic bandage. After that, the sensor was placed in the elbow joint and the electrical response of the device during the elbow motion was monitored. The sensor clearly detects the joint motion. The output current decreases when the elbow is bent, accordingly to the fact that bending the elbow induces a surface strain on OFET active layer. The elbow has been bent at different bending angles, and as shown in Figure 4.12, these different states can be clearly detected by the employed sensing system [19].

Very similar results were also obtained for the monitoring of the fingers' motions. In this case, the sensorized bend was placed in a finger and the electrical response of the device during the finger motion was monitored. In this case, the finger was bent over several cylinders with different radii, and the device output current was monitored during the whole experiment.

Interestingly, as can be observed from the plots reported in Figure 4.13, the sensor clearly detects the finger's joint motion. As already observed in the elbow experiments, the output current decreases when the finger is bent, according to the fact that bending induces a surface strain on the OTFT active layer. Moreover, it can be clearly noticed that by decreasing the bending radius, a more marked response to deformation is observed, according to the fact that the smaller is the bending radius, the higher is the induced surface strain on the OTFT structure (see Figure 4.13(c) inset). Also in this case, a reproducible response to mechanical deformation was obtained (see Figure 4.13(c)), and a clear dependence of the current variation vs. the applied bending radius was found, which make this system very promising for practical applications.

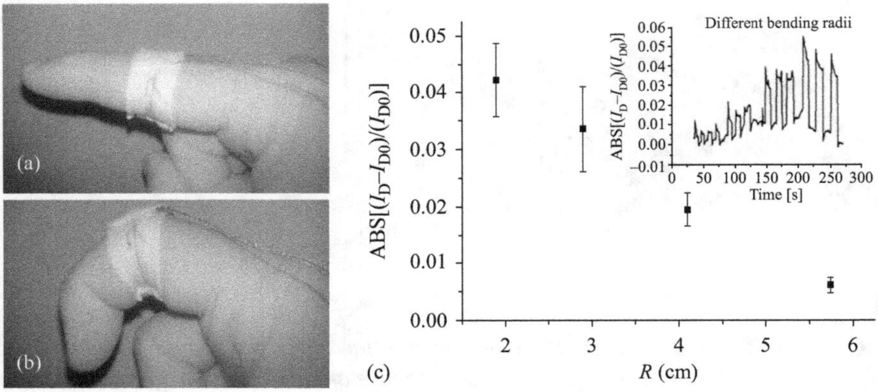

Figure 4.13 Images of the sensorized ribbon placed on a finger (a), (b); electrical response of the sensor during the finger motion (c)

A slightly different approach was recently employed for the fabrication of tactile transducers for the realization of artificial robotic skin. The fabrication of this system has been carried out by means of the inkjet printing technique using a silver nanoparticle-based ink, by means of a piezoelectric drop-on-demand Dimatix Materials Printer 2800 (DMP2800). Starting from a single OFET, with $W/L = 1{,}000$ and $C_{ins} = 1.86$ nF/cm^2, we fabricated matrices of 8×8 OFETs with a lateral pitch of 5 mm, with the layout reported in Figure 4.14. In particular, we employed a common source electrode, eight independent drain electrodes, one per column, and eight independent gate electrodes, one per row, so that each device of the matrix can be addressed independently. The most critical step of the fabrication process is the inkjet printing of the interdigitated source and drain electrodes, which represent the core of each device: the fine tuning of the printing parameters, calibrated in order to avoid short circuit and discontinuity problems, allowed us to obtain a yield, i.e. number of working devices, higher than 90%, and a very good reproducibility of the electrical performances between different devices. Organic semiconductor, TIPS-Pentacene, was deposited by means of drop-casting.

The output and the transfer characteristics of an inkjet-printed OTFT are shown in Figure 4.15, where a large field effect and a very small hysteresis can be noticed. We obtained a good mobility (0.07 ± 0.03 cm^2/(V·s), up to 0.1 cm^2/(V·s)), slightly negative threshold voltages (-1.5 ± 1.2 V) and high ION/IOFF ($\approx 10^5$).

It is important to underline that, in order to allow the devices to detect pressure, the devices structure was slightly modified. In particular, the fabricated matrix of mechanical sensors has been embedded into a 2-mm thick elastomeric film, namely Ecoflex® PDMS. In this way, when pressure is applied, it locally deforms the OTFT surface, thus leading to a change in the device output current.

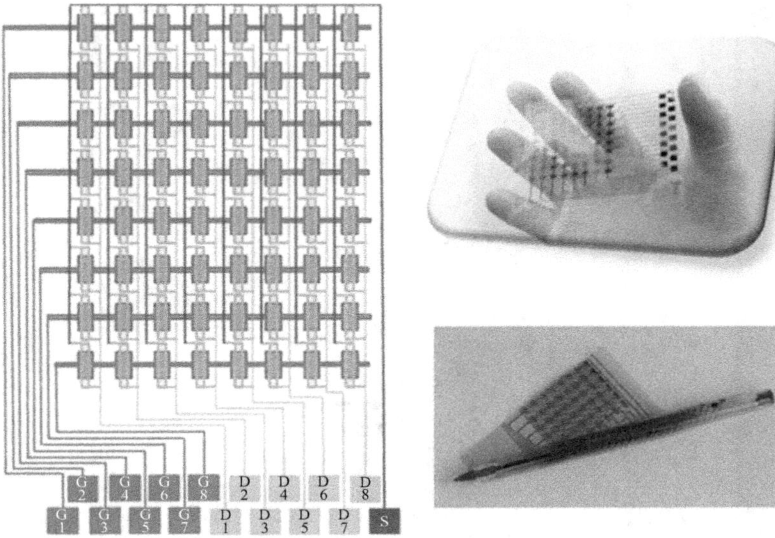

Figure 4.14 *Layout of the tactile transducers matrix (left), and two examples of fabricated matrices on different plastic substrates (right). A. Loi, L. Basiricò, P. Cosseddu, et al. "Matrices of inkjet printed OFETs for the realization of artificial robotic skin", Materials Research Society Symposia Proceedings, MRS Proceedings, mrsf111401t0402, reproduced with permission*

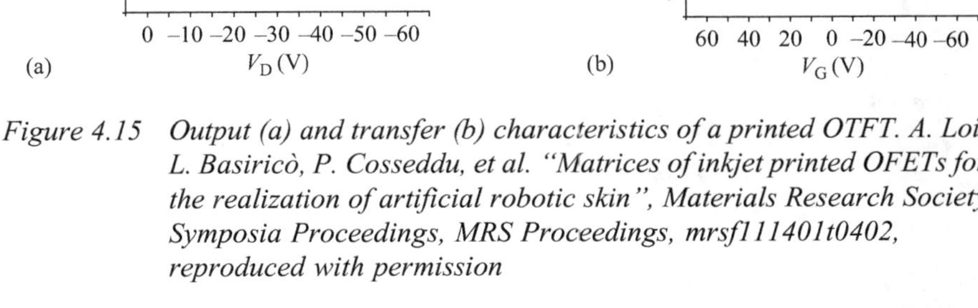

Figure 4.15 *Output (a) and transfer (b) characteristics of a printed OTFT. A. Loi, L. Basiricò, P. Cosseddu, et al. "Matrices of inkjet printed OFETs for the realization of artificial robotic skin", Materials Research Society Symposia Proceedings, MRS Proceedings, mrsf111401t0402, reproduced with permission*

The obtained results are better represented in Figure 4.16, where the response of the flexible robot skin is shown. When a certain pressure is applied on the device structure, a variation of the output current can be detected. In this case, an arbitrary pressure has been exerted with a finger on a single element of the matrix. It is

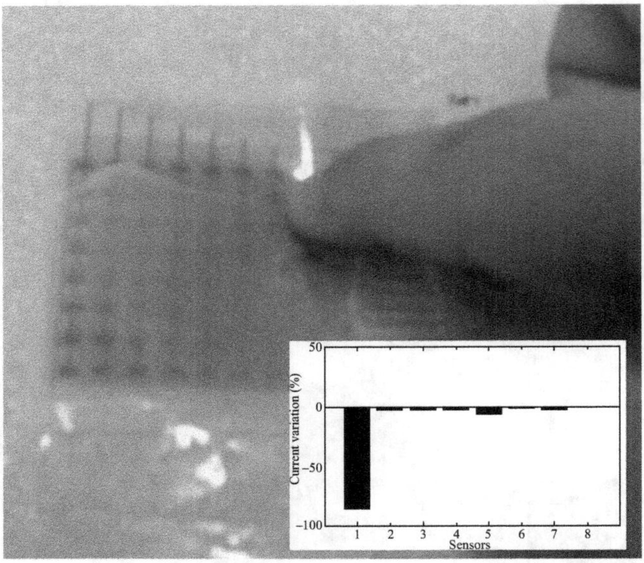

Figure 4.16 Electrical response of the flexible system fabricated for reproducing the sense of touch [18]

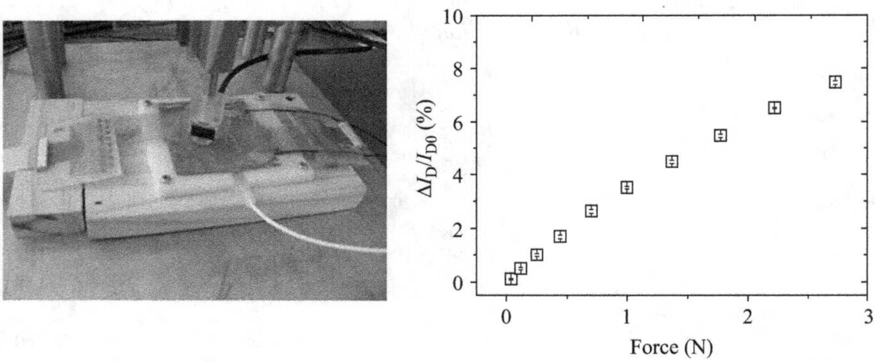

Figure 4.17 Mechanical indenter employed for the electromechanical characterization (left) of the tactile transducers, and the obtained calibration curve (right)

noteworthy that, when the finger is moved across the array surface each device responds to the applied mechanical deformation independently, and a very small cross-talking (both electrical and mechanical) can be detected. Moreover, the fabricated system was characterized using a mechanical indenter as shown in Figure 4.17 using this approach it was possible to apply controlled forces to each element of the matrix and simultaneously measure the devices response. Interestingly, the electromechanical characterization of the skin patch gave rise to a very reproducible response, which was found to be linear and fully reversible for applied pressures in the range of 0–3 N, with a sensitivity of 0.1 N.

4.4 OTFTs coupled with mechanical sensing components

A different possibility for achieving mechanical sensing using OTFTs is the employment of a different configuration in which the sensing element is not the semiconductor layer. In this case, OTFTs act only as part of the read-out electronics directly connected with the actual sensor, which modulates the OTFTs electrical response when an external stimulus is applied to the sensitive area. Therefore, the mechanical stimulus (force, pressure, strain, etc.) is sensed not directly by the organic semiconductor, but by the indirect modulation of the OTFTs performance. Despite the additional processing steps required to integrate the sensing element with the OTFT, which sometimes involve expensive techniques, the sensitivity obtained with indirect sensing is generally higher with respect to that achieved with direct sensing; moreover, indirect sensing is more versatile, since a great variety of sensing elements and techniques can be employed, which give rise to different sensitivities on different ranges of mechanical stimuli applied.

Using the first approach, Mannsfeld *et al.* [9] and later Schwartz *et al.* [21] studied and developed a microstructured dielectric layer of polydimethylsiloxane (PDMS). By using a silicon mold, they patterned a thin PDMS film with parallel rows of identically high pyramids, whose base and height was set from 6 to 3 μm and whose inter-pyramid space was eventually varied from 1.33 to 13.61 μm. The result is shown in Figure 4.18.

The first characterization was carried on the pressure-sensitive capacitor realized by laminating an *indium tin oxide* (ITO)-coated conductive flexible PET sheet

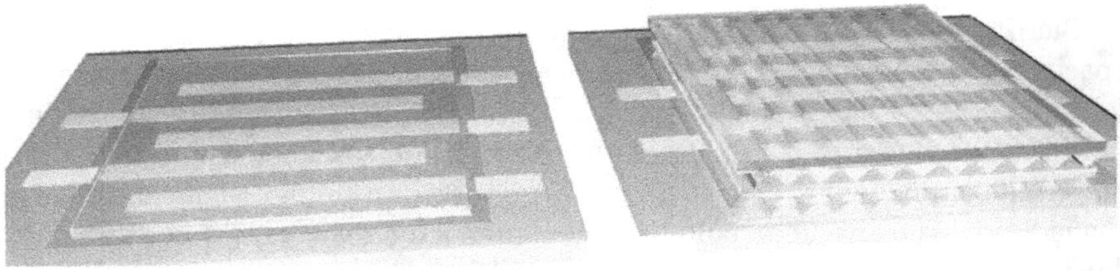

Figure 4.18 Schematic representation of the sensing structure of [21]. Adapted by permission from Macmillan Publishers Ltd: Nature Communications [21], © 2013

on the top of the microstructured PDMS film. When an external stimulus is applied, a very marked variation of the capacitance is observed also for very low pressures (<1 kPa). This is due to two main factors: (i) the overall capacitor is composed of PDMS and air gaps, which far decrease the film elastic resistance and (ii) during compression, the displaced volume is air, whose dielectric constant ($\varepsilon = 1.0$) is lower than that of PDMS ($\varepsilon \approx 3.0$). The pressure sensitivity of such a device is defined as the relative capacitance variation on unit pressure applied:

$$S = \frac{\partial\left(\frac{\Delta C}{C_0}\right)}{\partial P}\ (\mathrm{k/Pa}) \tag{4.1}$$

As can be seen in Figure 4.19, in the low pressure range (<1.5 kPa) the sensitivity is $S = 0.55\ \mathrm{kPa}^{-1}$, while for higher pressure range (>2 kPa) the sensitivity decreases to $0.15\ \mathrm{kPa}^{-1}$, probably due to the increasing elastic resistance with increasing compression. As a consequence of a variation of the gate capacitance, it is possible to detect pressure by means of the modulation of the drain current I_D. The sensitivity of the OTFT-based system is therefore defined as the relative I_D variation on unit pressure applied:

$$S = \frac{\partial\left(\frac{\Delta I_D}{I_{D0}}\right)}{\partial P}\ (\mathrm{k/Pa}) \tag{4.2}$$

In this case, as shown in Figure 4.19(b), the sensitivity in the low pressure range (<5 kPa) is even higher with respect to the previous example, reaching the value of $8.2\ \mathrm{kPa}^{-1}$, while for the medium and high-pressure range similar values can be obtained ($S = 0.38\ \mathrm{kPa}^{-1}$).

Moreover, microstructuring the PDMS film has a beneficial effect also on the time response. Figure 4.19 shows the comparison of the time response when a 15 kPa load is applied on the surface. While the unstructured layer, because of the visco-elastic properties of the PDMS, has a relaxation time of 10 s, the structured PDMS-films, patterned with triangular-shaped lines and with pyramids, show a faster time response, in the millisecond range. Finally, this structure exhibits good reliability also when cycling mechanical stress is applied, with *a* I_D degradation of 10% after more than 15,000 pressure cycles.

Starting from a similar structure, Kim *et al.* realized a OTFT where the gate capacitance is modulated by an external pressure and depends on the shape of the dielectric [22]. Using a block copolymer, polystyrene-block-poly(vinylpyridine) (PS-b-P2VP), they created a mold with a honeycomb structure with a very shape nano-needle array, whose ends had a radius of curvature around 10 nm and a height around 3 μm. They use this mold to create two polyurethane (PU) microstructured films: the first was obtained by pouring the PU on the nano-needle template and resulted with a inverse pattern (thus hemispherical), while for the second film the procedure was replied twice, therefore obtaining again the original nano-needle pattern transferred on the PU film.

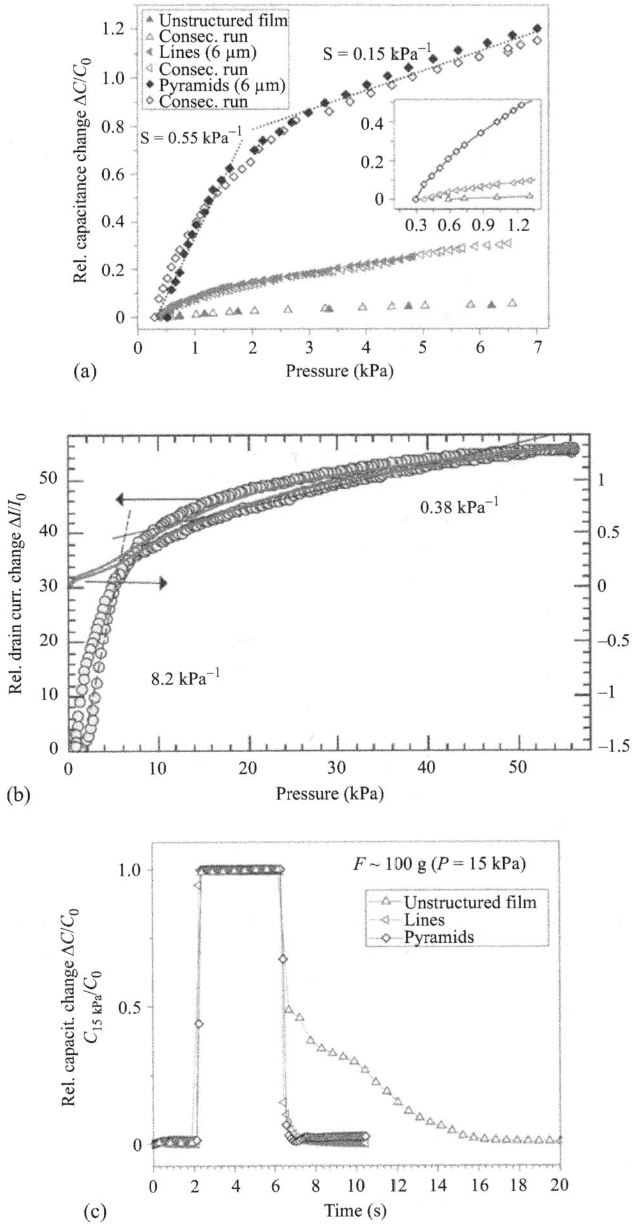

Figure 4.19 *Sensitivity of the PDMS capacitor (a) and of the OTFT with microstructured PDMS (b); time response for different patterns of the PDMS layer (c). Adapted by permission from Macmillan Publishers Ltd: Nature Materials [20], © 2010, and Nature Communications [21], © 2013*

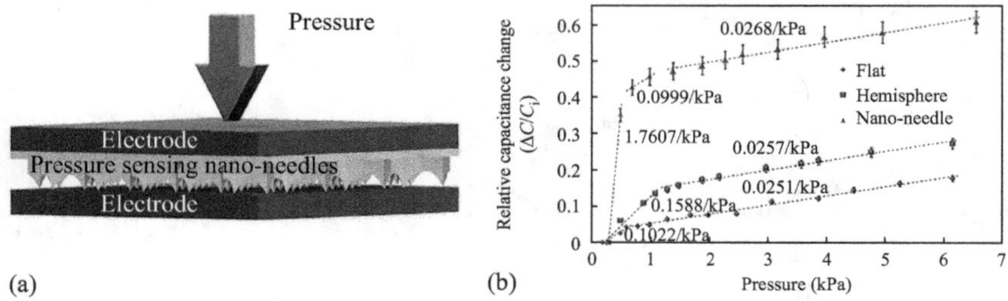

Figure 4.20 Structure of the nano-needle patterned capacitor (a) and comparison of sensitivity of the types of PU layer (b). Reprinted with permission from [22]. © 2012, AIP Publishing LLC

The first characterization was carried on by placing the second film (the nano-needle dielectric) between two metal electrodes, thus realizing a capacitive pressure sensor (Figure 4.20). For comparison, the same procedure was replied also for the first-type film (hemispherical pattern) and for an unstructured PU film. The results are shown in Figure 4.20(b), where the relative capacitance change is plotted vs. the applied pressure. In the medium pressure range, above 1 kPa, all the three types of film evidenced a very similar behavior, with sensitivity around 0.026 kPa^{-1}. The real advantage was observed in the very low pressure range (<1 kPa), where the unstructured, type 1 and type 2 dielectric layer had a sensitivity of 0.102, 0. 159 and 1.76 kPa^{-1}, respectively. Again, the difference in sensitivity of the three structures can be explained from the presence of air vacuum between needles or hemispheres, which reduces the elastic resistance to pressure and decreases the average dielectric constant of the layer when no mechanical stimulus is applied. Moreover, they demonstrated that the pressure sensitivity is independent of the density of nano-needles.

A different approach is based on the employment of poly(vinylidene-fluoride) (PVDF). PVDF, and its co-polymer poly(vinylidene-fluoride-trifluoroethylene) (PVDF-TrFE), is a piezoelectric polymer used in a wide range of applications [23]. Basically, when subjected to a mechanical deformation, it generates a voltage proportional to the applied stimulus: this can be used to detect a pressure event, and when it is coupled with an OTFT the external stimulus can be measured as a I_D variation. In this way, the most immediate integration with an OTFT consists of depositing the PVDF on the gate dielectric, so that the voltage variation caused by the pressure modulates the gate capacitance and hence the drain current.

Hsu *et al.* [23] realized a bottom-gate bottom-contact OTFT integrated with a PVDF sheet in the following way: a 52-µm PVDF film was metalized using 35 nm of unpatterned Al as a reference electrode for the measurement; Parylene C was deposited by thermal vapor deposition and Au source and drain electrodes were thermally evaporated; the structure was exposed to the UV-ozone treatment in order to shift the threshold voltage V_T to positive values and therefore allow the measurement of both compressive and tensile strain; finally, 25 nm of Pentacene were thermally evaporated and the whole structure was encapsulated with 200 nm of Parylene C and laminated to a PVC card.

The electromechanical characterization was carried out by means of a robotic actuator that induced a regular and precise bending on the whole structure. In this case, the sensitivity was defined as the I_D variation on unit strain (ε) induced:

$$S = \frac{\partial(\Delta I_D)}{\partial \varepsilon}(A/\%) \tag{4.3}$$

The authors found that, since V_T is positive, there was a non-zero current also in the initial state. When the device is bent, a large capacitive current, due to the capacitive coupling between the source/drain electrodes and the unpatterned aluminum gate electrode, is observed; after that, the current reaches a steady state in 50 s. From the measurements reported in their work, a retention time around 442 s is evaluated. Moreover, the calibration of the device, where three consecutive bending cycles were applied showed that, despite a hysteresis, caused by charge relaxation, a good reproducibility and a sensitivity $S = 0.182$ µA/strain is observed.

Adopting a similar device structure, Trung *et al.* [24] took advantage of the piezoelectricity and pyroelectricity of PVDF-TrFE to create a multimodal sensor which could detect, at the same time, thermal variations induced by infrared (IR) light and pressure/strain induced by a mechanical stimulus. In order to decouple the device response when both the two physical stimuli are applied together, they used an AC gate biasing and proposed the following model for I_D:

$$I_D = -\mu C\frac{W}{L}V_{G0}V_D\sin(2\pi ft) - \mu C\frac{W}{L}V_0V_D \tag{4.4}$$

where μ is the effective field-effect mobility of the device, C is the gate capacitance, W is the channel width, L is the channel length, V_D is the applied drain voltage, V_{G0} and f are the amplitude and the phase of the applied AC gate bias, and V_0 is the equivalent voltage corresponding to the remnant polarization P_r inside PVDF-TrFE ($P_r = \varepsilon_0\varepsilon_r V_0/d$).

Tests were carried out on varying the strain (0%, 0.2% and 0.31%) and the IR light intensity PIR (0, 500, 1,000, 1,500 and 2,000 mW/cm^2) and the results were plotted both as a function of PIR for different strains and as a function of strain for different PIRs. They found that, the sensitivity with respect to PIR changes at different strain, decreasing from 0.0024 to 0.001 Vm^{-1} W^{-1} cm^2 with increasing tensile strain from 0% to 0.31%. An Analogous trend is noticed for sensitivity with respect to strain at different PIR.

A major problem of all the previously reported structures is that the mechanical stimulus, i.e. the pressure, is directly applied to the OTFT: therefore, the electromechanical performances can be degraded if the sensor is over-stimulated. To overcome this problem, we have developed a new structure for the sensor, named pressure-modulated organic field-effect transistor (PMOFET), in which the

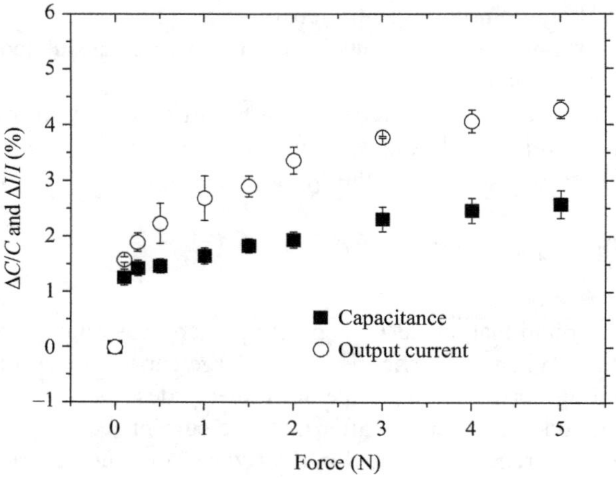

Figure 4.21 Response of the PMOFET

pressure-sensitive area is physically separated from the active area of the device [25]. Basically, it consists of a floating-gate transistor biased through a control capacitor C_{CF} and through a variable capacitor C_V, based on PDMS elastomer, which act as the sensing element. Since the total charge in the floating gate cannot vary, when a pressure is applied on the variable gate, C_V increases and the floating-gate voltage V_{FG} changes in order to maintain the total charge constant: the variation of V_{FG} modulates the OTFT drain current I_D and the pressure event can be detected. The separation between sensing area and active area, thanks to the presence of the floating gate, allows the implementation of large-area sensing sheets without actually damage the read-out active device. The sensitivity curves are reported in Figure 4.21, showing the relative variation of the capacitance and of the drain current as a function of the applied force. As expected, they both showed the same trend, but the sensitivity of the current is higher due to the transistor effect.

The second type of approach, which consists in modulating the effective V_{DS} applied to an OTFT, requires the employment of a conductive pressure sensor to be put in series with the source or drain electrode. The ideal candidates for this purpose are the conductive rubbers, which decrease their electrical resistance of several orders of magnitude when they are subjected to a pressure. Conductive rubbers are basically PDMS elastomers containing nanostructures, such as metal or graphite nanoparticles, that maintain the high flexibility of the organic devices and therefore are suitable for pressure sensing over large area.

Someya *et al.* [26, 27] adopted this solution to create a matrix composed of 32×32 sensors over 8×8 cm^2. On a polyimide substrate, Au electrodes were thermally evaporated, a polyimide precursor was spin coated as gate dielectric (thickness 500 nm) and Pentacene (mobility around 0.3 $cm^2\ V^{-1}\ s^{-1}$) was thermally evaporated to create a bottom-gate top-contact structure. The rubber sheet and a copper layer were laminated in the backside of the polyimide substrate, without any alignment. On the substrate some through-holes were made in order to obtain the electrical continuity between the source electrode and the backside rubber, which can indirectly modulate the I_D current through the modulation of the series resistance. As the pressure applied on the rubber ranges from 0 to 30 kPa, the series resistance decreases from 1 MΩ to 1 kΩ. The reported results show the OTFTs electrical response when various pressures are applied: while at 0 kPa the current is almost zero, for higher pressures it rises up to 9 μA for 30 kPa when biased at $V_{DS} = -20$ V and $V_{GS} = -20$ V. The conductive rubber can also be used as pressure sensor also when the device is bent, since upon a certain bending range the device performance are not dramatically affected. For example, when a rubber sheet is wrapped around a bar with a 10-mm radius, the related pressure is estimated to be around 6 kPa, which is below the sensitivity of the sensor (30 kPa). Therefore, bending experiments were carried on with different bending radii. Starting from a bending radius of 50 mm and scaling down to 10 mm, no remarkable current decrease can be observed (only 3% decrease is measured). Even at 2 mm of bending radius, the device is still working, showing a current decrease around 20%. The real damage that occurs for smaller bending radii (below 1 mm) is the breaking of gold electrodes: if it is avoided, even after a significant current decrease due to bending the device shows no relevant damage when it is measured in unbent condition.

Using the same architecture, Takei *et al.* [27] adopted Ge/Si nanowires (NWs) as semiconductor, and realized devices with mobility around 20 $cm^2\ V^{-1}\ s^{-1}$. The sensing element was a pressure-sensitive rubber with carbon nanoparticles embedded in it. From the base sensing cell, they realized a matrix of 18 × 19 sensors over 7 × 7 cm^2 flexible sheet that could be easily bent and rolled to small radii of curvature. Figure 4.22(a) shows the transfer characteristics for different pressure applied: a clear response to mechanical stimuli is noticed, since the actual VDS applied to the OTFT increases as the series variable resistance decreases. The sensitivity curve is shown in Figure 4.22(b). Here, two working regimes can be distinguished: below 2 kPa (inset) and over 2 kPa (main curve). The low-pressure regime shows an exponential dependence of the current on the applied stimulus, whereas for higher pressures the dependence is linear, with a sensitivity of around 11.5 μS/kPa. This reduced sensitivity for in the medium pressure range is due to the reduced elastic modulus of the rubber. Also, bending tests revealed a very poor dependence of this structure on the bending radius, down to 2.5 mm, at which the relative decrease of the output conductance is only 6%, in nice agreement with the previous work. Finally, cycling stress tests were performed, bending the substrate to a radius <10 mm and measuring its electric performances after relaxation. The results showed minimal response degradation even after 2,000 bending cycles and demonstrated the mechanical robustness of the devices.

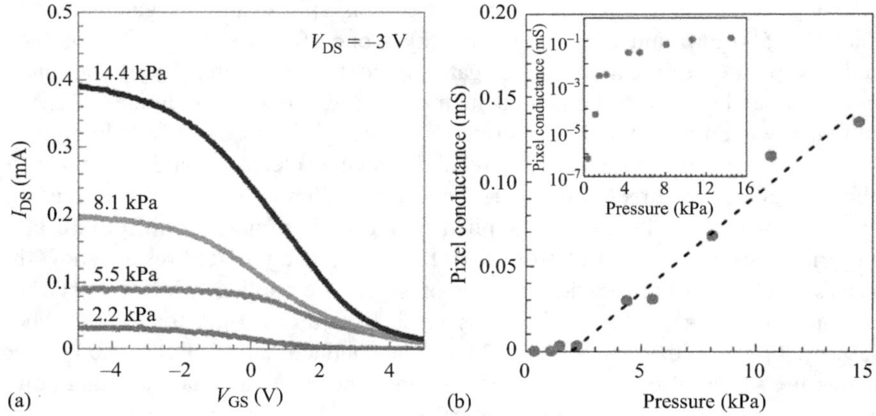

Figure 4.22 Transfer characteristics of the OTFT under different pressures applied (a) and sensitivity of the device in the low pressure range (<2 kPa) and in the middle pressure range (>2 kPa) (b). Reprinted by permission from Macmillan Publishers Ltd: Nature Materials [28], © 2010

References

[1] N. T. Kalyania and S. J. Dhobleb, "Organic light emitting diodes: energy saving lighting technology—a review", *Renewable and Sustainable Energy Reviews*, 16, 2696–2723, 2012. doi: 10.1016/j.rser.2012.02.021

[2] D. Braga and G. Horowitz, "High-performance organic field-effect transistors", *Advanced Materials*, 21, 1473–1486, 2009. doi: 10.1002/adma.200802733

[3] P. Lin and F. Yan "Organic thin-film transistors for chemical and biological sensing", *Advanced Materials*, 24, 34–51, 2012. doi: 10.1002/adma.201103334

[4] Y. W. Su, S.-C. Lan, and K.-H. Wei "Organic photovoltaics – review article", *Materials Today*, 15, 554–562, 2012. doi: 10.1016/S1369-7021(13)70013-0

[5] T. Sekitani, Y. Kato, S. Iba, *et al.* "Bending experiment on Pentacene field-effect transistors on plastic films", *Applied Physics Letters*, 86, 7, 073511, 2005. doi: 10.1063/1.1868868

[6] C. Yang, J. Yoon, S. H. Kim, *et al.* "Bending-stress-driven phase transitions in Pentacene thin films for flexible organic field-effect transistors", *Applied Physics Letters*, 92, 24, 243305, 2008. doi: 10.1063/1.2948862

[7] V. Scenev, P. Cosseddu, A. Bonfiglio, *et al.* "Origin of mechanical strain sensitivity of Pentacene thin-film transistors", *Organic Electronics*, 14, 1323–1329, 2013. doi: 10.1016/j.orgel.2013.02.030

[8] P. Cosseddu, G. Tiddia, S. Milita, and A. Bonfiglio, "Continuous tuning of the mechanical sensitivity of Pentacene OTFTs on flexible substrates: from strain sensors to deformable transistors", *Organic Electronics*, 14, 206–211, 2013. doi: 10.1016/j.orgel.2012.10.033

[9] P. Cosseddu, S. Milita, and A. Bonfiglio, "Strain sensitivity and transport properties in organic field effect transistors", *IEEE Electron Device Letters*, doi: 10.1109/LED.2011.2173898

[10] A. Jedaa and M. Halik, "Towards strain resistant flexible organic thin film transistors", *Applied Physics Letters*, 95, 10, 103309, 2009. doi: 10.1063/1.3216587

[11] Anatoliy N. Sokolov, Yadong Cao, Olasupo B. Johnson, and Zhenan Bao "Mechanistic considerations of bending-strain effects within organic semiconductors on polymer dielectrics", *Advanced Function Materials*, 22, 175–183, 2012. doi: 10.1002/adfm.201101418

[12] Zhenlin Rang, Marshall I. Nathan, P. Paul Ruden Vitaly Podzorov, Michael E. Gershenson Christopher R. Newman, and C. Daniel Frisbie, "Hydrostatic pressure dependence of charge carrier transport in single-crystal rubrene devices", *Applied Physics Letters*, 86, 123501, 2005. doi: 10.1063/1.1875761

[13] G. Darlinski, U. Böttger, R. Waser, *et al.* "Mechanical force sensors using organic thin-film transistors", *Journal of Applied Physics*, 97, 9, 093708, 2005. doi: 10.1063/1.1888046

[14] S. Jung, T. Ji, J. Xie, and V. K. Varadan, "Flexible strain sensors based on Pentacene-carbon nanotube composite thin films", in Proceedings of the Seventh IEEE International Conference on Nanotechnology, Hong Kong, August 2–5, 2007, pp. 375–378. doi: 10.1109/NANO.2007.4601212

[15] S. Jung, T. Ji, and V. K. Varadan, "Pentacene-based low-voltage strain sensors with PVP/Ta_2O_5 hybrid gate dielectrics", *IEEE Transactions on Electron Devices*, 57, 2, 391–396, 2010. doi: 10.1109/TED.2009.2036317

[16] I. Manunza, A. Sulis, and A. Bonfiglio, "Pressure sensing by flexible, organic, field effect transistors", *Applied Physics Letters*, 89, 143502, 2006. doi: 10.1063/1.2357924

[17] I. Manunza and A. Bonfiglio, "Pressure and strain sensing using a completely flexible organic transistor", *Biosensors & Bioelectronics*, 22, 2775–2779, 2007. doi: 10.1016/j.bios.2007.01.021

[18] A. Loi, L. Basiricò, P. Cosseddu, *et al.* "Organic bendable and stretchable field effect devices for sensing applications", *IEEE Sensors Journal*, 13, 4764–4772, 2013. doi: 10.1109/JSEN.2013.2273173

[19] A. Loi, L. Basiricò, P. Cosseddu, *et al.* "Matrices of inkjet printed OFETs for the realization of artificial robotic skin", Materials Research Society Symposia Proceedings, MRS Proceedings, mrsf11401t0402. doi: 10.1557/opl.2012.1300

[20] S. C. B. Mannsfeld, B. C.-K. Tee, R. M. Stoltenberg, *et al.* "Highly sensitive flexible pressure sensors with microstructured rubber dielectric layers", *Nature Materials*, 9, 859, 2010. doi: 10.1038/nmat2834

[21] G. Schwartz, B. C.-K. Tee, J. Mei, *et al.* "Flexible polymer transistors with high pressure sensitivity for application in electronic skin and health monitoring," *Nature Communications*, 4, 1859, 2013. doi: 10.1038/ncomms2832

[22] J. Kim, T. N. Ng, and W. S. Kim, "Highly sensitive tactile sensors integrated with organic transistors", *Applied Physics Letters*, 101, 103308, 2012. doi: 10.1063/1.4751354

[23] Y.-J. Hsu, Z. Jia, and I. Kymissis, "A locally amplified strain sensor based on a piezoelectric polymer and organic field-effect transistors", *IEEE Transactions on Electron Devices*, 58, 910, 2011. doi: 10.1109/TED.2010.2102631

[24] T. Q. Trung, N. T. Tien, Y. G. Seol, and N.-E. Lee, "Transparent and flexible organic field-effect transistor for multi-modal sensing", *Organic Electronics*, 13, 533, 2012. doi: 10.1016/j.orgel.2011.12.015

[25] S. Lai, P. Cosseddu, A. Bonfiglio, and M. Barbaro, "Ultralow voltage pressure sensors based on organic FETs and compressible capacitors", *IEEE Electronics Device Letters*, 34, 801, 2013. doi: 10.1109/LED.2013.2257660

[26] T. Someya, T. Sekitani, S. Iba, Y. Kato, H. Kawaguchi, and T. Sakurai, "A large-area, flexible pressure sensor matrix with organic field-effect transistors for artificial skin applications", *PNAS*, 101, 9966, 2004. doi: 10.1073/pnas.0401918101

[27] T. Someya, Y. Kato, T. Sekitani, *et al.* "Conformable, flexible, large-area networks of pressure and thermal sensors with organic transistors active matrixes", *PNAS*, 102, 12321, 2005. doi: 10.1073/pnas.0502392102

[28] K. Takei, T. Takahashi, J. C. Ho, *et al.* "Nanowire active-matrix circuitry for low-voltage macroscale artificial skin", *Nature Materials*, 9, 821, 2010. doi: 10.1038/nmat2835

Chapter 5

Smart fabric sensor architectures and technologies

Lina M. Castano[1] and Alison B. Flatau[1]

Abstract

This chapter presents a possible path towards fully wearable and fabric-based electronic platforms. It provides an overview of materials and methods in the field of smart fabrics and e-textiles with the perspective of achieving fabric-based stand-alone platforms. Electronic components, interconnects and sensors can be made out of fabric and polymeric materials. Smart fabric sensors can measure force, pressure, strain, temperature and chemicals, among other parameters. Fabric power elements compatible with textile platforms are found in textile capacitors, super-capacitors, solar cells and energy-harvesting components. All these elements can be integrated such that a complete sensing and electronics platform is primarily made of compliant materials. Flexible macroelectronics technologies have a direct impact on wearables and development of functionalized textiles.

5.1 Introduction

Fabric-based wearable devices have seen an increased market expansion and greater consumer acceptance over the last few years. Integration of flexible electronic devices into daily life requires development of new technologies which allow for portability while remaining fully functional. Wearable technologies started with small portable electronic devices such as glasses, watches, cameras and microphones and have since then opened a demand for more flexible and light-weight devices. Portability, however, remains one of the major challenges, along with resistance to mechanical wear and tear, high levels of moisture and chemicals present in washing cycles. One of the more promising realizations of wearable technologies entails the use of smart fabrics or e-textiles. The distinction between these two terms is found in their functional origins. Electronic textiles or e-textiles are textiles with embedded electronic functionalities, e.g. electricity conduction.

[1]Department of Aerospace Engineering, University of Maryland, College Park, MD 20740, USA

Smart fabrics encompass a wider set of functionalized fabrics, e.g. textile sensors and sensing fibers. In this chapter, these two terms will be used interchangeably, as they have become nearly synonymous due to technological developments.

As indicated by Castano and Flatau [1], fabrics can become fully stand-alone platforms (Figure 5.1). That is, a set of sensors, actuators, batteries, connectors, interconnects, circuit components, circuit traces, etc. which are based on textile materials entirely or to a large extent. All of these components can find analogues in textile materials, with a renovated perspective based on physical principles of operation. Smart fabric transducers and sensors, in particular, were shown to have different sensing modalities through a combination of fabric materials, sensing polymers and compliant substrates [1]. Conventional electronics, albeit not entirely replaceable due to reliability constraints, can be supplemented to a large extent by textile technologies used for wearable devices. The level of integration of fabric devices depends on intended usage and fabric construction. These are limited only by the availability and compatibility of materials required for wearable applications. Industrial-scale methods are still to be implemented for massive production of e-textiles devices and have not been fully developed. New fabrication methods and technologies need to be advanced for textile-based platforms. These will be driven mainly by civilian consumer demand, but also by the useful and promising applications in engineering, medicine and military operations [6], among other fields.

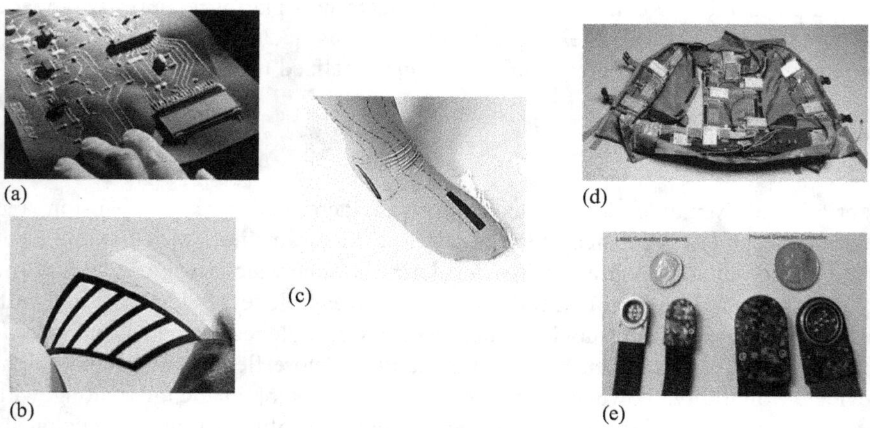

Figure 5.1 All-fabric-based wearable sensing platform: (a) Textile PCB [2], (b) Flexible solar cell [3], (c) Fabric sensors installed on smart garments (socks) [4], (d) Electrotextile antenna integrated into load carrying vest [5], Image courtesy of Dr. Carole A Winterhalter US Army NSRDEC, and (e) Textile data and power connectors [5], Image courtesy of Dr. Carole A Winterhalter US Army NSRDEC

Flexible macroelectronics developments and the personal electronics industry, i.e. microelectronics, are strongly associated with the development of e-textiles devices. Flexible macroelectronics, or large area electronics, e.g. thin-film solar cells, flexible displays and electronic skins, provide compliant solutions for e-textiles applications. Microelectronics developments produce miniaturized devices which have increased portability. Development in both microelectronics and macroelectronics are enablers and drivers of e-textile developments. At the same time, e-textile technologies have revealed a number of technological gaps and opportunities for development of new products and solutions.

The first part of this chapter will provide a layout of fabric architecture, followed by a second part where electronic textile elements will be presented as a basis for a fully fabric-based electronic system. The second part will also provide an overview of textile power elements and energy harvesting strategies. The third part presents possible flexible circuit solutions to be used with the textile electronic elements. It also includes an overview of possible connectors and interconnects technologies that can be used with textile-based devices.

5.1.1 Fabric origins and e-textile construction

Incorporation of electronic elements and functionality into textiles requires an understanding of fabric substrates. Fabrics are fibrous materials which are porous and have specific mechanical characteristics depending on the construction mechanism and materials used. They are composed of smaller units or fibers, which come in different lengths, widths and materials. Fiber origins have an impact on compatibility with materials, coatings and reliability. They also have an influence on effective coating methods and adhesion characteristics, as well as water absorption, thermal conductivity, melting point and chemical reactivity. These properties are of fundamental relevance to any given application.

Origins of fibers may be natural, synthetic or hybrid. Natural fibers are made of cellulosic materials found in plants, animals and minerals. They come mostly in staples, or clusters of fibers of a given origin-dependent length, and are water absorbent or hydrophilic. Fiber cross sections have specific characteristics depending on whether they are from animals or plants. These fibers are mostly flammable and have a higher melting point than synthetic fibers. Examples of cellulosic fibers include cotton, hemp, jute, flax and sisal; animal or protein fibers include silk and wool; mineral fiber examples are asbestos, ceramic fibers, glass filaments and natural rubbers. Synthetic fibers are composed of regenerated fibers and those based on synthetic polymers or metallic fibers. These fibers can be manufactured according to certain desired specifications and are mostly hydrophobic. Examples of regenerated fibers include acetate, rayon and azlon, which have a cellulosic origin; synthetic polymer fibers such as nylon, acrylic, polyester, spandex, nitrile and olefin; and metallic fibers which can be made of aluminium, silver, gold, copper, etc. Flame resistance, chemical inertness and sunlight resistance are additional factors to consider when fibers are exposed to heat expelled by electronics, or when continuously exposed to sunlight. Cotton, linen, silk and rayon have poor flame resistance, whereas nylon, polyester, acrylic, wool and spandex have good flame resistance. Fibers which have poor resistance to strong acids include cotton, silk, acetate, polyester, nylon and rayon. Similarly, those which are not resistant to alkalies include flax, jute, wool, acrylic and acetate. Sunlight resistance, or the ability to withstand degradation and damage from direct or prolonged sunlight exposure, is found to be strong for acrylic and metallic fibers and weak for nylon, rubber and spandex.

Aside from the inherent features of the materials used, the properties of a fabric depend on the construction method. Multiple fiber filaments form staples; staples are then morphed into thread of different characteristics. This thread is then used to form yarns of multiple plies or ply configurations by twisting the fibers at an angle or by holding them together by a film-forming polymer [7]. Yarns are woven, knitted, tufted, stitched or braided to form 2D and 3D textiles. Fibers and yarns can also be used with no structure to form non-wovens, or melted/knotted into nets. Each construction method has a number of variations depending on the intended use of the fabric. These fabric structures and construction mechanisms are the starting point for electronics-embedded devices, as the functionality can be incorporated at any step in the textile manufacturing process. Smart fabrics technologies make use of novel materials as well as customized adaptation methods.

Fabrics can be embedded with e-textiles elements at any point of the textile construction process, as shown in Figure 5.2. Fibers can be replaced or made out of materials with desired properties, i.e. electrical conductivity, gas sensing and thermochromic properties. Yarns can also be made of sensing materials or they can be constructed from a combination of conventional and functionalized fibers. Once the fabric has been woven, knitted, etc., a sensing or functionalized coating can be applied to the fabric surface. Coatings, however, can also be applied at the yarn and fiber level. A fabric composite can be made out of conventional materials or it can be made out of sensing or functionalized fabrics. Fabric composite or laminate behavior is a product of angle ply geometry, bonding agent and stacking sequence. Miniaturized conventional electronic components are typically introduced after the textile has been woven, or knitted, by replacing fibers, yarns or by electro-mechanical adaptation to the textile substrate. These electronic components can also be introduced into yarns, which are then inserted into knits, weaves, etc. Non-wovens and nets can be made directly from materials in fiber or yarn form.

Superficial attachment of conventional electronic elements, such as resistors or integrated circuit (IC) chips to the fabric, is another way of obtaining a functionalized textile. This method utilizes conventional technologies which have been miniaturized for ease of use and are adapted to textile substrates by various methods of encapsulation. Electronic elements and encapsulation must be of reduced size and weight. Circuit boards need to be either miniaturized or compliant enough to conform to the fabric application. Pre-packaged flexible electronics that can be applied to fabrics will be presented in a later section. Electrical and mechanical connectors and interconnects are also of importance to the success of a textile-based electronic device. This is an area of development for smart fabrics sensing technologies and functional realizations.

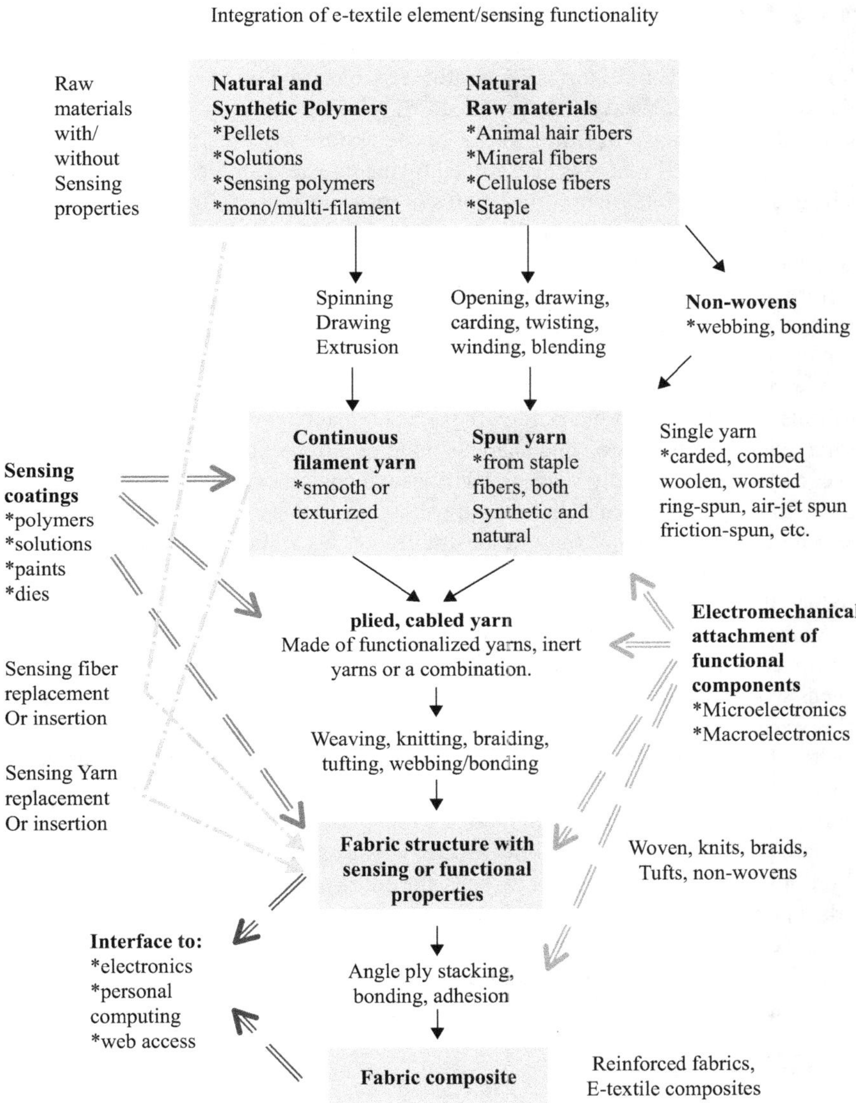

Figure 5.2 E-textile functionalization of fabrics, at the different steps in textile fabrication process

5.1.2 Fabric structure characteristics relevant to e-textile construction

Woven fabrics are produced by interlacing one or more warp and filling yarns (also known as length and cross yarns). These are usually made in a variety of looms, from home-made to industrial versions. The three basic weaves are plain weave, twill weave and satin weave. The plain weave is used in clothing, medical gauze and drapery, among other uses. The twill weave is used in garments such as denim, upholstery and industrial fabrics. The satin weave is used when surface smoothness is desired. Different weaves are produced by varying yarn ply configurations, the count and ratio of warp to filling yarns, or the texture of yarns. For e-textile applications, the choice of weave is determined by the intended application. Plain weaves can be useful when interlacing fiber optics or pressure sensing fibers for instance, as they provide a uniformly embedded pattern for the added sensing fibers. Plain weaves are commonly used as background for embroidery and printing and are therefore useful for e-broidery [8]. The basket weave, a variation of plain weaves, is porous, coarse looking, does not withstand friction or abrasion, and has poor shape retention as yarns tend to move and shift during use. When washed, these fabrics shrink, which is yet another consideration in the design of textile-based functional platforms. Given their porous nature, they can be placed for instance at points where thermal exchange is desired and shape is not a requirement. Twill weaves are more dense, compact and closely woven. Twill weave construction utilizes pied yarns or yarns with a high degree of twist. A twill weave may be useful for fabric applications where higher fiber density may be desirable, i.e. a conductive surface for textile switches, etc. Satin weaves have high luster and light reflection. These may be used to reduce the absorption of heat radiated from e-textile components [9].

As far as mechanical properties, the tensile behavior of woven fabrics usually has a strong correlation to the yarn mechanical properties and the weaving structure. Wovens will exhibit strain hardening. In addition, tensile strength is generally greater in the warp direction than in the weft direction, due to mechanical cycling during industrial fabrication [10]. Twill and satin woven fabrics usually have lower tensile strength as compared with plain woven fabrics, and a twill woven fabric exhibits larger elongation than plain woven fabrics. From geometric theories of woven fabrics, four main parameters describe fabric geometry: yarn diameter, thickness, cover factor and crimp. These parameters can be used for constructing models of warp and weft interlacing and therefore of textile sensors, interconnects and other structures which involve weaving or incorporation of sensing fibers. Fibers and yarns behave in a non-Hookean way during deformation and show hysteresis over time. The stress and strain relationship for a woven fabric can be modeled using the following equation [10]:

$$\sigma = \frac{e^{\alpha\varepsilon} - 1}{\beta} + e_r \tag{5.1}$$

In this equation, σ is the stress, ε is the strain, e_r is the error and α and β are parameters that can be estimated with regression analysis of experimental data. Bending rigidity in wovens can be described with different models, one of the simplest ones is shown in (5.2), where the bending rigidity B is a function of w, fabric weight per square unit and bending length c

$$B = wc^3 \tag{5.2}$$

Knitted fabrics have a very different method of construction. They are formed by interlocking loops with one or more yarns in preceding and succeeding rows.

Wales are the vertical loops and courses refer to horizontal loops. They can be made by hand or by using knitting machines, of which there are different types. The characteristics of a knit are defined by the type and content of fiber, yarn denier and type of yarn as is the case with wovens, but also the number, spacing and quality of needle sets, knitting bed displacement and geometry, latch kinematics, fabric take-up velocity and stitch size [11]. Knits are more stretchable than wovens and possess good thermal insulation properties. The main construction types of knit fabrics are the weft structure and the warp structure. In weft knits, the structure is built from a unique yarn, whereas a warp knit is made from a number of yarns. Therefore, weft knits can be unraveled and warp knits cannot. Weft structure includes the jersey, purl, interlock and rib knits, and the warp structure includes the milanese, ketten-raschel and tricot knits [9]. Weft knits have high elongation and elasticity properties as compared to warp knits. Weft knits may be useful when constructing strain sensors either out of sensing fibers or used as stretching substrates for strain-responsive sensing coatings. Warp knits may be useful for implementing sensors which need to conform to curved surfaces and have a moderate level of elongation.

Tensile properties for weft and warp knitted fabrics are a result of different mechanisms. Warp knits involve rotation and friction, while bending and friction are observed in weft knits. This is an important consideration for e-textiles constructed at the yarn level and the intended sensing mechanism. For weft knitted fabrics, stitch bending in the thickness direction decreases and inter-yarn friction and tension increase when the fabric is set in tension. In contrast to weft knitted fabrics, warp knitting has a number of parameters which include the number, position and movement of knitting guide bars, and stitch length of each bar. Many different approaches have been explored to approximate the tensile properties of knitted fabrics. In particular, a three-dimensional material model based on a strain energy function can be useful to predict non-linear characteristics of the knitted fabric. For instance, (5.3) represents the second Piola–Kirchoff stress S_{ij} as a Green strain derivative of strain energy function w, which is a function of three-dimensional knit fabric material parameters [12]:

$$S_{ij} = \frac{\partial w}{\partial E_{ij}} = \partial \frac{(Ce^{x})}{\partial E_{ij}} \tag{5.3}$$

The bending rigidity of knitted fabrics has a dependency on the modulus of elasticity of the material and the cross-sectional moment of area. For knitted fabrics, the relationship between bending moment and curvature is non-linear and is also non-constant, i.e. it increases with increasing curvature due to the increasing inter-yarn compression forces [10]. Knitting technology is useful for integration of functional yarns during a fabric manufacturing process. For instance, a compressive effect can be introduced by inserting elastic weft yarns of defined pre-tension into a knit structure. Yarns with thermal and conductive properties can also be inserted during the knitting process, as the inserted yarn material can be varied in each course [13].

In addition to woven and knitted fabric structures, non-wovens, braided fabrics, tufted fabrics and nets are also important substrates and construction configurations for e-textile technologies. Non-wovens are manufactured sheets or webs of randomly oriented fibers and/or non-fibrous elements bonded by friction, cohesion or adhesion [10]. The constitutive elements of non-wovens are not interlaced or interlooped bonded but are joined by a number of web-forming and bonding processes. Some of the most common web-forming processes include carding, cross-lapping, air laid, spunbond and meltblown. Common bonding methods include calendering and hydroentangling. The web-forming processes begin with polymer pellets and raw fiber materials which are melted and spun into webs. Bonding processes are then used to apply mechanical pressure and high temperatures to bond elements in these webs. The mechanical properties of the non-woven fabric and anisotropy of the constituent fibers are a result of the chemical, mechanical and thermal processes used. Non-woven processes can be applied to build sensing fabrics by using sensing materials to form the webs which are eventually bonded. Examples of non-woven fabrics include felt, filters and geocomposites.

Nets are very light-weight fabrics and can be constructed by twisting, knotting or fusing yarns together at intersection points. Nets are useful structures for sensors that require a high air passage, e.g. gas and chemical sensors, and that require yarn stability, i.e. photovoltaics. Tufted fabrics have only one hinging point and are particularly free to move with air currents, or external mechanical inputs. They can be manufactured in loop pile and cut pile constructions [9], and with varying degrees of yarn twist. Tufted fabrics can be applied to secure textile elements, to make touch sensors or to increase air surface interaction with chemical sensitive functionalized yarns. Braided textiles are constructed by intertwining yarns in diamond, regular and Hercules patterns. The braided fabric is typically made in tubular form and can conform to various shapes. For instance, braided conductor tracks can be wrapped around e-textile conductors for electromagnetic shielding. Braided fabrics can conform to curved surfaces, which may add wearability to textile sensors embedded in textile braids.

Mechanical characteristics of yarn and yarn construction methods also affect the final fabric construction. The twisting angle indicates the extent of the twisting effect. Higher twist or turns per inch produces larger yarn density but lower strength and modulus [14]. The amount of twist required to hold the fibers and yarns together depends on the yarn diameter. Thicker yarns will require less twist to hold them together and are referred to as low twist yarns, whereas fine yarns will require more twist and are referred to as high twist yarns. The degree of twist will affect the yarn's bending behavior, covering power, permeability, tensile strength, elastic characteristics, resistance to creases and abrasion, and fabric performance characteristics such as compressibility, porosity and stiffness. Twist also affect the texture of the finished fabric, i.e. lower twist will have a coarser feel and higher twist a smoother one. The yarn-ply characteristics also have an influence on resulting yarn and fabric. Plied yarns have increased strength, are thicker and heavier than single yarn. Yarns made for woven fabrics are denser than those made for knit fabrics, a direct result of the degree of twist, which is higher in wovens. Combed yarns are denser than carded yarns. Fiber migration, mobility and arrangement as well as yarn length, fineness and cohesion are other factors to be considered. In general, the finer the yarn becomes, the lower the degree the loose superficial fibers or "hairiness". All these parameters are design knobs for both yarn sensors and fabric sensors.

Sensing textile devices suffer macroscopic and microscopic deformation [11]. Macro-level deformation typically entails transverse and in-plane tension/compression, and out of plane bending. At the micro-level, a lot of different phenomena are present in yarn interactions, as shown in Figure 5.3. These include slip, shear, buckling and twist. Inter-yarn shear is common in many woven fabrics. Inter-yarn twist, slip and bending are influential in knits due to loop geometry. Micro-level considerations are important when manufacturing sensing fibers into fabric structures, or when introducing functionalized fibers into already manufactured structures [11].

A prime method for textile functionalization into e-textiles or smart fabrics is the application of sensing or functional coatings. These can be applied to all the fabric building blocks, i.e. fibers, yarns and fabric. The parameters necessary for consistency of coatings include enough substrate tension, appropriate viscosity of the coating material and substrate uniformity and porosity. One of the key parameters to consider before coating a fabric substrate is the coating factor, *C*, which gives a measure of the fabric openness, air permeability, degree of moisture resistance and adhesion [15]. This factor indicates the degree to which the coating can go through the fabric, having an impact on how well the coating will adhere to the substrate. For a woven fabric composed of a warp fiber and a filling yarn, the coating factor is given as follows:

$$C_{woven} = w * d_{w,y} + f * d_{f,y} - w * f * d_{w,y} * d_{f,y} \quad (5.4)$$

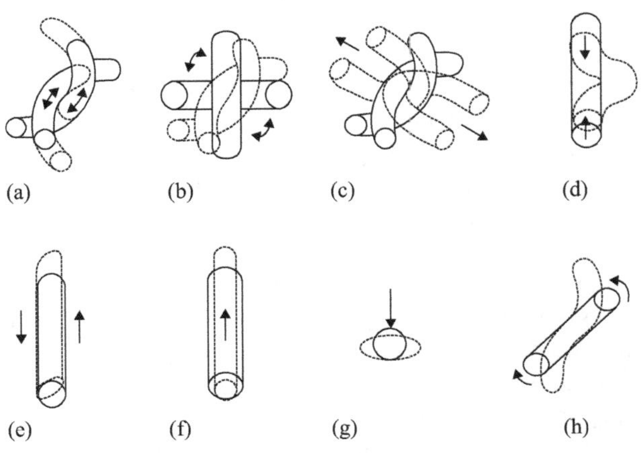

Figure 5.3 Micro-level fabric deformation modes: (a) inter-yarn slip, (b) inter-yarn shear, (c) yarn bending, (d) yarn buckling, (e) intra-yarn slip (inter-fiber friction), (f) yarn stretching, (g) yarn compression, (h) yarn twist. Excerpt from [11]

where w is the number of warp threads per inch, f is the filling threads per inch, $d_{w,y}$ is the warp yarn diameter and $d_{f,y}$ is the filling yarn diameter. For a knitted structure, the cover factor expresses the relation of yarn diameter to loop length. It correlates with characteristics of stability and some mechanical properties. To calculate the cover factor, the assumptions made are that the length of yarn in a stitch is the actual yarn length and that there is no interlacing between stitches [10]. The cover factor for a weft jersey knit composed of two yarns can be calculated as:

$$C_{knitted} = C_{yarn1} + C_{yarn2} = C_1 W_1 L_1 d_1 + C_2 W_2 L_2 d_2 \tag{5.5}$$

where C is the number of loop courses per unit length, W is the number of loop wales per unit length, L is the stitch length and d is the yarn diameter. If the fabric is knitted with several yarns in the same area, the total cover factor is the sum of the cover factors for each yarn. A variety of stitches produce different lengths, run in ratios and grammage, which need to be considered depending on the intended application.

It has been shown that a greater coating penetration depth prevents the yarns from rotating, stiffening the fabric and increasing flexural rigidity. Coating penetration can be modeled using Darcy's law for the flow through porous media, where the average velocity of the porous media is proportional to the vertical pressure gradient on the fabric substrate. The following calculation applies to the case of a coating blade of longitudinal dimension L, separated a distance $h(x)$ from the substrate, with upstream gap distance $h1$ and downstream distance $h2$. Assuming a Newtonian incompressible fluid coating flowing in the x-direction and no absorption of the liquid into the fibers, the square of the vertical penetration of the coating can be calculated as shown in the following equation [16]:

$$z(x)^2 = 12\frac{d_f^2}{16k_0}\frac{P^3}{(PF)^3}\frac{L}{P}\int_0^L \frac{(h1 - h(x))(h2 - h(x))}{(h2^2 - h1^2)h(x)^2}dx \tag{5.6}$$

where PF is the fiber packing factor and P is the porosity or the ratio of the volume of voids contained within the boundaries to the total volume, d_f is the fiber diameter and k_0 is the Kozeny constant, an empirical constant. From this expression, it can be observed that the penetration depth $z(x)$ increases by decreasing the gap distance and is independent of the coating speed and the viscosity of the coating. Depth, tension and viscosity have an impact on coating leveling, which is a critical step in achieving a uniform coating. If there is high surface tension, then the coating tends to crater generating micro-cracks [17] and if it has a very low viscosity it will sag; therefore, leveling needs to have a balance of these quantities. These dependencies can be quantified using the leveling half-time $T^{1/2}$ of the striation marks produced during the coating application process [18].

Coating techniques found in fabric sensors include screen printing, ink-jet printing, electrodeposition, electroless plating, sputtering of thin films, vapor deposition and thermoset coatings. Coatings can provide sensing qualities to a textile substrate at the cost of altering fabric mechanics such as their tensile, shear and bending properties [19]. This composite behavior is mainly due to the change in yarn mobility after the coating is applied [16]. The parameters that influence the consistency of coatings include the viscosity and uniformity of the coating material, and the tension, flexural rigidity [20], porosity and coating factor of the substrate [18].

5.2 Creating an all-fabric-based e-textile system

A stand-alone smart fabric sensing (or actuation) system would include fabric-based or flexible circuitry, miniaturized or fabric-based circuit elements, compliant connectors and interconnects, and fabric sensors (or actuators), and energy sources. Some of these components, if not all, need a level of mechanical and environmental encapsulation and potentially an interface to computing and web-enabled devices. As a first approach to incorporating electronic elements, conventional components that are miniaturized can be used in e-textiles platforms by proper adaptation. Textile circuit elements that are miniaturized can be adapted to the fabric structure. Small off the shelf components can be sewn to conductive interconnects directly on the fabric [8], or by using holders, sockets or sequins which can be attached to the fabric by different methods [21]. Sequins can be sewn into conductive thread lines while holder leads can be soldered onto flexible printed circuit boards (PCBs). These sewing tabs add a negligible amount of resistance to the circuit; a typical fabric tab to stitching joint has less than 1 Ω resistance. These tabs can also be created by twisting the leads of conventional components, e.g. resistors, capacitors, inductors, to create wire loops using crimp beads, and these loops are then sewn using conductive thread. The same techniques can be applied to microcontrollers and other circuit elements. For ICs, the corresponding IC socket is sewn into the fabric traces, one trace at a time. Fabric PCB breakout boards or thin-film bondable boards can also be used for each electronic component, with the added benefit of compliance. Conductive tabs for electrical components are composed of conductive traces to which wire or thread can be soldered or sewn. Another approach is to use mechanical force for establishing electrical contacts. This is achieved by using gripper snaps [21] or by stapling the components using pressure. This, however, causes reduced joint motion.

In essence, miniaturized electronic components can be attached to fabric if their lead connectors are adapted to fabric form factors. Component leads can be adapted by using crimp beads, loops, rivets, grommets, snap rings and pops, and conductive Velcro. Soldering the leads onto conductive fabric and stitching lead wires onto the fabric by using conductive thread [22] are other possible solutions. All of these adaptation options enable a convenient platform for textile circuitry [23]. Given a sufficiently large area of fabric, entire circuits of resistors, capacitors and inductors can be fabricated. In a fully fabric-based circuit, the density of components is highly reduced as compared to conventional counterparts and therefore might be considered a disadvantage. However, a lot of useful circuits for wearable applications are not composed of a large number of parts. For complex circuitry, or circuits with a high parts count, parts would need to be either installed on flexible PCB circuits or used in tandem through wireless or relayed transmission to processing units, i.e. cellphones and computers.

5.2.1 Passive and active circuit elements

Conventional resistors are typically composed of an electrically resistive material encapsulated in an insulating case composed of ceramics, glass and fiberglass. The resistive material is connected to connector caps and leads. Resistive materials include carbon composites and granules, metal-oxide films such as tin oxide, wire-shaped alloys such as nichrome and manganin, among others. The power dissipated is determined by the current and voltage at which the resistor is operated and is typically proportional to the physical size of the resistor. A similar principle of operation is found in resistors created directly onto fabrics. If a non-conductive textile is used as the resistor substrate, the electrically resistive material can be manufactured using polymer composites and conductive fibrous materials. These materials include thread, yarn and fabrics which have been made conductive or have been made out of conductive materials. The degree of electrical conductivity will depend on the properties of the conductive polymers, in the case of polymer coatings. Similarly, the resistive properties of fabric materials will depend on the properties of the conductive materials used to construct the fabric structures.

Passive and active electronic circuit elements can be made out of conductive yarn [17] and other soft materials [8] by embroidering, weaving knitting, printing, metal ink sputtering and silk screening of conductive materials, among other methods. Passive elements include resistors, capacitors, inductors, transformers and active elements include transistors, ICs and light-emitting diodes. Electronic elements can be made out of conductive thread [8] by sewing thread fibers in patterns of multiple crossings to achieve desired electrical properties. This is applicable to a variety of available conducting thread of various degrees of conductivity, e.g. copper [24], silver [25], gold and steel [24] wire/thread. Passive elements can also be formed with conductive paint [26], inks [27] and polymers by applying them on textile substrates. If combinations of these materials are used, the resulting resistivity of combined materials can be calculated using the basic formulation of resistivity for a continuous material. Resistors, capacitors and inductors can be made by planar printing techniques such as screen printing or sputtering metal inks onto fabric substrates [27,28].

Resistors can also be made from a stretch of fabric of different shapes and lengths [29]. Resistive elements can be constructed by adjusting the dimension of an already coated conductive polymer fabric [30]. Fabrics can be made conductive for construction of passive elements or sensing elements. Their conductivity can be enabled during manufacturing or after manufacturing [31]. Before fabric construction, thread can be made conductive by applying conductive coatings, or twisting a conductive fiber or yarn with a non-conductive one. Extrusion of conductive polymer materials can also be used to form conductive thread. After manufacturing, fabrics can be altered by metal or conductive polymer coatings or by inserting conductive thread or yarn by sewing or embroidery into the manufactured textile. Laminating methods are also applicable to enable conductive fabrics; these include techniques adapted from conventional and flexible electronics [28]. Different fabrication methods for fabric-based resistors are shown in Figure 5.4.

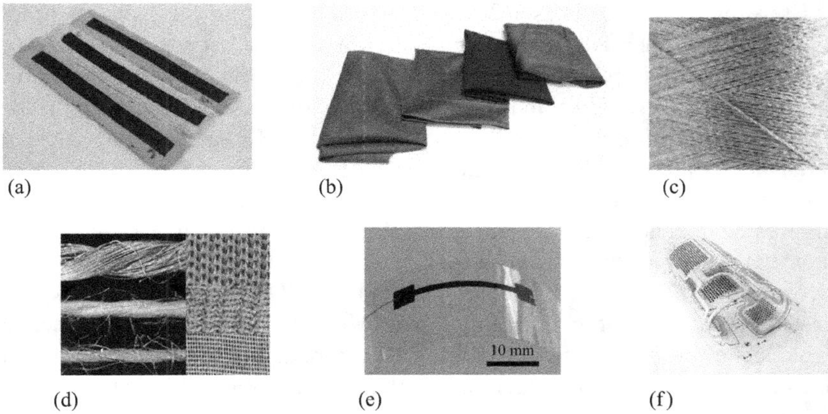

Figure 5.4 Fabric-based resistors: (a) Carbon-based screen-printed resistor on cotton-polyester knits [4], (b) Polypyrrole-based conductive fabrics [32], (c) Conductive yarn [34], (d) SEM images (×200) of three conductive fabrics made of silver-plated nylon-based elastic yarns [35], (e) Graphite strain gauge printed on PEN poly (ethylene naphthalate) [36], (f) Printed resistors on plastic film [37], reproduced with permission from DuPont Electronics & Communications

Fabric elements can be made out of compliant and polymeric materials [38]. Due to increasing interest in miniaturization and compliance of components, fibers turn out to be a very appropriate target for standalone element construction. A capacitor is made for instance out of a hollow core fiber with the first electrode lining the inside of the hollow core and a second plastic electrode wrapping the fiber from outside. The conductive layer is a polyethylene (PE)-based carbon black, and the dielectric layers are made of low-density PE [39]. Conductive polymers can be used to make fibers with sensing properties and use them to build fabric structures. Such is the case for soft capacitor fibers [40] where multiple layers of carbon-based polymer/PE are drawn over an air/copper core.

Construction of capacitors and inductors relies to a greater extent upon the specific geometry of the conductive elements. Attaining specific values with fabric strips is more challenging than for woven resistors. In the case of transistors, the core of a metalized yarn can be used as gate, while source and drain contacts can be made using evaporation or soft lithography methods for deposition of metals or polymers [41–44]. Transistors can also be fabricated on strips of kapton which can later be interlaced into a textile substrate [45]. Other examples include building thin-film transistors on Kapton® thin flat fibers [45] and piezoelectric films on textiles [46]. Organic transistors have also been fabricated on cotton fibers [47].

Fibers themselves can be made of sensing materials by conventional and unconventional methods (Figure 5.4). Electrospinning, wet-spinning, self-assembly, carding, combing and die extrusion are some of the methods used to make fibers out of sensing materials. These fiber and yarn construction mechanisms need to also incorporate other fabric parameters such as crimp, degree of fabric cover, density, flexural rigidity and thickness in order to suitably integrate the element into the fabric. These fiber and yarn fabrication options can be customized as needed for applications with specific requirements. Mechanical properties of a material are limited by defects during synthesis and fabrication. In the case of polymer fibers, these defects consist of chain ends and topological defects such as chain entanglements which prevent good alignment and chain packaging. Other phenomena that limit the strength of a fiber-based material are the inclusions or voids within the fiber and surface defects. Figure 5.5(a) shows a feedstock which consists of a liquid hydrocarbon solution in which ferrocene (catalyst precursor) and thiophene (promoter) are dissolved. This compound is injected into a hot furnace along with a hydrogen carrier gas. With heat, the compound breaks down and

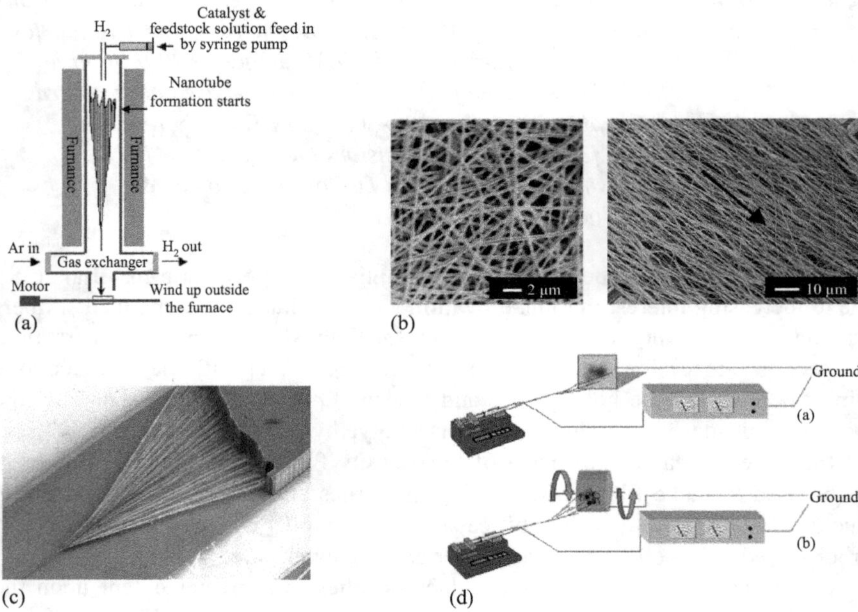

Figure 5.5 Examples of yarn fabrication methods to construct yarns from polymeric materials with sensing properties: (a) Schematic of the direct spinning process of carbon nanotube fibers and films [48], (b) SEM image of amorphous entangled polyacrylonitrile (PAN) fibers made by electrospinning onto a stationary target (left) and those collected onto a take-up rotating wheel [51], (c) Carbon nanotube (CNT) yarn formation by drawing from CNT forest and insertion of twist [50], (d) Electrospinning apparatus of polyacrylonitrile nanofibers with (a) a stationary, grounded target and (b) a rotating grounded target [51]

Table 5.1 Carbon-based conductive yarns manufactured from sensing materials

Yarns	Method	Conductivity/ resistivity	Max. tensile strength
CNT [14]	Twisting/shrinking	~2.5×10^{-8} Ω m	1.1 GPa
CNT-PVA [52]	Solution treated/ furnace dried	9.2×10^{4} S/m	2.0 GPa
SWNT/MWNT [55]	Direct spinning by CVD	8.3×10^{5} Ω^{-1}/m	1.0 GPA
SWNT/MWNT [57] Cotton core	Solution dipping	20 Ω/cm	87.8 MPa

reacts to form carbon nanotubes (CNTs) which then interact to form a continous aerogel that ends up being collected outside the furnace as a fiber or film. Similarly, other spinning processes draw the fibers from films [49] or from CNT forests and then twisted into yarns [50], or are drawn by charged targets [51] which may be stationary or rotating, to produce aligned fibers [49].

Carbon nanotube (CNT) yarns have been made by direct drawing from CNT bulk solutions [52,53], directly spun by chemical vapor deposition (CVD) from solutions, suspensions [54], gels or aerogels [55], coagulation-spinning, these are examples of wet-spinning processes. Dry spinning methods are composed of direct synthesis, growth from super aligned arrays [52], twisting from CNT films [14,49]. In addition to nanotubes, carbon nanoparticles, nanofibers, nanowires, nanosphere chains can be applied to change properties of textile substrates [56]. Conducting fibers can also be intrinsically grown by electrophoretic deposition into forming yarns. These can subsequently be woven into textiles to either increase their mechanical strength or to increase the substrate conductivity if they are used as conductive paths. Resistance has units of Ohms and conductance (reciprocal of resistance) has units of Siemens (S). Similarly, resistivity has units of Ohms-meter (Omega.m) and conductivity (reciprocal of resistivity) has units of Siemens per meter (S/m). Table 5.1 shows some examples of fibers made directly from sensing materials. Combinations of CNTs and inherently conductive polymers can have increased conductivities, a mixture of PANi-CNT composite fiber with 750 S/cm has been reported. Other combinations such as polyaniline single-walled nanotubes (PANi-SWNT) [58] and polyaniline multi-walled nanotubes (PANi-MWNT) [59] have also been reported to increase the overall conductivity. Other nano-fillers used to increase conductivity besides CNTs and S/MWNTs are metal-oxide nano-particles (i.e. TiO_2, ZnO) which also have antimicrobial qualities and carbon black nanoparticles which have a high chemical resistance. Furthermore, fibers can be made such that they are given conventional electronic features, e.g. antennas [60], batteries [61], powering elements [62] and other electronic functional units.

5.2.2 Fabric sensors

The design process for fabric sensors begins with the intended application. This will determine requirements in the choice of materials and how flexible, stable or impermeable the fabric substrate or yarns need to be. Depending on these mechanical constraints, a sensor type and polymer materials or functional textile materials can be chosen. Once the sensor type, sensing materials are chosen the connectors can be crafted, adapted and the bonding agents and encapsulation materials determined, based on environmental requirements. The intended application will also determine whether telemetry or inductive power is needed, if the textile device needs to transmit data continuously or if it gets to be locally stored and then transmitted after a period of time. Personal area networks, Wi-Fi and Bluetooth technologies can be incorporated into the functionality of the standalone all-fabric-based sensing device. Fabric textile boards can aid in making these non-invasive sensing devices. Flexible circuit technologies are particularly promising for housing complex circuitry, and for being used in tandem with fabric textile boards if needed.

The following sensing technologies can also be used as stand-alone elements if the sensing mechanism is truncated or simply not enabled. For instance, a fabric capacitor may change a compressible spacer fabric dielectric layer for a non-compressible one, etc. Fabric textile sensors use textile-based materials to sense changes in force, pressure, strain, chemicals, gases, ions, analytes, humidity and temperature. Mechanical sensing is achieved by capacitive, resistive, electromagnetic, optical and piezoelectric parameter variations. Sensing elements for these mechanisms include woven or knitted electrodes, plates coated with inherently conductive polymers or particle-doped polymers, piezoelectric and piezoresistive-coated fibers/fabrics and coatings of polymers with numerous sensing possibilities [63]. These can be made from semiconductors, semiconducting metal oxides, solid electrolytes, metals, insulators, organic semiconductors, electrolytes and catalytic materials. Gases and chemicals can be detected by using chemoresistive polymers, as well as amperometric, and optical sensing mechanisms of fibers attached to the fabric or coatings on textile substrates. Temperature sensing can be achieved by means of resistive, optical, and shape memory polymers, as well as polymers with pyroelectric and ferroelectric response [6]. Electrically conductive polymers have a wide range of sensing properties as their electrical resistance changes with mechanical force, pressure, strain, light, chemicals and moisture. Electroactive polymers are a subcategory of sensing polymers which also possess actuator qualities and have been applied to textile sensors. These polymers can have piezoresistivity, thermoresistivity, magnetoresistivity, chemoresistivity and photoresistivity as passive sensing mechanisms, and can use piezoelectricity, thermoelectricity, pyroelectricity and photoelectricity to do work on the environment, i.e. exerting actuator properties [64]. Coating methods that are used for enabling sensing features on fabrics include sputtering, electroless plating, electrospinning, electrochemical deposition, CVD, dip coating, chemical polymerization and screen printing. These methods can be applied to fibers, yarns, wovens, knits, braids, non-wovens, etc.

For resistance-based sensors, all of the methods mentioned in the previous section are applicable. Strain and pressure sensors are formed when these conductive compliant materials are coated on stretchable substrates. Candidate materials include polymers with inherent conductivity and those that have been doped with conductive particles, nanoparticles, nanotubes, nanowires, etc.

Nano-composites and a number of polymers have been shown to have a wide range of reactivity to various types of stimuli. In Resistance-based sensors can be used to measure contact resistance, electrical resistance changes with mechanical inputs, i.e. strain, pressure, tension; chemical inputs, i.e. analytes, gaseous inputs; thermal and moisture inputs; and optical stimuli. In the case of mechanically generated resistance changes, elastic recovery will depend on the elasticity of the fabric substrate and the content of spandex and other resilient synthetic fibers. Polypyrrole [65], polyaniline [66] and PEDOT [67] are inherently conductive piezoresistive polymers that have found prolific application to fabric sensors in pure form and mixtures. This is also the case for a number of carbon composites, e.g. graphite, carbon black, and carbon nanocomposites, i.e. SWNT and MWNT.

Capacitive sensors can be constructed following the conventional two plate and spacer configuration, however, with fabric materials. The plates can be made of the variety of conductive materials mentioned earlier, e.g. woven/knitted/embroidered conductive yarns, electroplated fabrics, Intrinsically Conductive Polymer (ICP) and Extrinsically Conductive Polymer (ECP) coatings, CNT-based polymers, conductive inks, paints, glues, coatings, conductive fabrics used for Electromagnetic field (EMF) shielding, etc. The spacer or dielectrics in capacitors can be made of a number of materials that have a specific dielectric constant and have a measurable and consistent compressive response. Possible dielectric materials include Porous 3D fabrics, spacer fabrics, and synthetic foams, as well as porous PE, styrene, butadiene, silicone rubber, etc. Electrochemical and electrostatic capacitors are also possible. These will be discussed in the textile power section. Capacitive fiber sensors have also been constructed using textile and polymer materials [40,68,69]. Their construction resembles that of planar capacitors only with cylindrical geometry. Switches, tactile sensors and human body proximity sensors are possible using these capacitive fibers.

Optical fibers can be woven to form a textile substrate or they can be inserted into textile structures. Interweaving [70], interlacing and fiber bonding [71] are some of the insertion methods that can be used. Optical fiber sensors can be used in multiple sensing applications. They can measure strain, pressure, temperature and humidity, among other quantities. Similarly, fabrics with electroluminescent [72] and chromic properties [73] can be inserted into textile structures. Temperature and humidity fabrics follow similar construction in that polymers with corresponding sensitivity are used for coating yarns and other fabric substrates. Other humidity sensing devices entail flexible transistors [44], and changes in capacitive spacer dielectrics [74], and organic sensors that react to humidity changes.

In summary, using the process described in Figure 5.2, any textile substrate can be modified for sensing purposes. Sensing coatings are the prime method for fiber, yarn or fabric functionalization. The second best method entails insertion of functionalized yarns into textile substrates. The sensitized yarns can be either manufactured from sensing materials or they can be coated or twisted with other sensitized yarns to form a ply or cabled yarn. Miniaturized sensors can also be adapted and embedded at any point during fabrication of the textile. Passive and active electronic components can be constructed out of textile materials, possibly with reduced precision. Albeit enabling the option of completely fabric-based systems.

5.2.3 Connectors and interconnects

Electrical connections between the different fabric components for a fully textile-based platform can be realized with a variety of techniques. Many of these techniques are young technologies which are still in early development. These connection methods, interconnects and connectors all depend on the targeted sensing application and what it implies and demands of the textile substrate. Wearables [75] have a particular set of requirements as compared with smart textiles used for, e.g. building monitoring applications [76], ship and aviation industries [2,6] and medicine, to name a few. Therefore, connectors and interconnects are to be chosen accordingly.

Conductive fibers, yarns and fabrics can be used as interconnects. Fibers and yarns can come in different configurations where either the wrapping fibers or the core fiber are conductive, generating a conductive path overall. These configurations depend on the nature of the fibers or coatings used to make them conductive. Intrinsically or particle-doped polymers can be used to achieve conductive properties. Table 5.2 shows the types of fabric materials used for textile circuitry connections. Interconnects and traces can be made out of compliant fabric and polymers, pastes, glues and adhesives with added functionality. Fabric electrodes and connectors are also a way to connect two electronic elements together. For instance, when two electrodes that are separated by a hollow spacer contact one another by exertion of mechanical pressure, an electrical connection is established [21]. Fabric USB and chord connectors are a practical way to connect textile devices to conventional connectors. These can then connect to personal computing devices. Specialty connectors can also be developed based on materials used and intended usage, e.g. Electromyography (EMG) sensor connector [78], data connector [88] and buckle connector [90].

Bonding and attachment of component parts of a textile circuit typically require the bonding material to have very low resistivity. Connections among the different circuit components strongly depend on the materials being connected and are typically a combination of an electrical connection and a mechanical connection. Conductive fibers and conductive thread may be electrically bonded and then mechanically restrained by casting polymers around the joint [22]. These may also be connected to other circuit components using conductive polymers [4]. A metallic snap may be soldered onto a PCB board or breakout board, to enable electrical connection with a matching snap connector on the fabric [22]. Connecting mechanisms for textile circuit components entail modification of lead connectors in the case of miniaturized components to adapt to fabric connectors, e.g. crimp beads, pops, rivets, rings. This also includes bonding wires to conductive fabrics by utilizing conductive paints, polymers, glues or adhesives.

Adhesion can be achieved by interfacial forces, mechanical interlocking or chemical reactions. Mechanical adhesion to plastics can be achieved mainly by using thermoplastic and thermoset bonding agents. Solvent cements, hot melts, silicones, urethanes, cyanoacrylates, methalcrylates, acrylics and epoxies [91] are some of the materials that can be used for mechanical bonding of textile elements

Table 5.2 Conductive traces, interconnects, and connection methods

Traces and interconnects	Connection methods	Connectors and electrodes
• Conventional wire and thread: copper [24], gold, silver [8], stainless steel [8] • Embedded metal wire paths [31] • Woven, knitted, embroidered conductive yarn paths [77,78] • Conductive fabric: metal plated (Ag, Ni, Cu) and polymer coated (PPy, PEDOT/PSS, PANi, CNTs) [9,32,79] • Conductive composites: epoxies [80], paints [26], pastes [31], glues [74], adhesives [31], inherently conductive and doped polymers [81] • Thin-film sheets: Kapton [45]	• Mechanical attachment by gripping, crimping, mold casting, snapping [17] • Joint soldering [82] and electroplating [83], micro-welding [8], thermoplastic adhesion [84], conductive polymer adhesion [33] • Conductive adhesives [85] • Conductive thread lead and loop stitching [22] • Conductive polymers: doped and inherently conductive [9] • Crimp beads, snaps, pops, rings, rivets [8,22] • Copper tape/thread soldering [22] • Soldering onto conductive yarn or fabric [8] • Conductive tape [22] • Chemical bonding through oxidation, covalent bonding, hydrogen bonding and plasma pre-treatments [86] • Strips of conductive fabrics: metal plated and polymer coated	**Connectors:** • Thin and flexible conventional electronic connectors [21] • Fabric snap fastener [87] • Fabric USB connector [88] • Polyurethane mechanical connector [78] • Magnetic snaps [22] • Electrically enabled textile mechanical fastener [22] **Electrodes:** • Conductive fabrics: metal plated and polymer coated [32] • Embroidered, woven or knitted conductive thread surfaces [24] • Conductive particle-doped or functionalized polymers [89] • Conductive Velcro [22]

which may be fibers/yarns, plastic or metal. Associated conductive composites used for electrical connections include technologies used in electronics packaging [92] as well as non-conventional materials, e.g. inherently conductive polymers, and polymers with conductive particle fillers. Principle influences in joint adhesive behavior are particle size, particle geometry, filler content on percolation threshold, influence of surface films on contact resistance, oxidation and galvanic corrosion. Prior to applying any of the adhesive agents, the fiber surfaces to be bonded need to be cleaned from any impurities acquired during manufacturing, such as dirt, oil and stains. Wetting, spreading and wicking of adhesives on fibers and fabrics depend primarily upon fiber surface, and in particular the surface energy of the fiber surface [93]. Surface tension is a result of chemical treatments, bleaching, dyeing, printing or chemical finishing. Coating, lamination and bonding of adhesives are also affected by fabric shrinkage. This is a result of thermal relaxation of the internal molecular structure of, e.g. synthetic fibers. Aside from treatments of the textile surfaces, if adhesives are to be applied onto metals, i.e. wire leads, then proper surface treatment is required for strong adhesion. Surface pre-treatments of metals include abrasion, priming, solvent and chemical cleaning and conversion.

5.2.4 Thermal considerations

An important design consideration for fabric-based electronic components is heat transfer and insulation. In fabrics, heat transfer is composed of three mechanisms: conduction, convection and radiation. It is generally considered that the first two mechanisms are more significant than the third. Thermal conduction is related to the thermal conductivity of each component material as well as that of air. It entails the transfer of heat by phonons (quantized vibrational energy) and free electrons. In metals, thermal current is carried by both electrons and phonons, in insulators only by phonons. Thermal conduction is described by Fourier's law, which relates heat transfer to temperature gradient. The second mechanism, thermal convection, entails heat transfer by means of macroscopic movement of fluids [94]. The fluid motion can be laminar or it can have random fluctuations, i.e. turbulent flow. Convective heat transfer in textiles occurs mostly through direct contact with air and water and is affected by the fabric surface geometries and air permeability. Thermal resistance is reduced with wind velocity for fabrics of low permeability. Radiation, the third mechanism, is the transfer of energy by electromagnetic waves with frequencies ranging from infrared (10^{-2} cm) to ultraviolet (10^{-5} cm). The rate of emission of radiant energy from a solid surface is given by the Stefan–Boltzmann law, which is a function of temperature and surface emissivity. Radiative heat transfer depends on the surface emissivity which is a result of the material properties of constituent fibers of the fabric and any coatings or dies applied. Light-colored materials have low thermal emissivity and dark-colored materials have high thermal emissivity. Rough surfaces, i.e. made of yarns with low twist, possess higher emissivities than corresponding smooth surfaces [95], as the effective area is increased. Effective heat absorption or emission is achieved by rough surfaces and dark colors, whereas smooth and shiny surfaces have reduced reception and emission of heat. Surface emissivity is shown to be of major importance for thin fabrics, with a reduced effect for thicker materials [96]. Plastics, ceramic fibers and paint act as radiation heat insulators.

Fabrics are porous materials (i.e. woven, knitted) which are composed of fibers and entrapped air, and therefore have a thermal conductivity that results from their combined thermal properties. Thermal properties of textiles are influenced by microscopic properties of component fibers, mesoscopic yarn structure, and macroscopic fabric characteristics [97]. Fiber specific heat, material density and morphology are parameters associated with the microscopic characteristics of fibers. Fiber arrangement within yarns or yarn packing density is an example of the yarn structure parameters. Cover factor, porosity, surface density, permeability and compressibility are instances of macroscopic fabric characteristics. All of these affect heat transfer. In particular, macroscopic fabric thickness seems to have the most influence on thermal resistance and heat transfer properties. It has been shown that the reciprocal of thermal conductance is linearly related to fabric thickness and weight [96]. Thermal conductivity can be calculated using the fabric volume porosity, and the thermal conductivity of each fabric component and air. The thermal conductivity can be calculated as [7]

$$\lambda_{fabric} = (1 - P)\lambda_{fiber} + P * \lambda_{air} \tag{5.7}$$

where λ_{fabric} is the thermal conductivity of the fabric, λ_{fiber} represents the thermal conductivity of the component fiber, which can be cotton or polyester, and λ_{air} which is the thermal conductivity of air. Thermal conductivity gives the speed of heat transmission through textile fabric. If two regions are separated by a distance d, i.e. clothing thickness, the thermal conductivity K will describe how the temperature varies spatially due to the heat flow between the two regions per unit area A, from the hotter region to the colder region:

$$\frac{1}{A}\frac{dQ}{dt} = -\frac{K}{d}\Delta T \tag{5.8}$$

The left-hand side of (5.8) represents the heat flux, which is a function of heat capacity, or $Q = mc\Delta T = C\Delta T$. The heat capacity C of an object describes how its temperature changes with time due to heat flow. It is calculated as the specific heat (per unit mass) c times the total object mass. The temperature rise ΔT of an object with a heat flow Q to the body is the ratio of the heat flow over the heat capacity [98]:

$$\frac{C}{A}\frac{d\Delta T}{dt} = -\frac{K}{d}\Delta T \tag{5.9}$$

where the temperature becomes an exponential function of the thermal conductivity, as well as fabric thickness, and heat capacity. The thermal properties of a textile also depend on the fabric's construction method, thermal absorptive capacity, surface thermal resistance, thermal emission characteristics [99], air permeability and air volume distribution [97]. Other thermal characteristics of textile materials are described as softening, melting, decomposition and flammability [7]. All these factors need to be considered when designing fabric-based devices with embedded electronic components. For instance, if an electronic component is mounted near a high-temperature source, which may be the sun in the case of sports applications, the heat absorbed by the electronic components may be reduced by using materials and textile surfaces with low thermal emissivity. Thermal conductivity and thermal resistance increase with fabric density [99]. Higher fabric thickness produces good thermal insulation and fabric temperature variation decreases with increasing fabric thickness. Therefore, if a particular component is expected to dissipate a certain amount of heat, the fabric that hosts this component needs to be of lower thickness and lower material density.

5.2.5 Power dissipation

In terms of power requirements, components can be classified into three categories. The first corresponds to components that dissipate such low levels of power that they do not require any thermal analysis, and rarely exceed the maximum operating characteristics. This category includes passive elements, logic devices and other components that do not consume/dissipate much if any power during operation. The second category corresponds to components that consume/dissipate some power during operation, and might reach their maximum operating temperatures if not in a properly designed system. In general, these are components which would have a temperature rise of at least 10 °C above ambient temperature, or components which will operate in environments with high temperatures, or those which are in proximity of parts that add to the overall temperature rise of the PCB. They require being able to dissipate heat through a specific thermal path. The third category is of components that consume/dissipate enough power that they require specific thermal management solutions, such as heat sinks, conduction paths or forced airflow. In general, it can be considered that these components would have a high-temperature rise above ambient. In smart fabric wearables, most of the devices are of low power and medium power consumption/dissipation, depending on the components used.

Electronic components of conventional devices have power dissipation and required thermal resistance [100]. For wearable and textile-based devices, thermal transfer between the human body and the electronic device is a factor to be considered. Thermal insulation is desirable to shield the electronics from the heat expelled by the human body. Conversely, electronics need a thermally conductive platform to disperse heat and avoid its transference to the skin. Humans regulate their body core temperatures within a range of 36.5 and 37.5 °C at rest, and up to 41.5 °C during exercise. The normal average human skin temperature is 33 °C [101]. Human skin can tolerate a wider range of temperatures, from −0.6 to 45 °C; however, these are extremes which should be avoided. The ways in which the human body loses heat to the surroundings are conduction, convection, radiation, evaporation [102] and vasomotor flow [103]. The human body convects heat to surrounding air. Warm air then moves away and cooler air receives heat again, this is a cyclical process. Convective heat loss rises as an exponential function of the speed at which air moves across the body [101]. The human body is also cooled by perspiration, or evaporation of water on the skin but it is not very effective if the humidity of air is very high, as this slows down the heat dissipation rate. Heat exchange may also vary depending on whether the human is inside a confined space or outdoors. In indoor environments, air is usually stratified, which means that there are continuous temperature fluctuations depending on air conditioning settings. All these factors need to be considered for wearable textile devices in the context of thermal management.

Fabric-based electronic devices must be robust to not only temperature changes but also to localized heat transfer, which may entail varying degrees of humidity, especially if in close proximity to the skin. Heat is a byproduct of metabolism; approximately 75% of the energy consumed is lost as heat. Fabric-based electronic devices must not hinder the ability of the human body to carry out its thermoregulatory process. This is especially true for full or partial-body e-textile suits [75] and e-textile applications that cover large body surfaces. Heat transmission through low-density materials is primarily dependent upon the resistivity of the air layer. In practice, the function of garments is to trap a layer of air, to function as insulation to prevent circulation of air currents around the subject. If there are no convection currents, then the insulation is that of a layer of air equivalent in thickness to the textile material. This layer of air then determines heat flow [96]. Power dissipation from circuit components made of textile materials follow similar principles to those governing conventional electronics. For a fabric resistor made of conductive fabric, every square meter could dissipate several hundred watts of power. The configuration of the fabric resistor as a very thin broad sheet is ideal for efficient heat dissipation. Similarly, for capacitors and inductors, the power dissipation is theoretically small.

Another thermal factor is matching the thermal expansion coefficient of the materials involved. In the case of textile materials, similar values of thermal expansion coefficients albeit desirable are not entirely necessary. The coefficient of thermal expansion is a material property which allows for prediction of linear or volumetric change when the material is exposed to a change in temperature. All materials undergo thermal cycling and therefore expand or contract based on their material properties. When two materials are bonded or secured together, the closer their coefficient of thermal expansion is, the lesser the chance for bending, fracture or degradation at the joining points or surfaces is. This is also important at the locations in the fabric where textile yarns or polymeric fibers connect to conventional electronics, i.e. connector interfaces. Textiles, however, have the advantage of being slightly more resilient than plastic, metallic or PCB platforms. Some fibers have a higher coefficient of thermal expansion than others and these effects have the potential of causing the fabric to crimp, fabric sensors to fail, or of causing breakage of mechanical connections with other fibers. The position of components is also affected by thermal expansion, i.e. fabric contact switches may not properly function if thermal expansion causes contact area displacements. Other instances of thermal effects influencing e-textile performance are that changes in the dimensions of fabric-based inductors, resistors and capacitors alter their circuit characteristics. This is particularly noticeable in the case of capacitance- or inductance-based resonators. Additionally, electrical resistance is generally temperature sensitive; therefore, it is also necessary that the variances in electrical resistance of the materials chosen for electrical components connections are satisfactory over the intended temperature range.

5.3 Fabric power elements

Non-conventional energy sources such as fabric voltaics are necessary for fabric device portability, i.e. one that generally is not hard wired to a stationary source of power. A voltaic textile entails mainly two categories: either a flexible battery mechanically attached to the textile or a voltaic textile structure built out of polymers, dyes and other textile compatible compliant materials. Other conventional elements can be made out of fabric-based materials. Such is the case for embroidered stainless steel inductors [60], which can be woven onto fabric substrates. Conductive polymers, i.e. PEDOT, can be used as charge storage devices on textile substrates. PEDOT coatings are placed over three, silver-coated polyamide yarns interwoven in the substrate weft [104].

Flexible solar cells, fiber-optic based batteries, inductive radio frequency identification tags (RFIDs) and mechanical storage elements are some of the flexible battery devices which can be attached to a fabric substrate. Even miniaturized conventional batteries can be placed in fabric holders and then stitched to a fabric conductive path [21]. Battery films and fibers can be interwoven into a non-conductive fabric [45]. Voltaic textile structures that can be grown on fabric substrates use layers of sensing dyes to build structures similar to those of conventional batteries. Such is the case of FlexEl [105], a thin-film battery cloth that uses ruthenium oxide to achieve ultra-high charge storage capacity. Paper battery and battery fibers are other possibilities for light-weight compliant power [106]. Two examples of alternative energy sources entails wireless recharging [107] and continuous wireless power supply [108] by inductive coupling.

Flexible solar technologies on fabrics are an example of the first category. Photovoltaic textiles are those which can generate energy upon exposure to light. These are being used in many applications such as jackets, backpacks, soldier uniforms, usually to power up portable electronic devices. Flexible solar cells have been made out of copper indium gallium selenide on steel foils, cadmium telluride cells and silicon (Si) thin films. These last ones are very promising due to their balance between the required flexibility and performance; polycrystalline silicon cells are ready for direct lamination onto textiles and high-efficiency single crystalline silicon transfer cells are suitable to be used as add on accessories [109]. Organic solar cells such as polymer solar cells have promising features due to their compliant nature. Conjugated polymers with fullerene derivatives made by solution-based techniques and cooperating small molecular materials made by thermal evaporation techniques are some of the approaches currently at the forefront of research [110]. Advantages of organic solar cells include the possibility of being manufactured in small as well as large areas and being potentially suitable for mass production at low cost using, for instance, printing or coating techniques [111]. They are also flexible given they use substrates such as polyethylene terephthalate (PET), poly(ethylene naphtalate), polyimide (PI), kapton and poly(ethersuphone). Fabrics can become the flexible substrate [112] by applying an interfacing layer of a very flexible polymer such as PE to the fabric in addition to other physical treatments to improve the adhesion of the bottom layer to the subsequent layers. Organic flexible solar cells, however, pose challenges with respect to the rapid degradation and polymer permeability.

In the second category, cloth fibers, polymer fibers, compliant materials, dyes and other chemicals have been used to construct fibers with voltaic characteristics that can be used to produce a complete fabric, or which can be introduced into textiles as one of the constituent elements. Fabric capacitors can be created with soft materials that follow a typical plate configuration. A flexible conductive polymer-based fiber with high electric capacitance has been reported by Gu *et al.* [39]. This fiber is fabricated using fiber drawing, where a macroscopic preform is drawn into a sub-millimeter capacitor fiber. These fibers can be applied in distributed sensing and energy storage. The measured capacitance of these fibers is in the order of 60–100 nF/m.

5.3.1 Textile supercapacitors

There are Faradic and non-Faradic type electrochemical energy storage devices. The former one entails charge storage which is achieved by electron transfer that produces a redox reaction. In a non-Faradic device, there is ideally no electron transfer and the storage of electric charge and energy is electrostatic. This latter type corresponds to the mechanism found in supercapacitors, where the chemical and electrode potentials are a continuous function of degree of charge. Structural and electrochemical properties of electrode materials as well as packaging designs determine the characteristics of both batteries and supercapacitors [113].

It has been suggested that textile supercapacitors can be built out of PEDOT nanofibers with polyacrilonyte membranes as separators and carbon fabric as collectors [114]. In this approach, alternating layers of fiber mats (or flat clusters of fibers) and separating layers are joined and embedded into a PVDF-based electrolyte which when evaporated produces a flexible textile supercapacitor. A rechargeable polymeric battery can be made of a solid PEDOT/PSS electrolytic layer deposited between two woven silver-coated polyamide yarns [115]. Similarly, the supercapacitor fibers made of single walled nanotubes or SWNTs have an electrolytic separating layer between two fibers, which are later re-coated into a single unit [116] and then sewn into fabrics. These can also be used as interconnects, electromagnetic shields and antennas. Furthermore, cotton and polyester substrates can be coated with SWNT inks to form plate capacitors. This type of supercapacitor can have an area capacitance up to 0.48 F/cm^2 [117]. Flexible energy storage devices can be manufactured in fiber or sheet form. They can be based on separated thin electrode and spacer layers, albeit with non-optimal performance. A variation of this basic structure entails a combination of porous cellulose as the spacer and multi-walled CNTs as electrodes, and an electrolyte [113]. This supercapacitor–battery hybrid combination was demonstrated, showing that the nanocomposite paper has great mechanical flexibility, wide temperature range and works for a wide variety of electrolytes.

5.3.2 Photovoltaic cells

Photovoltaic cells can be made with alternating layers of coatings, using the same construction principle used for design of fabrics that have embedded optical fibers [118,119]. In this case, the electrodes are wires but the wire core undergoes a series of coatings, all of which have a different function in the photo cell. Other developments in this area use the same concept of dye-sensitized solar cells but using different dyes such as photosensitized nanomatrix materials for conversion of the photons into electric potential [120]. The fiber core includes a flexible polymeric material, e.g. PET, and flax, cotton, wool silk, nylon and/or combinations of these, and the photosensitized nanomaterial includes nanoparticles or a heterojunction composite material. The photosensitizing agent may be a dye or an organic molecule such as xanthine, cyanine, merocyanine, pthalocyanine or pyrrole. In this configuration, the photo-conversion nanomatrix material lays over the fiber core, with an electrical conductor circumferentially covering the photo-conversion material. Other fiber substrate materials include optical fibers, PI-coated silica fibers, polypropylene fibers, tapes and stainless steel fibers [110]. Photovoltaic fibers may have advantages over planar solar cells due to their cylindrical structure which may lead to a greater power generation due to a more constant illumination given its symmetric structure. However, fiber architecture and suitable manufacturing processes need to be optimized given they need to meet the requirements for full functional fabric integration. Some of the different types of textiles capacitors that can be used for energy storage in fabrics are shown in Table 5.3. Figure 5.6 illustrates the usage of fiber capacitors woven into a textile weave.

Table 5.3 Textile capacitors and batteries

Type	Elements	Fabrication method	Electrolyte
Textile supercapacitor [114]	Active material: PEDOT fiber mats. Separators: polyacrilonite membranes	Alternating layers of textile materials. Collector: carbon cloth	Flexible PVDF – co-HFP and EMIBF4
SWNTs – single-walled nanotubes [116]	Two nanotube composite fibers	The fibers are coated with electrolyte, twisted and re-coated	Polyvinylalcohol/ phosphoric acid
Konarka's photovoltaic cell [118]	Dye-sensitized titanium, metal wire core	Sequential coating	Electrolyte
Polymer-based rechargeable battery [115]	Silver yarn electrodes/ PEDOT on woven fabric	Multiple coatings	PEDOT/PSS

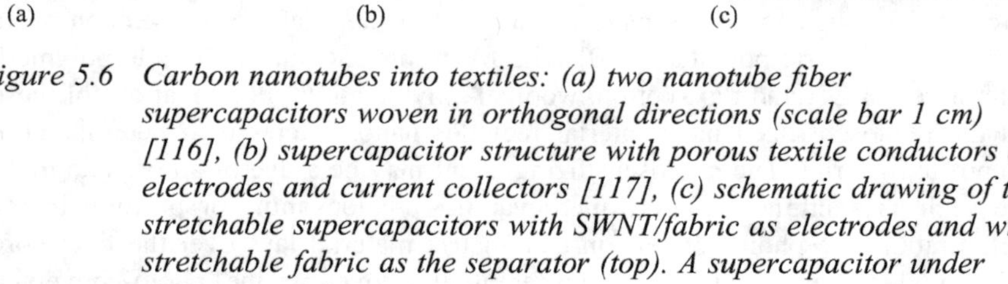

Figure 5.6 Carbon nanotubes into textiles: (a) two nanotube fiber supercapacitors woven in orthogonal directions (scale bar 1 cm) [116], (b) supercapacitor structure with porous textile conductors as electrodes and current collectors [117], (c) schematic drawing of the stretchable supercapacitors with SWNT/fabric as electrodes and with stretchable fabric as the separator (top). A supercapacitor under 120% strain (bottom) [117]

5.3.3 Energy-harvesting technologies embedded in fabrics

Exhaustible energy sources are of common use in today's portable electronics industry, they are also used to power up many existing smart fabric devices. This is particularly true as the total personal power consumption has increased due to the large number of miniaturized integrated devices that continuously appear in the market [121]. A growing number of opportunities exist for fabric devices that use ubiquitous energy sources such as energy harvesting systems applied to fabric sensors. These systems need to be compliant to the human body and of reduced size. They need to be as portable as possible for the wearer's convenience and for reduced user input. The main technologies being considered for energy harvesting from the human body and its surroundings are; inertial kinetic energy-harvesting devices, low frequency vibration, solar energy, photovoltaics [121], and thermoelectric devices [122]. Inertial kinetic devices harvest energy from human motion by converting mechanical energy into electrical energy, including low frequency vibrations. Solar energy and photovoltaics harvest natural and artificial electromagnetic radiation sources. Thermoelectric devices use the heat that the human body produces by converting thermal energy into electrical energy. Another possibility for recharging energy units entail the use of inductive coupling charging or wireless charging.

Figure 5.7 shows a power generating tactile sensor array where an energy harvesting piezoelectric film is woven and integrated with elastic tubes. This device converts mechanical stretch into electrical energy and also works as a capacitive tactile sensor. Another example of bio-mechanic energy harvesting entails piezoelectric shell structures can be placed around the finger and elbow joints to harvest energy from mechanical flexing of these joints [124]. It has been reported that a single shell structure produced a power density of 2.18 mW/cm^2 [125]. This output is on par with shoe-based harvesters (1.6 mW/cm^2) and backpack-based harvesters (0.58 mW/cm^2). Piezoelectric generators can be integrated into the fabric structures at low temperatures [129], this represents a manufacturing advantage. Thermoelectric harvesters use thermal gradients to produce power. Miniaturized thermoelectric generators are an example of this type of energy harvester [127]. The human body produces several tens of watts of heat which can be captured and reused. Thermoelectric harvesters can harness part of this energy and convert it into electrical power for use in powering smart fabric sensors. These generators are built in a silicon chip and consist of a large number of thermocouples, electrically connected in series and arranged to make best use of a given area. These devices operate by using the Seedbeck effect or thermoelectric effect, whereby a temperature difference between both ends of a closed circuit composed of two different metals generates a voltage and causes an electrical current to pass through the circuit. An output power of 1.6 mW can be obtained from a 7.0-mm^2 chip containing 16,000 micro-thermocouples, which is enough to power devices such as a wrist watch. Another example of a flexible thermoelectric harvester is one where the thermoelectric materials are flexible. These are dispensed into a thin silicone matrix, generating 2.1 mW in a 50 $\times$ 50-mm area [128].

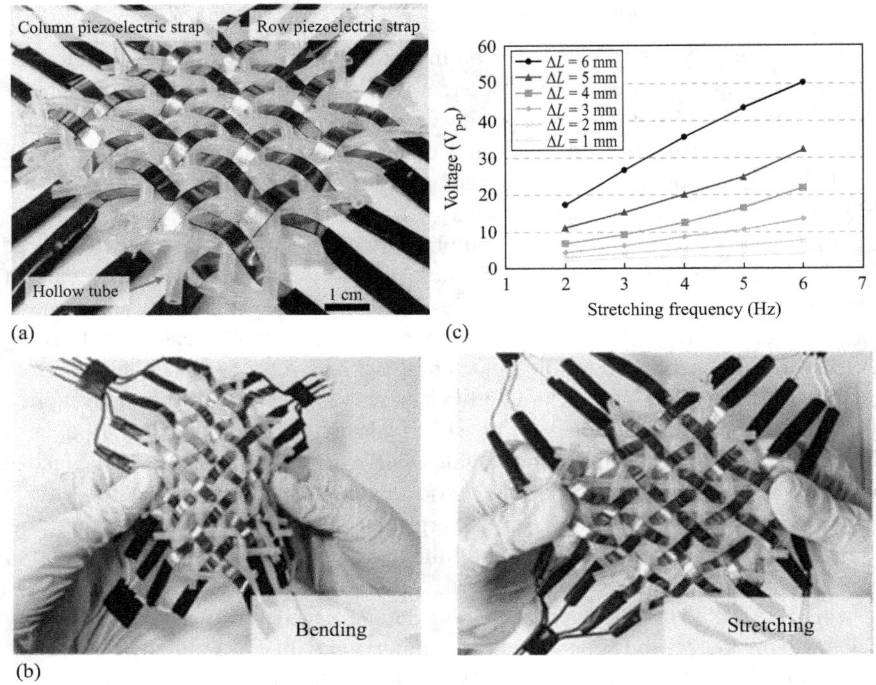

Figure 5.7 Piezoelectric energy harvester: (a) Interwoven PVDF straps and elastic tubes, (b) bending and stretching of tactile array, (c) Peak-to-peak output voltage from 10 PVDF straps connected in parallel for various stretching frequencies [123]

Increasing energy and power densities in energy storage devices can also be obtained by developing new chemistries and materials or by the optimization of current systems. Some of the parameters that can be enhanced include: conversion efficiency, recharge ability, safety and environmental properties [121]. Nanomaterials play a major role in this development as they allow granular control of material architecture and therefore of charge transport phenomena at the electrode interfaces. Graphene, for instance, could be used for a new generation of energy harvesting nanodevices. This material could convert energy available in the environment, i.e. noise, mechanical vibrations, electromagnetic radiation, into usable energy to power electronic devices [121]. Low-frequency vibration or friction energy can be converted into electricity using piezoelectric zinc oxide nanowires grown radially around textile fibers [129] or in arrays to maximize the harvested power [130].

Fabrics that contain embedded energy harvesters have special interconnect and packaging technologies such as those in Table 5.2 and Table 5.4. For instance, in the case of the thermoelectric chip the connections are formed by using a narrow strip of polyester fabric in which several warp threads have been replaced by copper wires that have been coated with silver and polyester, for connection endurance and electrical insulation. The electrical connections are created by removing the surrounding textile material with a laser and soldering the exposed areas with tiny contact plates and thin bonding wires. The complete unit is then molded and encapsulated, forming a hermetically sealed casing protecting it against mechanical and chemical stresses caused by wearing and cleaning the textile [127]. For a thermoelectric harvester that has been printed on PDMS, flexible printed circuit boards with Cu-patterned lines are attached to establish electrical connections [128]. These connections allow for increased user mobility.

5.4 Technologies related to the development of smart fabrics

5.4.1 Fabric circuitry and macroelectronics

Fabric circuit boards are textile analogues to conventional PCBs or printed circuit boards. Conductive traces are drawn in a similar fashion as conventional circuit boards. Some of the techniques to draw the circuit traces include metal sputtering, silk printing [28], screen printing [27], fabrics with metallic fibers/yarns [131], laser-cut/ironed conductive/solderable fabric [21], machine weaving [132], embroidery [8], conductive inks [133], conductive paint [22], conductive polymers [134], strips of conductive fabrics and conductive yarns [76]. All of the conductive materials mentioned in the previous section can all be used to form electric traces on fabric. One important observation, however, is that the electrical resistance of the traces must not vary too much under the expected bending/torsion wear and tear.

Flexible electronics circuits are technologies that also allow for a flexible substrate. The substrate materials are of critical importance, as they can dominate the mechanical properties of the flexible system. Substrates can be plastic films, metal foils and fibrous materials [135]. These technologies often use flexible plastics such as polyester, polyimide (PI), polyethylene terephthalate (PET), polyethylene naphthalate (PEN), polycarbonate (PC), polyethersulfone (PES), and polyarylate (PAR). Other substrates include elastomer sheets, polydimethylsiloxane (PDMS) silicon slabs [136], bucky gels and stretchable conductors on stretchable polymeric substrates. As for conductors, fine metallic foils are used, primarily copper in electrodeposited and rolled-annealed forms. Other conductive metals that can be sprayed, sputtered and electrodeposited can also be used [137]. Adhesives used in flex circuits include acrylics, epoxies and phenolics.

Equally important are all of the manufacturing technologies associated with flexible electronics such as lithography, transfer printing, roll-to-roll printing and all the processes that enable the mass production of macroelectronics. These technologies can be adapted to complement smart fabrics technologies due to their unique compliant characteristics. Novel fabrication techniques have allowed to harness potential of 1D semiconductors such as carbon nanotubes, nanowires and quantum dots, as well as 2D semiconductors and dielectrics such as graphene, molybdenum disulfide, and hexagonal boron nitride monolayers [121]. However, some of these materials still need to overcome issues of cost, scalability, and stability when subjected to prolonged electrical or optical bias.

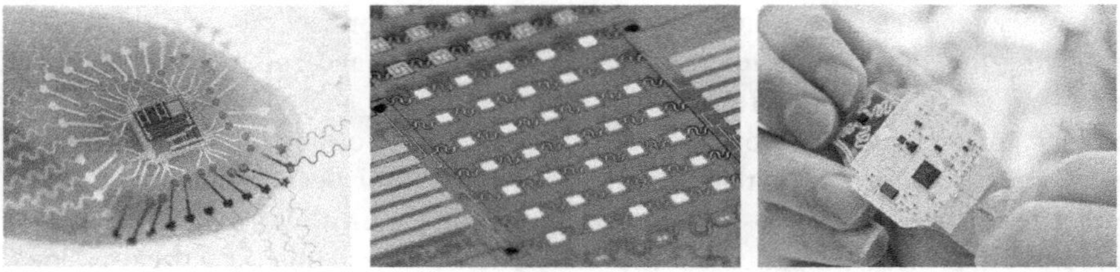

(a) (b) (c)

Figure 5.8 Stretchable and flexible electronics: (a) Ultra-thin chip package with fan out interconnections, (b) Stretchable LED microdisplay with thermoplastic polyurethane lamination, and (c) Bluetooth flexible sensor patch with accelerometer, temperature and humidity sensors [138]

Figure 5.8 shows a flexible electronics ultra thin chip package on polyimide film and fan out interconnections. It also shows a stretchable LED where the LEDs have been interconnected to each other by meandering printed conductors in order to make the display stretchable. This feature makes it especially useful for implementation into garments. Figure 5.8c shows a Bluetooth sensor patch that includes an accelerometer, as well as temperature and humidity sensors. Conductive adhesives were used to assemble components on a screen printed circuit pattern.

The design of smart fabrics transducers has benefited from developments in the area of flexible or stretchable electronics. These technologies provide flexibility and functionality to fabric-based transducers. Some of these technologies include smart skins, electronic paper (e-paper), i.e. Gyricon, organic light-emitting diodes, electronic inks (e-ink), microelectromechanical systems, RFIDs [139], flexible solar cells, flexible PCBs, printable flexible liquid crystal displays (LCDs) and other flexible displays [140]. E-paper consists of millions of spherical beads with optically contrasting hemispheres contained in a matrix. Dipole moments cause the beads to rotate giving the possibility to display either of the two colors. E-ink works on a similar principle but the charged particles are much smaller and color independent. Both these technologies are used for flexible displays as well as cholesteric inks, which consist of modified liquid crystals suited for reflective displays removing the need for color filters and polarizing layers. Printable LCDs can be manufactured by a sequence of simple coating and UV curing processes on plastic substrates resulting in ultra-thin and flexible displays [141]. Another set of technologies that greatly improve fabric transducers is wireless technologies and nanotechnology. Nanotextiles have emerged from the application of nanotechnology to fabric substrates and materials. These developments will lead to new smart textile materials [58]. Wireless transmitters can be attached to fabrics in the form of wearable antennas [142,143] and bluetooth devices [144] and can be used in wireless local area networks (WLANs) and personal networs (WPANs), in near-field communication (NFC), or with devices that have access to a Global Positioning System (GPS) [121].

5.4.2 Encapsulation and environmental considerations

Connectors as well as encapsulation mechanisms for e-textile devices need to be protected from chemicals that may deteriorate their function. Cleaning fluids, detergents, fluxes and other chemicals encountered in daily or industrial usage may affect the function of fabric interconnects and encapsulation mechanisms. These interactions can be prevented by proper choice of materials. Material property descriptions for e-textile devices should include a chemical resistance chart. If this information is not available, then it may be necessary to determine these chemical interactions prior to fabrication. Temperature may also be a factor in determination of chemical interactions. Chemical deterioration may be induced by ionizing radiation, ultraviolet radiation, and may cause elastomers to become brittle. Deterioration by humidity exposure and corrosion is another important factor. A lot of cellulose-based fibers are hydrophilic and may decompose faster than synthetic fibers. For metals, intergranular corrosion may produce cracking in the presence of certain acids or bases. Galvanic corrosion, or the electrochemical reaction in the presence of electrolyte, i.e. salt water, may cause perforation or delamination of electrical connection points by anodic migration [95]. Electrical contacts in switches are not permanently engaged and connectors need to be permanently engaged. Requirements for both types are slightly different for fabric-based devices.

Components in electronics modules require thermal, mechanical and chemical protection. This includes limiting temperatures caused by internal heat generation, avoiding excessive mechanical stresses, reducing dynamic shock and vibration loads to avoid reaching fragility levels, avoidance of chemical and corrosive agents and proper attachment to the intended platform. These basic requirements are applied both at the fabric-grown electronics level and at the electronics-adapted level. Electronic components which are grown on the fabrics themselves or which are adapted to textiles substrates require proper encapsulation for protection from the environment and the human wearer. Encapsulation barriers for single components or entire electronic systems constitute a mechanical and thermal barrier between the internal component environment and the external environment. Human and atmospheric factors are included in the latter one. The internal components in electronics-embedded fabric devices include electronic modules, components, soft interconnects and conventional interconnects. These encapsulated devices require heat dissipation in greater or lesser proportion depending on the level of integration into the fabric [95,145].

Encapsulation of electronic components should allow the encapsulating material to flow freely, not capture air bubbles, be lightweight, have good electrical, thermal and chemical properties and resist cutting and abrasion after cure. During cure, the material should not produce high temperatures or excessive stresses. After cure, the material should not absorb moisture and not produce forces during temperature cycles, be stiff enough to support components and connections and be able to be removed for purposes of repair. Three primary groups of materials are used for encapsulation: epoxy-resin-based materials, silicon-based materials and polyurethane-based materials. Some of the plastics used in electronic packaging which are applicable to e-textiles components are described in Table 5.4. Epoxy resins

Table 5.4 Plastics used in electronic packaging, applicable to e-textiles [95,145]

Thermoplastics	Typical applications	Characteristics
ABS	Enclosures, insulators, molded parts	Temperature range is limited
Acrylics	Conferral coating	Temperature sensitive, flammable
Fluoropolymers	Wire insulation, electrical insulation, dielectric purposes	Heat resistant, inert to most chemicals zero water absorption, low dielectric losses
Nylon	Connectors, wire jackets, low pressure bearings	Tough, abrasion resistant, good mechanical strength, absorbs water, dielectric strength dependent on humidity Will degrade with sunlight and will creep
Polyetherimide	Fuse blocks, circuit board insulators, chip carriers, circuit breaker housings	Wide temperature range, high radiation resistance
Polyethylene	Molded form for discrete component enclosures, battery cases. Insulation sheath for wires and cables	Abrasion resistant Will burn if exposed to flame
Polyimide plastic	PCBs, connectors, films. Bearings, seals, insulator, flexible cable, tapes, wire enamel, chip carriers	High radiation resistance, low coefficient of thermal expansion, high electrical resistance, resist high temperatures. Degraded by hot caustic solutions and highly polar solvents
Polystyrene	Foam insulation	Degraded by many solvents and chemicals, modestly elevated temperatures
PVC	Wire jackets, tubing and insulation	Good resistance to most chemicals. Sensitive to UV radiation. Temps from 0 to 100 °C
Allyl polymer	Connector inserts, electronic parts and housings	Can withstand high temperatures, good moisture resistance
Epoxy	Conformal coating, microcircuit encapsulation, adhesives. Potting, casting	Can be filled with different particles to modify characteristics
Phenolic	Knobs, handles, connectors chip carriers in laminate form	High temperature, high mechanical strength
Polyurethane	Paints, wire enamels, conformal coatings, foam dielectric, thermal insulators	Tear resistance, abrasion resistance, and unusual toughness. Can be degraded by acids and bases. Temperature should not exceed 120 °C
Silicon-based materials	Sealants, adhesives, paints, encapsulants	Broad temperature range −40 to +250 °C Non burning Water resistant when fully cured
Elastomers	Sealing gaskets, drive belts	Much more vulnerable to environmental degradation, loss of elasticity at low temperatures

are used when strength and rigidity are required. Silicon materials tend to remain flexible and are water resistant. Polyurethane-based materials tend to remain flexible and are of a rubber-like consistency; however, they are not as adhesive as silicones. Polyurethanes and silicones, however, have a useful temperature range. Moisture absorption is another necessary consideration for fabric-based electronics. Absorbed moisture may cause corrosion and lead to failure by expanding within the material at cold temperatures. For materials used as electrical insulators, the insulator may become electrically conductive when moist. Therefore, hydrophobic materials are preferred for insulator applications. For organic materials, moisture absorption leads to a loss of mechanical strength and physical deterioration.

Environmental considerations include potential for exposure to High temperature, high corrosive chemicals, solar radiation and ultraviolet radiation. Considerations for selecting good contact materials are electrical conductivity, chemical inertness, and galvanic corrosion and wear resistance. This also depends on the mechanical requirements for the intended e-textile application. Wearable devices may require conductivity at a percentage of elongation; by contrast, furniture devices may not have such requirements. Contact requirements may also depend on whether the electrical contact will sustain high or low electrical currents. For low currents, films and tarnishes that can form on some materials, i.e. silver and lead, are enough to cause connection issues. In this case, a strong removable mechanical connection can get rid of these insulating films, albeit at the expense of faster connector erosion. Higher electrical currents allow for a larger material selection, which may include copper alloys, silver alloys, lead–tin alloys and nickel. Similar considerations are applicable to switch contacts, however, with particular attention to wear resistance and chemical resistance.

Temperature range is important when temperatures are very high or very low. Metal and non-metal encapsulation materials as well as e-textile components will become brittle and stiff at lower temperatures and may soften, lose strength and creep or flow at elevated temperatures. If conventional electronic units are embedded in fabrics, they must be kept at temperatures within prescribed limits. Considerations for encapsulation material selection are similar to those discussed for the e-textile components, and include thermal conductivity, chemical inertness, resistance to corrosion, wear resistance, sublimation, thermal expansion, and ultraviolet and ionizing radiation resistance. One of the design goals for thermal conditioning and reliability is to ensure heat flow between the heat source and the heat sink. A design for good heat convection in fabrics entails increasing the area of contact between the heat source and the air or fluid in charge of the heat exchange. When considering space applications, radiation is the primary mode of heat transfer. Plating steel with nickel or zinc for corrosion protection, as well as black oxide, and phosphate coating. Metal electroplated with another metal can cause delamination if the coefficient of thermal expansion is different. The further apart materials are listed on the galvanic series, the greater is the opportunity for galvanic corrosion to occur.

Electromagnetic and electrostatic shielding is required to avoid interference in electronics devices on fabric. Electrostatic shielding helps prevent charge formation on the surface of an electronic enclosure which may cause damage to electronic components. Electromagnetic shielding entails shielding from electromagnetic waves from the environment on circuit elements. Magnetic shielding protects a circuit from fluctuating magnetic field energy which may induce currents in wires and metallic elements, i.e. metalized thread. Shielding can be made out of conductive fabrics or can be made out of plastic cases which are coated internally with a conductive paint. This paint may contain nickel or silver particles. Plastic and fabric enclosures need tests for combustibility, as many will burn and emit toxic fumes [95].

5.5 Conclusions

Fabric sensor architectures for creation of all-fabric-based devices were explored. All these components exist and can be interfaced. Mechanical properties of yarns, fabrics and coatings, along with deformation modes, were presented and discussed to provide a framework for fabric sensor design. Sensing properties can be imparted to textiles at scales that range from modifications to fibers and yarns to modifications of built textile structures. Coatings can be applied to all the fabric structures, as well as miniaturized electronic elements. Passive and active circuit elements can be constructed from textile elements and circuit boards can be fabric based or can be based on flexible plastic substrates. Textile sensors have an ample range of sensing options using polymeric and fabric materials. Power textile elements, i.e. photovoltaics and supercapacitors, are also possible by using textile materials as well as plastic fibers. Connectors, interconnects and encapsulation technologies are also presented and discussed. These components constitute the textile sensing platform architecture. Interfacing all these elements yields an all-fabric-based sensing system. Advances in Flexible macroelectronics [146] technologies and nanomaterials, are expected to be a key driver in the widespread development of these textile platforms.

References

[1] Castano LM, Flatau AB. Smart fabric sensors and e-textile technologies: a review. Smart Materials and Structures. 2014;23(5):053001.

[2] TITV Greiz, The Institute for Special Textiles and Flexible Materials (Cham, Switzerland). Available at: http://www.titv-greiz.de/. Last accessed: Oct 18, 2016, 121–146.

[3] Kim BJ, Kim DH, Lee YY, *et al.* Highly efficient and bending durable perovskite solar cells: toward a wearable power source. *Energy & Environmental Science*. 2015;8(3):916–21.

[4] Castano L M, Winkelmann A E, Flatau A B. A first approach to foot motion monitoring using conductive polymer sensors. *Proceedings of the SPIE Sensors and Smart Structures Technologies for Civil, Mechanical, and Aerospace Systems (San Diego, CA)*; 2009.

[5] United States Army Natick Soldier Research, Development and Engineering Center (NSRDEC). Available from: https://www.army.mil/info/organization/natick. Last accessed: Sept 28 2016.

[6] Castano LM, Flatau AB. Smart textile transducers: design, techniques, and applications. In: Hosseini M, Makhlouf AS, eds. *Industrial Applications for Intelligent Polymers and Coatings*. (Springer); 2016.

[7] Colinart T, Pierre T. Radiative and hygrothermal properties of flax/viscose spunlaced nonwovens. *Journal of Industrial Textiles*. (Cambridge, England) 2015:1528083715606104.

[8] Post E R, Orth M, Russo P R, Gershenfeld N. E-broidery: design and fabrication of textile-based computing. IBM Systems Journal. 2000;39:840–60.

[9] Gioello DA. *Understanding Fabrics: From Fiber to Finished Cloth*. 1st edition (New York: Fairchild); 1982.

[10] Schwartz P, editor. Structure and mechanics of textile fibre assemblies. (Elsevier); 2008.

[11] Duhovic M, Bhattacharyya D. Simulating the deformation mechanisms of knitted fabric composites. Composites Part A: Applied Science and Manufacturing. 2006;37(11):1897–915.

[12] Yeoman MS, Reddy D, Bowles HC, Bezuidenhout D, Zilla P, Franz T. A constitutive model for the warp-weft coupled non-linear behavior of knitted biomedical textiles. Biomaterials. 2010;31(32):8484–93.

[13] Cherif C, ed. *Textile Material for Lightweight Constructions*. (Berlin: Springer); 2015.

[14] Liu K, Sun Y, Zhou R, *et al.* Carbon nanotube yarns with high tensile strength made by a twisting and shrinking method. Nanotechnology. 2009;21(4):045708.

[15] Adanur S. *Wellington Sears Handbook of Industrial Textiles* (Lancaster, PA: Technomic Publishing Company, Inc.); 1995, pp 625–40.

[16] Farboodmanesh S, Chen J, Mead J L, *et al.* Effect of coating thickness and penetration on shear behaviour of coated fabrics. Journal of Elastomers and Plastics. 2005;37:197–227.

[17] Kallmayer C, Simon E 2012 Large area sensor integration in textiles. *International Multi-Conference on Systems, Signals and Devices SSD* (Chemnitz, Germany).

[18] Tracton A. *Coatings Technology Handbook*. 3rd edition (Boca Raton, FL: CRC Press); 2006.

[19] Chen Y, Lloyd D W, Harlock S C. Mechanical characteristics of coated fabrics. Journal of the Textile Institute. 1995;86:690–700.

[20] Dubrovski P D, Cebasek P F. Analysis of the mechanical properties of woven and nonwoven fabrics as an integral part of compound fabrics. *Fibres and Textiles in Eastern Europe*. 2005;13:50–3.

[21] Buechley L, Eisenberg M. Fabric PCBs, electronic sequins, and socket buttons: techniques for e-textile craft. Personal and Ubiquitous Computing. 2009;13:133–50.

[22] Hard/Soft Fabric Electrical Connections. Available at: http://www.kobakant.at. Last accessed: Oct 18, 2016.

[23] Buechley L. A Construction Kit for Electronic Textiles. *10th IEEE International Symposium Wearable Computers (Montreux, Switzerland)*; 2006.

[24] Cottet D, Grzyb J, Kirstein T, Troster G. Electrical characterization of textile transmission lines. IEEE Transactions on Advanced Packaging. 2003;26: 182–90.

[25] Meyer J, Arnrich B, Schumm J, Troster G. Design and modeling of a textile pressure sensor for sitting posture classification. Sensors Journal, IEEE. 2010;10:1391–8.

[26] Calvert P, Patra P, Sawhney A, Agrawal A, Duggal D. Printed conducting polymer strain sensors for textiles. *23rd International Conference on Digital Printing Technologies (Anchorage, AK)*; 2007.

[27] Kim Y, Kim H, Yoo H J. Electrical characterization of screen-printed circuits on the fabric. IEEE Transactions on Advanced Packaging. 2010;33:196–205.

[28] Kim H, Kim Y, Kim B, Yoo H J. A wearable fabric computer by planar-fashionable circuit board technique. *Sixth International Workshop on Wearable and Implantable Body Sensor Networks (Berkeley, CA)*; 2009.

[29] Mattila H, editor. *Intelligent Textiles and Clothing*. (Cambridge, England: Woodhead Publishing); 2006.

[30] Lilja J and Salonen P. Making flexible resistors out of conductive polymer fabrics. Electronics Letters. 2011;47:602–4.

[31] Locher I. *Technologies for System-on-Textile Integration.* Dissertation, Swiss Federal Institute of Technology, Zurich, Switzerland; 2006.

[32] Conductive and Heat Resistant Fabrics, Foams and Textiles, Eeonyx Corporation. Available at: http://www.eeonyx.com. Last accessed: Oct 18, 2016.

[33] Jin G, Norrish J, Too C, Wallace G. Polypyrrole filament sensors for gases and vapours. Current Applied Physics. 2004;4:366–69.

[34] Conductive Yarn. Available at: http://www.sparkfun.com. Last accessed Oct 18, 2016.

[35] Bera TK, Mohamadou Y, Lee K, *et al.* Electrical impedance spectroscopy for electro-mechanical characterization of conductive fabrics. Sensors. 2014;14(6):9738–54.

[36] Bessonov A, Kirikova M, Haque S, Gartseev I, Bailey MJ. Highly reproducible printable graphite strain gauges for flexible devices. Sensors and Actuators A: Physical. 2014;206:75–80.

[37] Flexible PTC heater; Printed materials for wearable electronics and smart fabrics. DuPont Electronics & Communications. Available at: www.dupont.com. Last accessed Oct 18, 2016.

[38] Conductive carbon resistive coating. Conductive Compounds. Available at: http://www.conductivecompounds.com/. Last accessed: Oct 18, 2016.

[39] Gu J F, Gorgutsa S, Skorobogatiy M. Soft capacitor fibers using conductive polymers for electronic textiles. Smart Materials Structure 2010;19:115006.

[40] Gu J F, Gorgutsa S, Skorobogatiy M. Soft capacitor fibers for electronic textiles. Applied Physics Letters. 2010;97:13.

[41] Lee J B, Subramanian V. Weave patterned organic transistors on fibre for E-textiles. IEEE Transactions on Electron Devices. 2005;52:269–75.

[42] Bonfiglio A, De Rossi D, Kirstein T, *et al.* Organic field effect transistors for textile applications. IEEE Transactions on Information Technology in Biomedicine. 2005;9:319–24.
[43] Maccioni M, Orgiu E, Cosseddu P, Locci S, Bonfiglio A. Towards the textile transistor: assembly and characterization of an organic field effect transistor with a cylindrical geometry. Applied Physics Letters. 2006;89: 143515.
[44] Nilsson D, Kugler T, Svensson P O, Berggren M. An all-organic sensor–transistor based on a novel electrochemical transducer concept printed electrochemical sensors on paper. Sensors and Actuators B: Chemical. 2002;86:193–7.
[45] Bonderover E, Wagner S. A woven inverter circuit for e-textile applications. Electron Device Letters, IEEE. 2004;25:295-97.
[46] Edmison J, Jones M, Nakad Z, Martin T. Using piezoelectric materials for wearable electronic textiles. *Proceedings of the Sixth International Symposium on Wearable Computers ISWC (Seattle, WA)*; 2002.
[47] Mattana G, Cosseddu P, Fraboni B, Malliaras G G, Hinestroza J P, Bonfiglio A. Organic electronics on natural cotton fibres. Organic Electronics. 2011;12:2033–9.
[48] Motta M, Kinloch I, Moisala A, Premnath V, Pick M, Windle A. The parameter space for the direct spinning of fibres and films of carbon nanotubes. Physica E: Low-dimensional Systems and Nanostructures. 2007 Mar;37 (1):40–3.
[49] Ma W, Liu L, Yang R, *et al.* Monitoring a micromechanical process in macroscale carbon nanotube films and fibers. Advanced Materials. 2009;21(5):603–8.
[50] Miao M. Yarn spun from carbon nanotube forests: Production, structure, properties and applications. Particuology. 2013 Aug;11(4):378–93.
[51] Fennessey SF, Farris RJ. Fabrication of aligned and molecularly oriented electrospun polyacrylonitrile nanofibers and the mechanical behavior of their twisted yarns. Polymer. 2004;45(12):4217–25.
[52] Jiang K, Wang J, Li Q, Liu L, Liu C, Fan S. Superaligned carbon nanotube arrays, films, and yarns: a road to applications. Advanced Materials. 2011; 23(9):1154–61.
[53] Xie X, Ye M, Hu L, *et al.* Carbon nanotube-coated macroporous sponge for microbial fuel cell electrodes. Energy & Environmental Science. 2012; 5(1):5265–70.
[54] Koziol K, Vilatela J, Moisala A, *et al.* High-performance carbon nanotube fiber. Science. 2007;318(5858):1892–5.
[55] Li YL, Kinloch IA, Windle AH. Direct spinning of carbon nanotube fibers from chemical vapor deposition synthesis. Science. 2004;304(5668): 276–8.
[56] Maheshwari G, Mallik N, Abot J, *et al.* Sensing & measurement powering up nanoparticles: versatile carbon materials for engineering and medicine. 2008; SPIE Newsroom. doi: 10.1117/2.1200809.1305

[57] Shim B S, Chen W, Doty C, Xu C L, Kotov N A. Smart electronic yarns and wearable fabrics for human biomonitoring made by carbon nanotube coating with polyelectrolytes. Nano Letters. 2008;8:4151.

[58] Coyle S, Wu Y, Lau K T, De Rossi D, Wallace G, Diamond D. Smart nanotextiles: a review of materials and applications. MRS Bulletin. 2007;32:434–42.

[59] Devaux E, Koncar V, Kim B, *et al.* Processing and characterization of conductive yarns by coating or bulk treatment for smart textile applications. Transactions of the Institute of Measurement and Control. 2007;29:355–76.

[60] Catrysse M, Puers R, Hertleer C, Langenhove L V, Van Egmondc H, Matthys D. Towards the integration of textile sensors in a wireless monitoring suit. Sensors and Actuators A: Physical. 2004;114:302–11.

[61] BatteryCloth by Flexel-Innovative Flexible Power Solutions. Available at: http://www.flexelinc.com. Last accessed Oct 18, 2016.

[62] Yan L, Yoo J, Kim B, Yoo H-J. A 0.5 micro-Vrms 12 micro-W patch type fabric sensor for wearable body sensor network. IEEE Journal of Solid State Circuits. 2010;45:2356–65.

[63] Hu J, Meng H, Li G, Ibekwe S. A review of stimuli-responsive polymers for smart textile applications. Smart Materials and Structures. 2012; 21:053001.

[64] Carpi F, De Rossi D. Electroactive polymer-based devices for e-textiles in biomedicine. *IEEE* Transactions on Information Technology in Biomedicine. 2005;9:574.

[65] Collins G E, Buckley L J. Conductive polymer-coated fabrics for chemical sensing. *Synthetic* Metals. 1996;78:93–101.

[66] Wu J. *Synthesis, Characterization and Applications of Conducting Polymer Coated Textiles*. Dissertation, University of Wollolong; 2004.

[67] Lang U, Rust P, Schoberle B, Dual J. Piezoresistive properties of PEDOT: PSS. *Microelectronic Engineering*. 2009;86:330–4.

[68] Hasegawa Y, Shikida M, Ogura D, Sato K. Novel type of fabric tactile sensor made from artificial hollow fiber. *Proceedings of 20th International Conference on Micro Electro Mechanical Systems (Hyogo, Japan)*; 2007.

[69] Takamatsu S, Kobayashi T, Shibayama N, Miyake K, Itoh T. Meter-scale surface capacitive type of touch sensors fabricated by weaving conductive-polymer-coated fibres. *Symposium on Design, Test, Integration and Packaging of MEMS/MOEMS DTIP (Bernin, France)*; 2011.

[70] Rothmaier M, Luong M, Clemens F. Textile pressure sensor made of flexible plastic optical fibres. Sensors. 2008;8:4318–29.

[71] El-Sherif M A, Yuan J, MacDiarmid A. Fiber optic sensors and smart fabrics. Journal of Intelligent Material Systems and Structures. 2000;11:407–14.

[72] Akcelrud L. Electroluminescent polymers. *Progress in Polymer Science*. 2003 Jun;28(6):875–962.

[73] Dashti M, Mokhtari J, Nouri M, Shirini F. Imparting conductivity and chromic behaviour on polyester fibres by means of poly(3-methylthiophene) nanocoating. Journal of Applied Polymer Science. 2012;124:3007–12.

[74] Kim H, Kim Y, Kwon Y, Yoo H. A 1.12mW continuous healthcare monitor chip integrated on a planar-fashionable circuit board. *Digital Technical Papers IEEE International Solid-State Circuits Conference (ISSCC) (Philadelphia, PA)*; 2008.

[75] Antelope suit. Available at: http://www.antelope.club/en/. Last accessed: Oct 18, 2016.

[76] Fraunhofer Institute for Reliability and Microintegration IZM. Available at: http://www.izm.fraunhofer.de/en/institute.html Last accessed: Oct 18, 2016.

[77] Axisa F, Schmitt P M, Gehin C, Delhomme G, McAdams E, Dittmar A. Flexible technologies and smart clothing for citizen medicine, home healthcare, and disease prevention. IEEE Transactions on Information Technology in Biomedicine. 2005;9:325–36.

[78] Linz T. *Analysis of Failure Mechanisms of Machine Embroidered Electrical Contacts and Solutions for Improved Reliability*. Dissertation, Department of Electronics and Information Systems, Ghent University; 2011.

[79] Inaba M, Hoshino Y, Nagasaka K, Ninomiya T, Kagami S, Inoue H. A full-body tactile sensor suit using electrically conductive fabric and strings. *International Conference on Intelligent Robots and Systems* (Vol. 2, pp. 450–7); 1996.

[80] Mercier P P, Chandrakasan A P. A supply-rail-coupled eTextiles transceiver for body-area networks. IEEE Journal of Solid-State Circuits (Shropshire, United Kingdom). 2011;46:1284–95.

[81] Scilingo E P, Lorussi F, Mazzoldi A, De Rossi D. Strain-sensing fabrics for wearable kinaesthetic-like systems (eds. Ephraim Suhir, Y C Lee, C P Wong). Sensors Journal, IEEE (New York, United States). 2003;3:460–67.

[82] Jung S, Lauterback C, Strasser M, Weber W. Enabling technologies for disappearing electronics in smart textiles. *Digest of Technical Papers – IEEE International Solid-State Circuits Conference (Philadelphia, PA)*; 2003.

[83] Bhattacharya R, Van Pieterson, L, Van Os K. Improving conduction and mechanical reliability of woven metal interconnects. IEEE Transactions on Components, Packaging and Manufacturing Technology. 2012;2:165–8.

[84] Linz T, von Krshiwoblozki M, Walter H. Novel packaging technology for body sensor networks based on adhesive bonding: a low cost, mass producible and high reliability solution. *International Conference on Body Sensor Networks BSN (Singapore)*; 2010.

[85] Locher I, Troster G. Fundamental building blocks for circuits on textiles. IEEE Transactions on Advanced Packaging. 2007;30:541–50.

[86] Netravali A N, Bahners T. Adhesion promotion in fibers and textiles using photonic surface modifications. Journal of Adhesion Science and Technology. 2010;24:45–75.

[87] Winterhalter C A, Teverovsky J, Horowitz W, Sharma V, Lee K. Wearable electro-textiles for battlefield awareness. *Army Natick Soldier Center MA Report ADA431955*; 2004.

[88] Universal Serial Fabric Bus, U.S. Army Natick Soldier Systems Center. Available at: http://www.natick.army.mil/about/pao/pubs/warrior/02/janfeb/smartclo.htm. Last accessed: Oct 18, 2016.

[89] Araki T, Nogi M, Suganuma K, Kogure M and Kirihara O 2011 Printable and stretchable conductive wirings comprising silver flakes and elastomers IEEE Electron Device Letters. 2011;32:1424–6.
[90] Slade J. Mechanical testing of electrotextile cables and connectors. *MRS Proceedings, Symposium D Electronics on Unconventional Substrates* (Vol. 736, pp. 73–8); 2002.
[91] Dunn DJ. Engineering and structural adhesives. (iSmithers Rapra Publishing); 2004.
[92] Morris JE, Liu J. Electrically conductive adhesives: a research status review. *InMicro- and Opto-Electronic Materials and Structures: Physics, Mechanics, Design, Reliability, Packaging* (pp. B527–B570) (Springer US); 2007.
[93] Holme I. Adhesion to textile fibres and fabrics. International Journal of Adhesion and Adhesives. 1999;19(6):455–63.
[94] Kaviany, M. *Principles of Heat Transfer* (New York, United States: John Wiley & Sons); 2002.
[95] Fackler, W C, Kutz, M, eds. Electronic materials and packaging. *Mechanical Engineers' Handbook, Volume 1: Materials and Engineering Mechanics*. (Hoboken, New Jersey, United States: John Wiley & Sons); 2015.
[96] Winston G, Backer S. *Measurement of the Thermal Transmission of Textile Fabrics*. Quartermaster CORPS Washington DC, 1948.
[97] Stanković, S B., Popović D, Poparić G B. Thermal properties of textile fabrics made of natural and regenerated cellulose fibers. Polymer Testing. 2008;27.1: 41–8.
[98] Herman, I P. *Physics of the Human Body*. (Cham, Switzerland: Springer); 2016.
[99] Abdel-Rehim, Z S., Saad M M, El-Shakankery M, Hanafy I. Textile fabrics as thermal insulators. AUTEX Research Journal. 2006;6.3:148–61.
[100] Gurrum S, Romig M. Using Thermal Calculation Tools for Analog Components. *TI Report SLUA566*; 2010.
[101] MacAuley, D, ed. *Oxford Handbook of Sport and Exercise Medicine*. (New Delhi, India: OUP, Oxford); 2012.
[102] Rathore, M M. *Thermal Engineering*. (London, England: Tata McGraw-Hill Education); 2010.
[103] Karwowski, W, ed. *International Encyclopedia of Ergonomics and Human Factors*. Vol. 3 (CRC Press); 2001.
[104] Bhattacharya R, de Kok M M, Zhou J. Rechargeable electronic textile battery. Applied Physics Letters. 2009;95:223305.
[105] Flexel-Innovative flexible power solutions. Available at: http://www.flexelinc.com. Last accessed Oct 18, 2016.
[106] Axisa F, Schmitt P M, Gehin C, Delhomme G, McAdams E, Dittmar A. Flexible technologies and smart clothing for citizen medicine, home healthcare, and disease prevention. IEEE Transactions on Information Technology in Biomedicine. 2005;9:325–36.

[107] Catrysse M, Puers R, Hertleer C, Langenhove L V, Van Egmondc H, Matthys D. Towards the integration of textile sensors in a wireless monitoring suit. Sensors and Actuators A: Physical. 2004;114:302–11.
[108] Yan L, Yoo J, Kim B, Yoo H-J. A 0.5 micro-Vrms 12 micro-W patch type fabric sensor for wearable body sensor network. IEEE Journal of Solid State Circuits. 2010;45:2356–65.
[109] Schubert M B, Werner H J. Flexible solar cells for clothing. Materials Today. 2006;9:42–50.
[110] Bedeloglu A. *Progress in Organic Photovoltaic Fibers Research, Solar Cells – New Aspects and Solutions*. 2011:255–86.
[111] Krebs F C. Fabrication and processing of polymer solar cells: a review of printing and coating techniques. Solar Energy Materials and Solar Cells. 2009;93:394–412.
[112] Krebs F C, Biancardo M, Winther-Jensen B, Spanggard H, Alstrup J. Strategies for incorporation of polymer photovoltaics into garments and textiles. Solar Energy Materials and Solar Cells. 2006;90:1058–67.
[113] Pushparaj, V L., *et al.* Flexible energy storage devices based on nanocomposite paper. Proceedings of the National Academy of Sciences. 2007;104(34): 13574–7.
[114] Laforgue A. All-textile flexible supercapacitors using electrospun poly (3,4-ethylenedioxythiophene) nanofibres. Journal of Power Sources. 2011; 196: 1.
[115] Bhattacharya R, de Kok M M, Zhou J. Rechargeable electronic textile battery. Applied Physics Letters 2009;95:223305.
[116] Dalton A B, *et al.* Super tough carbon nanotube fibres. Nature. 2003; 423(6941):703.
[117] Hu L, *et al.* Stretchable, porous, and conductive energy textiles. Nano Letters. 2010;10(2):708–14.
[118] Gaudiana R, Eckert R, Cardone J, Ryan J, Montello A. Photovoltaic Fibres. *Proceedings of SPIE Organic Photovoltaics VII* (*San Diego, CA*); 2006.
[119] Bedeloglu A, Demir A, Bozkurt Y, Serdar S N. A photovoltaic fibre design for smart textiles. Textile Research Journal. 2010;80:1065–74.
[120] Chittibabu K, *et al.* Photovoltaic Fibers. US Patent US 6,913,713; 2005.
[121] Nathan A, *et al.* Flexible electronics: the next ubiquitous platform. Proceedings of the IEEE. 2012 May;100(Special Centennial Issue):1486–517.
[122] Mitcheson P D. Energy harvesting for human wearable and implantable bio-sensor. *Annual International Conference of the IEEE Engineering in Medicine and Biology Society (Buenos Aires, Argentina)*; 2010.
[123] Ahn Y, Song S, Yun KS. Woven flexible textile structure for wearable power-generating tactile sensor array. Smart Materials and Structures. 2015 May;24(7):075002.
[124] Yang J H, Cho H S, Park S H, Song S H, Yun K S, Lee J H. Effect of garment design on piezoelectricity harvesting from joint movement. Smart Materials and Structures. 2016 Feb;25(3):035012.

[125] Yang B, Yun K S. Piezoelectric shell structures as wearable energy harvesters for effective power generation at low-frequency movement. Sensors and Actuators A: Physical. 2012;188:427–33.

[126] Hackworth R, Moriera J R, Maxwell R, Kotha R, Ayon A A. Piezoelectric charging for smart fabric applications. *Symposium on Design, Test, Integration and Packaging of MEMS/MOEMS DTIP (Bernin, France)*; 2011.

[127] Jung S, Lauterback C, Strasser M, Weber W. Enabling technologies for disappearing electronics in smart textiles. *Digest of Technical Papers – IEEE International Solid-State Circuits Conference (Philadelphia, PA)*; 2003.

[128] Jo S E, Kim M K, Kim M S, Kim Y J. Flexible thermoelectric generator for human body heat energy harvesting. Electronics Letters. 2012;48:1013–15.

[129] Qin Y, Wang X, Wang Z L. Microfibre-nanowire hybrid structure for energy scavenging. Nature London. 2008;451:809.

[130] Xu S, Qin Y, Xu C, Wei Y, Yang R, Wang Z L. Self-powered nanowire devices. Nature Nanotechnology. 2010;5:366–73.

[131] Post E R, Orth M. Smart fabric, or "wearable clothing". *IEEE 1st International Symposium on Wearable Computers*, Digest of Papers *(Cambridge, MA)*; 1997.

[132] Sergio M, Manaresi N, Tartagni M, Guerrieri R, Canegallo R. A textile based capacitive pressure sensor. *Proceedings of the First IEEE International Conference on Sensors (Orlando, FL)*; 2002.

[133] Karaguzel B, Merritt CR, Kang T, *et al.* Flexible, durable printed electrical circuits. The Journal of The Textile Institute. 2009;100(1):1–9.

[134] Calvert P, Duggal D, Patra P, Agrawal A, Sawhney A. Conducting polymer and conducting composite strain sensors on textiles. Molecular Crystals and Liquid Crystals. 2008;484:291–302.

[135] Harris K D, Elias A L, Chung H J. Flexible electronics under strain: a review of mechanical characterization and durability enhancement strategies. Journal of Materials Science. 2016 Mar;51(6):2771–805.

[136] Rogers A J. Materials and mechanics for stretchable electronics. Science. 2010;327:1603–7.

[137] Macleod P. A Review of Flexible Circuit Technology and its Applications. PRIME Faraday Partnership; 2002.

[138] van den Brand J, *et al.* Flexible and stretchable electronics for wearable health devices. Solid-State Electronics. 2015 Nov;113:116–20.

[139] Sayhan I, Bauersfeld M L, Woellenstein J, Becker T. Dye and polymer based light sensor for tag integration. Microsystem Technologies. 2008;14:659–64.

[140] Tsai C C. Recent development in flexible electronics. *16th Opto-Electronics and Communications Conference*, OECC *(Kaoshiung, Taiwan)*; 2011.

[141] Vogels, J P A, Klink S I, Penterman R, De Koning H, Huitema E E A, Broer D. Robust flexible LCDs with paintable technology. Journal of the Society for Information Display. 2004;12:411–6.

[142] Zhang L, Wang Z, Volakis J L. Textile antennas and sensors for body-worn applications. IEEE Antennas and Wireless Propagation Letters. 2012; 11:1690–93.

[143] Wang Z, Lee LZ, Psychoudakis D, Volakis JL. Embroidered multiband body-worn antenna for GSM/PCS/WLAN communications. IEEE Transactions on Antennas and Propagation. 2014;62(6):3321–9.

[144] Lee J, Kwon H, Seo J, *et al.*. Conductive fiber-based ultrasensitive textile pressure sensor for wearable electronics. Advanced Materials. 2015;27 (15):2433–9.

[145] Linz T, Vieroth R, Dils C, *et al.* Embroidered interconnections and encapsulation for electronics in textiles for wearable electronics applications. In: Advances in Science and Technology (Vol. 60, pp. 85–94). (Trans Tech Publications); 2008.

[146] Reuss RH, Chalamala BR, Moussessian A, *et al.*. Macroelectronics: Perspectives on technology and applications. *Proceedings of the IEEE*. 2005 Jul;93(7):1239–56.

Chapter 6

Organic gas sensors and electronic noses

Juan Soto Camino[1,2] *and Román Bataller Prats*[1]

Abstract

Smell is one of the five human senses. It is fundamental to people and to many living creatures and by extension it is of great importance for the quality control of online production processes in many mass-production industries.

G. Dodd and K. Persaud published the first work regarding the artificial sensing of smells and odours and a few years later Gardner defined the technological solution to an electronic nose as: 'an instrument, which comprises an array of electronic chemical sensors with partial specificity and an appropriate pattern-recognition system, capable of recognising simple or complex odours.'

In agreement with the definition, a competent electronic nose should be capable of working continuously providing useful information of the samples under test and presenting no fatigue over the measurements.

Two key elements should be the base of every electronic nose: a sensor array of cross-selective electrodes and a pattern recognition tool to statistically analyze the samples, which has to be able to point out the similarities and dissimilarities among all the measured samples. These statistical methods are known as chemometrics, which combined with sensors arrays has been recognized by many as a method to enhance the results of sensor systems.

One of the most classic sensor arrays with which the first electronic noses were built, and that are still in use, is based on metal oxides. However they present some problems and so it has been necessary to investigate and develop new technologies. These will permit the characterisation of gases, vapours and aromas in a fast, economic and, if possible, online manner, in order to generalise the use of electronic noses in many distinct areas and applied technologies.

Sensors using semiconducting polymers have been used since the 1970s due to their favourable features. Recently, there have been a number of further

[1]Centro de Reconocimiento Molecular y Desarrollo Tecnológico (IDM), Departamento de Química, Universitat Politècnica de València, Camí de Vera, s/n. 46022 Valencia, Spain
[2]Departamento de Química, Universitat Politècnica de València, Camí de Vera, s/n. 46022 Valencia, Spain

developments in the field of organic semiconductors, which suggests that devices based on these may become even more relevant in the near future. This chapter discusses about semiconducting polymers, organic polymers used in gas sensors, their recognition principles and about adsorbent organic conducting polymer composites.

Three configurations are used for organic sensors: chemiresistors, transistors (and diodes) and optodes. Some applications in which these sensors are used are also described, such as humidity, ammonia, nitrogen oxide, hydrogen chloride and nitroaromatic explosives.

6.1 Introduction

Of the five traditionally recognised senses that human beings possess, smell might be the least appreciated. However, its utility is fundamental for many people and essential to other living creatures. For example, for many living beings the sense of sight might play a secondary role in their perception, whereas hearing, smell or touch result in perception of greater importance. In many species, olfaction, or the sense of smell, is fundamental to guaranteeing their survival (food search, scent marking, litter recognition, avoidance of the ingestion of spoiled food, etc.) [1].

Some domestic mammals have an especially well-developed olfactory system. This makes them suitable for finding specific types of food, which is why they are trained by humans to search for fungi with high culinary value, such as truffles.

The olfactory sensitivity of dogs (*Canis familiaris*) is 10,000–100,000 times greater than that of humans [2,3], which has lead this species to be used for tracking smells. They are able to detect explosives [4] and drugs [5], to find the location of lost or injured people and even to detect specific diseases in people [6].

Moreover, to evaluate the quality of certain foods, beverages or perfumes, the need to train a panel of experts in sensory testing has arisen. These panels are used extensively in the food and perfume industries, although there are some drawbacks. The main obstacle for specialised testers is the very long training process required for their olfactory education. Another important issue is connected to the olfactory fatigue of the panellists due to saturation of their olfactory receptors. This happens to every expert, even after a relatively short period of work, necessitating a recovery time for the panellists to regain their perceptive capabilities. These factors make sensory panels quite unfeasible for the quality control of online production processes usually required in many mass-production industries.

The human sense of smell is a very subjective attribute that depends on the factors mentioned in the previous paragraph and on many others: the genetic predisposition of the individual, which is strongly influenced by age; the state of health in that specific moment; the type and amount of food recently ingested; the level of physical exhaustion; whether or not the individual is a smoker; the acquired olfactory education; etc. Because of this, when a specific product is assessed, not just one tester's sensory experiences are relied upon but rather the combined analyses of an entire group of sensory panellists are needed. This slows down production and raises the price of the final product, thus making the use of sensory panels a hindrance rather than a benefit in mass-production processes.

The technological solution to the problems brought up by the use of sensory panels is the use of an electronic nose. In 1982, G. Dodd and K. Persaud published an article in the journal *Nature*, in which they announced the birth of this new technology [7]. A few years later, Gardner defined the electronic nose as 'an instrument, which comprises an array of electronic chemical sensors with partial specificity and an appropriate pattern-recognition system, capable of recognising simple or complex odours' [8].

According to this definition, a properly built and tuned electronic nose should be a versatile instrument capable of working online in a quasi-continuous mode without presenting fatigue symptoms and providing objective information about the physical and chemical features of the system under study. Its usefulness depends, then, on technical factors such as the price of the instrument itself, the cost of maintenance and consumables, the time needed to measure one single sample, the speed at which that information is processed and the repeatability, accuracy and precision of the instrument.

That first proposed definition of an electronic nose was slightly modified over the years as new methodologies were introduced into the procedure, such as data sampling and information analysis [9]. In any case, these definitions became somewhat outdated as soon as new technologies such as mass spectroscopy and gas chromatography were introduced as sensor elements of the electronic nose.

6.2 Chemometrics and sensor arrays

Gardner's definition proposes that the electronic nose consists of two basic elements: first, a sensor array with cross-selective response and second, a pattern-recognition system. The first one is generally formed by a series of sensors of low specificity, while the second one is a statistical analysis tool that allows the differences and similarities among all the measured samples to be identified in an objective way (chemometrics). This combination of sensors and mathematical analysis has been recognised as a method to improve the performance of sensors [10]. Some of these mathematical processing procedures, for instance principal components analysis, have been developed for the identification and classification of samples, while others, such as partial least square (PLS) regression, are used to quantify and predict the value of specific parameters in complex systems [11].

The analysis techniques described above may present limitations when non-linear tendencies are considered between the signal acquired from the sensor and the concentration or partial pressure of some components (vapour or gas) to which the sensor responds. In these situations, the PLS method does not operate properly and it is necessary to utilise some other techniques such as artificial neural networks, which are techniques that are especially suitable for the analysis of non-linear response systems.

In some cases, as has happened in studies conducted using sensors based on conducting polymers, if a direct analysis is done on the electrical resistance variations of the sensor as a function of the analyte's concentration, the prediction results are simply wrong. When the relationship between the concentration and the acquired signal is known, in which case the concentration is proportional to the inverse of the electrical resistance, the analysis of the information must be done based on the electrical conductivity. If this is taken into consideration and the information used to calculate the electrical conductance comes from the measured electrical resistance, then appropriate statistical fittings can be obtained.

6.2.1 Hybrid sensor systems

The mammalian sensor system is formed by a collection of organs (eyes, ears, nose, tongue, skin) which contain a series of sensory cells that collect information from the surroundings by means of specific perception mechanisms: visible radiation (sight), air vibrations (hearing), the aromatic composition of volatiles (smell), non-volatile chemical composition (taste) and thermal radiation, hardness, etc. (touch: thermal and pressure sensors). The sensory system captures the information and sends it through the nervous system in the form of many electrical impulses to the brain. The analysis of the information coming from these different sources is performed by the neural network and depends on the series of patterns acquired during the gradual development of that specific individual. This analysis provides fairly complete information about the characteristics of their surroundings and helps them make the most appropriate decision in each specific moment.

It can be said that a sensor system is hybrid when the data collected by the central information processor (the brain or a computer) come from different sources or receptor structures and/or by means of various transduction systems (eyes, ears, skin, etc.), capable of interaction with the surroundings through several channels (visible radiation, infrared radiation, ultraviolet radiation, vibration, chemical interactions with volatile agents, etc.). Nature has classified species based on their abilities and, though they have different versions, the hybrid sensor system is the most widespread among living beings. Data collection coming from different sources provides a more complete vision of the physical and chemical conditions of the surroundings or of a specific object, such as its shape, mass, smell, temperature and chemical components.

Generally, an optimal hybrid system has to be particularised for each specific application. A hybrid sensor system is considered useful when it is comprised of the sensors necessary to provide information that serves to reliably characterise the system under study.

Beneficially, the use of hybrid sensor systems may lead to significant savings in resources when assembling advanced control equipment.

6.2.2 Redundant sensor systems

It can be said that a sensor is redundant when another sensor that performs the same measuring task by means of the same technique exists in the same sensor array.

In nature, redundant sensors and redundant sensor systems are extensively common. For example:

1. The vast amount of colour and brightness sensors (approximately 6 million cone cells and 120 million rod cells) existing in the retina of the human eye allow the human brain to create a precise image of the surroundings, their brightness and their colour distribution. As a consequence of this, the eye is able to continue working for many years even if some of the cells and integrating elements of the lens are damaged by ageing over time. The retina of the eye is a redundant sensor system because it is formed by hundreds of millions of sensitive cells. The duplicity of the eye (redundancy of the perceptive organ) allows the brain to perceive the tridimensional composition of space based on the analysis of the non-overlapping image from each eye, which contributes to the perception of distance between objects [12]. The results are very favourable; stereoscopic vision is beneficial and evolution has selected this duplicity (or multiplicity) of the sensory organ of vision because it has proved to be very useful for survival.
2. The sense of smell of snakes is located in the nose and in Jakobson's organ, which is located near the front of the mouth. Their forked tongue allows them to collect volatile aromatic particles and bring them to Jakobson's organ in order to obtain details on the differences between local concentrations so they are able to feel where the source of the smell is coming from. In addition, the intensity of the smell provides information on its proximity. So by combining these features they can obtain a tridimensional position of the prey.
3. The Loreal pits where the infrared sensor system is located in reptiles is also redundant (there are four to six pits depending on the species). They are located near the front of the reptile's face and allow them to obtain a thermal picture of their prey in the dark [13] and to estimate their position (3D night 'vision') prior to initiating attack.
4. Echolocation by means of ultrasound [14] is the 3D location mechanism used by the order Chiroptera. It is tremendously efficient and is based on echo analysis of the ultrasounds emitted during animal flight, which allows them to avoid obstacles, locate prey and distinguish other individuals of their own species in completely dark environments. The echo signal processing system is not significantly affected by ultrasounds emitted by other bats.

It is interesting to mention that redundant sensor systems not only allow more reliable signal measurements to be obtained and anomalous data to be identified and discarded in the case of a system defect, but they also allow, through the brain's data processing, additional information to be obtained regarding the characteristics of the observed system. All of this cannot be obtained with a single sensor.

Many papers have been written about studies that imitate systems originated in nature, where a single sensor array combines different measuring techniques (hybrid sensor system) or where the duplicity or multiplicity of equivalent sensors (redundancy) is observed. Regardless of the system, they seek the same objective: to increase the resolution of the measurements and to enhance the discrimination capacity.

The hybridisation of techniques has been applied to the field of electronic tongues in a very generalised manner. Potentiometric sensors, voltammetric sensors and conductometric sensors have been combined to form a hybrid electronic tongue which obtains information using different measuring techniques. Winquist [15] used this hybrid sensor system for monitoring fermentation processes in milk. Gutiérrez and his collaborators, on the other hand, combined potentiometry, voltammetry and optical absorbance measurements to quantify the varietal composition of grapes in different wines [16]. In these cases in which a hybrid signal was used to analyse the system's information, the discrimination results and the prediction capabilities improved noticeably.

Hybrid sensor systems are also formed by measuring devices that can perform measurements on properties in different physical phases (such as gas–liquid). Thus, a system integrating an electronic nose and an electronic tongue can be considered a hybrid sensor system as well [17].

Petrei *et al.* [18] combined three measuring devices to evaluate the quality of the analysed samples: an electronic nose, an electronic tongue and what can be called an electronic eye, which performed a wavelength scan to obtain the transmittance spectra of the sample. This hybrid system was used to quantify the polyphenolic content, the acidity and the bitterness of Spanish olive oils.

6.3 Electronic noses and probes based on the use of metal oxides

Perhaps the most classic sensor arrays with which the first electronic noses were built, and that are still in use, are those based on metal oxides [19,20]. These sensors are generally formed by a semiconducting layer of a metal oxide (for instance SnO_2, TiO_2, ZnO or ZrO_2), which coats a ceramic substrate (for instance alumina), and to which a built-in heater is sometimes attached, which is essential for the desorption of the sample and the subsequent regeneration of the surface activity of the semiconductor or metal oxide. The use of this type of sensor is due to its great capacity to absorb vapours and volatiles, which is normally exhibited by large variations in the electrical conductivity of the material when the absorbed species are very polar gases or vapours and to a lesser degree depending on their semiconducting properties [21,22].

Gas-FET devices are formed by a transistor (field-effect transistor – FET) whose gate is a thin layer of platinum or palladium and will be exposed to the gas or mixture of gases to be analysed. Pure hydrogen or gas hydrogen contained in a gaseous mixture dissociates when it reaches the surface of the metal by diffusing through the layer; once it reaches the semiconducting phase, it oxidises to form protons at the metal/gate interface, which changes its potential and affects the semiconducting behaviour of the sensing device. The temperature range for operating this kind of device is from 200 to 300 °C when the substrate is made of silicon, while the working temperature increases to 600 °C when the transistor is made of silicon carbide. This kind of sensor is successfully used in the production of the so-called lambda sensors, currently used to monitor the emission of gases in the exhaust systems of the automotive industry [23,24].

The fundamental problems that have been described in the use of electronic noses based on metal-oxide semiconductors (MOSs) are the following:

1. It takes a long time for the vapours and gases to be absorbed onto the active surface of the sensing elements or the electronic nose.
2. The required time to reach a dynamic equilibrium in order to permit a reproducible measurement which each one of the sensing elements of the electronic nose is very long.
3. The time needed to regenerate the sensors is long; it usually entails the desorption of the adsorbate at moderate or high temperatures to cause its pyrolytic oxidation. This operation requires high energy consumption.
4. There is an observable drift in the signals as the sensing system of an electronic nose ages. This usually necessitates the recalibration of the system periodically by employing a series of absorption patterns.
5. Some chemical compounds which make up the matrix of the system under study, such as water vapour (present in all fresh foods) or ethylic alcohol (present in every fermented beverage), interfere with the measurements.

6.3.1 Working protocol of electronic noses (MOS)

In order to mitigate the problems described in the previous paragraphs, a series of working protocols have been developed to regenerate the sensing system. The aim is to desorb the remains of adsorbate molecules held by the sensor by heating them or to destroy them using oxidation (pyrolysis). In the basic operation of a typical electronic nose, this procedure depends essentially on the type of sensor integrating the array and on the chemical composition of the sample. However, a typical methodology can be described for working with this type of electronic nose (Figure 6.1), and is usually composed of the following steps:

1. Pumping of the gas or vapour sample through a filtering system, which retains small particles such as dust and occasionally vapours that may interfere with the measurement, such as water vapour.
2. Insertion of the gas or vapour sample into the adsorption headspace of the electronic nose and subsequent recirculation through the sensors for the time required to reach a dynamic equilibrium state in some cases, or a quasi-stationary situation in others.
3. Measurement of the physicochemical changes in each of the sensing elements of the electronic nose.
4. Purging of gases and vapours from the measuring headspace of the electronic nose and regeneration by heating or pyrolysis of the sensors.
5. Reestablishment of the temperature of the sensing system and insertion of the following sample.

Despite these efforts, the results obtained by electronic noses based on sensor arrays made up of oxides and/or inorganic semiconductors are still far from those

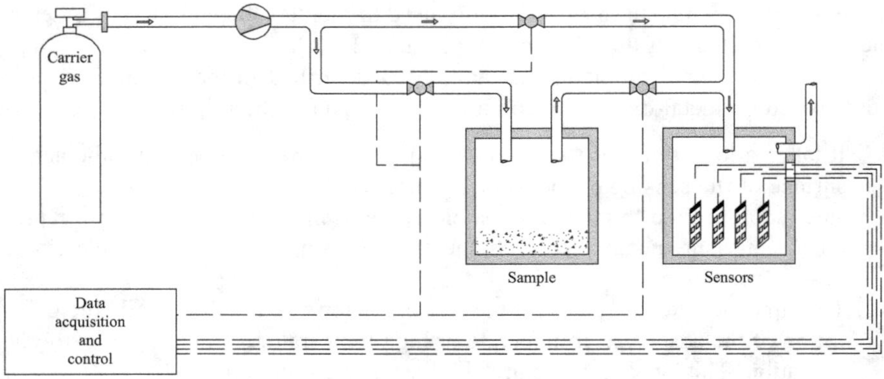

Figure 6.1 Diagram of an electronic nose

required for their application as useful monitoring devices. The main drawbacks to their working operations are the long amount of time required to measure consecutive samples and the high power consumption required in the cleaning process.

Moreover, the change in the sensors caused by ageing and doping seems hard to avoid despite the regeneration of the sensors. This process has been attributed, among other things, to structural changes in the metal oxide, to changes in the chemical composition of the oxide due to contamination from doping agents in the absorbate and to changes in the oxidation state of the metal oxide. The heating elements and metallic contacts of the sensors also suffer from high operating temperatures and regeneration of the sensor system, indicating that as the hours of use increase, the oxidation of the metallic contacts themselves increases as well and so does the electrical resistance of the system. Working at high temperatures also requires high power consumption and excessively long absorption times and regeneration intervals.

Although very effective specific applications exist (for instance probes for monitoring gas concentrations in combustion engines) and many commercial electronic noses [25] (based on sensor arrays of MOSs and metal-oxide semiconductor field-effect transistors (MOSFETs)) exhibit optimal performance and reliability, it is still necessary to investigate and develop new technologies. These will permit the characterisation of gases, vapours and aromas in a fast, economic and if possible online manner, in order to generalise the use of electronic noses in many distinct areas and applied technologies.

6.4 Organic conducting polymers for gas probes and electronic noses

The synthesis and general properties of organic conducting polymers have been described in the previous chapters of this book. Although the first descriptions of the organic semiconductor were made in the 1970s, the importance of organic transistors was highlighted for the first time by Tsumura and Koezuka between 1986 and 1987 [26,27] when they were able to form a semiconducting active layer through the electrolysis of polythiophene (PT).

As a result of the discovery of organic conducting polymers, a great deal of effort has been dedicated to studying the modification of their chemical structures as well as the electrical properties of devices based on this kind of polymer. One of the objectives pursued has been to improve the performance of the sensors made with MOSs [28,29].

Conducting polymers seem to be very suitable for use in the manufacturing of devices that detect volatile compounds [30] and it has been demonstrated that they have several very favourable features for their application in the manufacturing of gas probes and electronic noses, the most noteworthy being:

1. Organic conducting polymers exhibit generally fast adsorption–desorption kinetics at room temperature.
2. Unlike metal-oxide sensors, they do not require a heating system to induce desorption, resulting in a low-power consumption system.
3. Polymers may be structurally modified and they can be designed specifically for the recognition of the specific chemical species which need to be monitored.
4. Interference by specific vapours and volatile compounds, such as substances containing sulphur, may be inhibited.

Recently, there have been a number of further developments in the field of organic semiconductors, which suggests that devices based on these may become even more relevant in the near future.

6.4.1 Semiconducting polymers are hybrid sensors

The detection of different volatile substances by means of semiconducting polymers has been carried out using several measuring techniques to monitor the changes in the electrical properties or other physical magnitudes that occur to the polymer when the recognition process of the analyte takes place. Among them, the following are worthy of mention:

1. Systems in which analyte activity is determined by changes in impedance [31], electrical resistance [32] or even the interfacial capacitance [33] of the sensor array.
2. Systems in which the detection is accomplished by measuring the changes in the semiconducting properties of the sensing polymer [34]. For example, the variations in the gate potential of a transistor, the threshold potential between the gate and the source of a transistor, its current intensity, the drift in diode characteristics, etc.
3. Models in which the detection is achieved by using the sensors as an electrochemical cell. In this way, the potential, current or conductivity of the sensory membrane can be correlated with the content in gas or vapour phase of some of the compounds integrated into the fluid mixture [35].

4. Models in which spectroscopic techniques are used to monitor sensor changes. In these systems, the recognition of the substance by the polymer is associated with significant variations in the absorption spectra or in fluorescence [36,37].

Electronic noses constructed based on organic semiconductors should then be considered hybrid systems from which the interaction of the sensors with the gas or vapour samples may be analysed by means of different techniques [38]. The answer to making use of both the electrical and optical signals is a combined array with optoelectronic sensors (a combined electronic nose and electronic eye system) [39].

Nevertheless, the combination of measuring techniques does not seem to be sufficiently utilised in the field of electronic noses based on conducting polymers and this is an aspect to take into consideration if the intention is to improve sensitivity, reduce noise and enhance repeatability in this type of device.

Moreover, it is important to keep in mind that making a measurement might significantly alter the physicochemical state of a sensing element. For example, it is known that performing amperometric measurements significantly alters the sensor–electrode interface because the faradic current flow causes a series of redox processes that generally leave the device out of a natural equilibrium state for what might be a very long time interval for solid-state systems [40].

The recommended methodology for working with hybrid sensors is to start by taking measurements which involve the absence of external current flow, such as the open-circuit potential) of the sensor [41]. This potential can be correlated with the amount of reagent adsorbed by the polymer and then, using this as a new parameter, it can be associated with the gas or vapour pressure to which it has been subjected. After this, on the same sensor, optical measurements can be performed on the polymer, such as IR (infra-red) spectroscopy, ultraviolet–visible (UV–Vis) spectroscopy, spectrophotometry or spectrofluorometry. Next, the measurements of the conductivity and impedance spectroscopy can be taken by using low amplitude alternating current (AC, e.g. 10 mV) and in a frequency range in which the redox reactions cannot take place (for instance from 100 Hz to 1 MHz). Finally, measurements that involve the application of direct current (DC) signals can be performed, such as voltammetry or transistor characterisation.

The most advisable way of working with hybrid sensor systems is to introduce a series of additional redundant sensors so that if their physical properties are altered while measurements with different techniques are taking place, they can easily be replaced.

6.4.2 Most commonly used organic polymers for sensing devices

How organic semiconductors work and the reasons behind their molecular recognition capabilities are still current aims in material science research. Trying to address the connections between both molecular and crystalline structures of polymers and their coordination capacity and electrochemical reactivity is essential to discovering why charge transfer is enhanced or inhibited depending on the circumstances or particularities of the chemical surroundings. This and other questions of interest can be analysed by means of a series of examples which can explain the working fundamentals of this type of sensor.

In 1976, Heeger *et al.* discovered conducting polymers and the way to dope these polymers in order to transform them into conductors as if they were metals [42]. They demonstrated that the conductive characteristics of this type of compound are based on the alternation of simple and double carbon bonds, which is repeated periodically along the polymeric structure (bond conjugation), permitting highly delocalised electronic states to be established.

The first systematic studies were realised on the cis and trans isomers of polyacetylene (PA) (Figure 6.2) by studying the doping effects of N and P types with oxidising and reducing agents [43]. Through ordinary chemical synthesis, its linear form precipitates as a black powder, is sensitive to contact with air and is not easy to work with or mechanise. By using the synthesis procedure proposed by Shirakawa, polymers of the desired size and thickness can be obtained but they still have chemical stability problems in the presence of oxygen. An additional and important counterpoint in the use of these polymers is their limited solubility. As a consequence of these problems, PA is rarely used in the manufacturing of devices that require reliable and stable measurements.

The stability of polymeric membranes can be improved by either introducing a heterocyclic compound (or ring structure) into the conducting chain or stabilising them with carbon aromatic rings. The polymers obtained from aniline [29], pyrrole [44] or thiophene [45] are stable enough to be used under normal working conditions (Figure 6.3).

6.4.2.1 Polyaniline

Out of all the conducting polymers, those obtained from polyaniline (PANI) are of great interest due to their low cost, their chemical stability in air, their controllable electrical conductivity and their redox properties that are attributable to nitrogen in the polymeric structure. The chemical synthesis of PANI is normally done in aqueous acidic media in which an oxidising agent is added, such as ferric chloride [46]. The

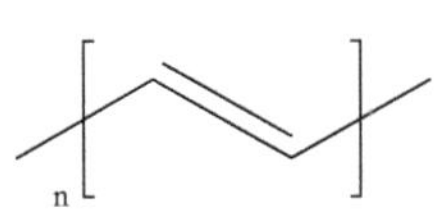

Figure 6.2 Polyacetylene

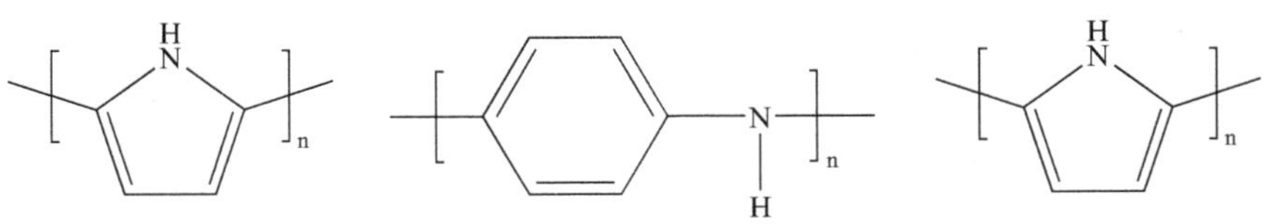

Figure 6.3 Polypyrrole, polyaniline, polythiophene

synthesis is performed at a low temperature (around 0 °C), which provides better results than at moderate or high temperatures. The resulting product is an insoluble dark powder that, after rinsing, may be compacted into pellets.

The electrochemical synthesis of this polymer can also be carried out with cyclic voltammetry in the range of 0–1 V versus a normal hydrogen electrode at a scan rate between 50 and 100 mV/s. In this way, conducting polymeric films with good characteristics and a controlled thickness can be obtained [47] and in addition, they can easily be processed in dissolution. The electrochemical behaviour of PANI has been studied thoroughly [48] and it is probably one of the most widely used polymers for study in different applications. It has several intrinsic redox states which determine its conductivity and its behaviour as a chemical sensor [49]. Figure 6.4 shows the polymers obtained by P-doping PANI.

These forms can be either protonated or deprotonated by interaction with acidic or basic gases based on the mechanisms shown in Figure 6.5.

Of all the possible forms that PANI may appear in, emeraldine salt is the most conductive form. The detection response time when sensors fabricated with this polymer are used is in the range of minutes, and depends on the thickness of the membrane employed.

Nanostructures produced with PANI (nanofibres, nanowires, nanotubes and nanorods) [50] exhibit high specific surfaces and shortened times required for gas adsorption and for the regeneration of the material.

It has been demonstrated that long periods of exposure to ambient air result in polymer deterioration, as revealed by the appearance of peaks in the IR zone associated with the formation of C=O bonds.

6.4.2.2 Polypyrrole

One of the aspects that makes polypyrrole (PPy) so attractive is that its chemical synthesis and the cost of the raw materials are very cheap. PPy prepared by the oxidation of its monomer with $FeCl_3$ and other Fe(II) or Cu(II) salts is also a black

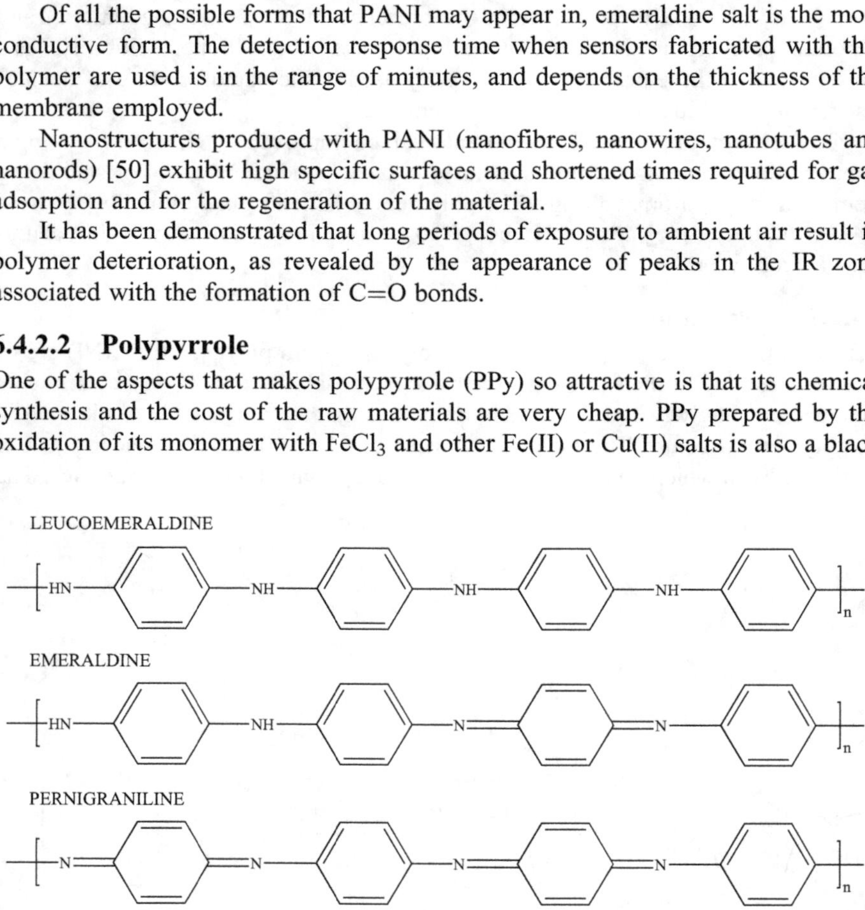

Figure 6.4 Polyaniline oxidation states

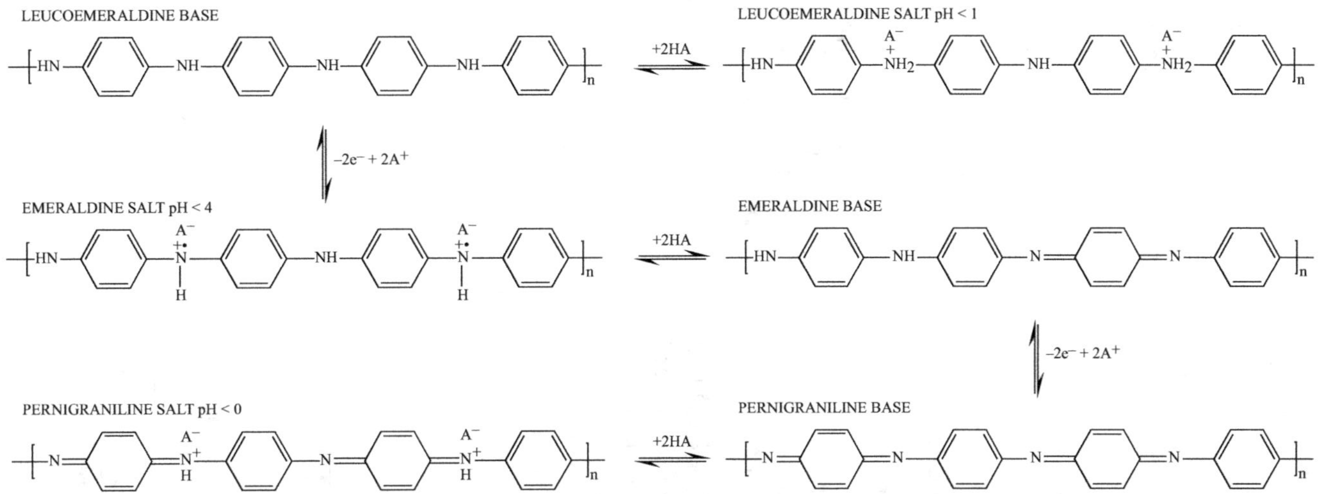

Figure 6.5 Conversion between different states of PANI

conducting powder, and the yield of the synthesis process is very high (close to 100%). For the preparation of highly crystalline films, electrochemical synthesis provides better results [51]. Electrochemically obtained PPy polymers in aqueous solutions are flexible films with very high conductivity values and are stable at air temperatures up to 150 °C [52].

Several studies have been done on devices fabricated with this conductive polymer. As case in point is the conductometric detection of carbon dioxide with an upper detection limit of 700 ppm [53]. The influence of temperature and relative humidity on PPy sensors has also been studied by controlling the amount of methanol in the sample with reproducible results [54]. In order to improve the sensing properties of PPy for the detection of volatile organic compounds (VOCs), several modifications have been made to the polymer by manufacturing composites with polystyrene fibres, which has improved the repeatability and shortened the response times noticeably [55].

6.4.2.3 Polythiophene

PTs are a family of conducting polymers that result from the polymerisation of the monomer thiophene. Their chemical synthesis can be done by working under similar conditions to those described for PPy. The proposed mechanism is again similar to the synthesis mechanism of PPy and obeys the reaction scheme shown in Figure 6.6.

However, the synthesis yield rates are inferior to those obtained by PPy and PANI. Also, while the unsubstituted chain of PT is conductive after the doping process and exhibits good stability in air, it is only soluble in arsenic trifluoride and arsenic pentafluoride. The use of substituted PTs has resulted in products of certain solubility which may be employed in the fabrication of polymeric films.

The in situ polymerisation of aniline and pyrrole is a simple method that is extensively used in the fabrication of conducting polymeric membranes, although

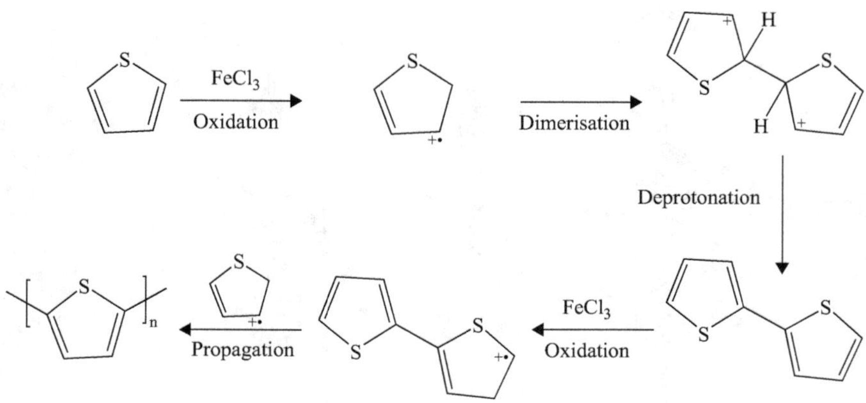

Figure 6.6 Chemical polymerisation mechanism of tiophene

the preparation of PT at ambient temperature with this procedure is much more complex due to the elevated oxidation potential of the monomer.

In 2005, Mazur described an electropolymerisation method that allowed creating membranes of good quality and of the proper size for their application in gas sensing [56]. In another study, electrodeposition was performed on quartz microbalance sensors by means of cyclic voltammetry, which resulted in the fabrication of thin films of thiophene derivatives [57]. The behaviour of these sensors in the presence of VOCs has been studied and they exhibit good predictive responses to the measured signal. The studies on vapour binary systems, performed with the same electronic nose, exhibited good results when the measured vapours had noticeable differences in polarity.

Several studies have been performed on VOCs by means of resistive and optical transduction [58,59], leading to similar results and conclusions. The detection of ammonia compared to other amines and other vapours has also been studied, by using amperometric transduction, thus demonstrating the reversibility of the process and its repeatability in situ on polymerised membranes from different pyrrole derivatives [60].

Using a similar methodology to detect food spoilage, the behaviour of several PT derivatives with volatile amines has been studied by working with sensors with thin-film transistor configuration [61]. The substituent type inserted into the PT chain depends on its size and polarity and on the structure and functional groups already in the polymeric chain. These substituents and also the thickness of the microtransistors affect the discrimination capacity of the sensors towards the substances under study.

6.4.3 Recognition principles

Conducting polymers do not present much conductivity in their neutral charge states. Their intrinsic conductivity is the result of the appearance of their own transport charge mechanisms created by oxidation (positive doping) or reduction (negative doping) processes of the polymer itself.

To explain this conduction phenomenon, it is convenient to resort to the band theory, which is used to describe the conducting properties of materials. Conduction basically depends on the space between the band containing the bonding electrons (the highest occupied molecular orbital) and the superior energy levels (the lowest unoccupied molecular orbital) that are associated with the free electrons in the structure and to the conduction process. In the limiting case, considering a partial overlapping of both bands, we would have a system with metallic conduction. In the opposite case, where the bands are separated by an interval higher to 3–4 eV, the material behaves as an isolator (diamond 5.33 eV, AgCl 3.20 eV). The intermediate situation is where the semiconducting materials fit, and is characterised by those materials in which a minimal thermal or radiating energy permits the flow of electrons between bands and thus either conducting processes appear in the material or absorption processes and electromagnetic radiation emissions occur in easily accessible areas of the electromagnetic spectrum.

The gap between the valence band and conduction band in the most common conducting polymers is approximately 1.7 eV for PA [62], 3.16 eV for PPy [37] and 3.6 eV for PANI [63]. The last two polymers mentioned are considered non-conducting when they are not intrinsically doped; nevertheless, the material is transformed into a conductor when it is auto-doped by a chemical reaction. This drastically reduces its band gap due to the appearance of the levels that are associated with the formation of bipolarons.

Table 6.1 summarises the basic interaction mechanisms of conducting polymers with different adsorbates. It is suggested that the detection of gases and vapours by means of this type of device is due to the combination of physical interactions, such as superficial adsorption or absorption at the polymeric substrate, and chemical interactions, such as reduction–oxidation reactions, protonation and deprotonation processes of constituent elements of the polymeric membrane or coordination processes in the polymeric chain.

The physico-chemical interaction generated between the sensing polymer and the analyte is transformed by means of a suitable transductor into an electronic signal that will be useful for characterising the system. The sensor–analyte interactions may be of different types as stated in Table 6.1. Among the most significant mechanisms of detection and quantification described and acknowledged nowadays, the following can be found:

1. *Systems undergoing an electrochemical reaction* in the presence of the analyte. The electrochemical properties of a polymer change due to this reaction. For example, the conductivity of many semiconducting polymers increases significantly because of doping processes caused by redox reactions. The electrons transferred from the valence band or the conduction band of the polymer cause important variations in its conductive properties and work function. This kind of process occurs in films formed with polymeric structures such as PPy, poly (ethylenedioxythiophene) (PEDOT) (Figure 6.7) and in some cases also with PANI when they are exposed to the action of gases such as NH_3, NO_2, I_2, H_2S and other active redox gases.

   Electron acceptors such as NO_2 and I_2, for instance, may remove electrons from the aromatic rings of polymeric conductors. When this happens to a P-type semiconductor, its conductivity varies significantly. The opposite process can also occur with gas or vapour molecules of donor electrons, which

Table 6.1 Interaction mechanisms of several gases and vapours with polymeric conductors

Physical interactions		**Chemical interactions**	
Absorption	***Oxidation***	***Reduction***	***Protonation***
$CHCl_3$, CH_2Cl_2, acetonitrile, alkane, benzene, toluene, acetone, alcohols, etc.	NO_2, SO_2, O_3, O_2	H_2, N_2H_4, NH_3, H_2S	HCl, NH_3

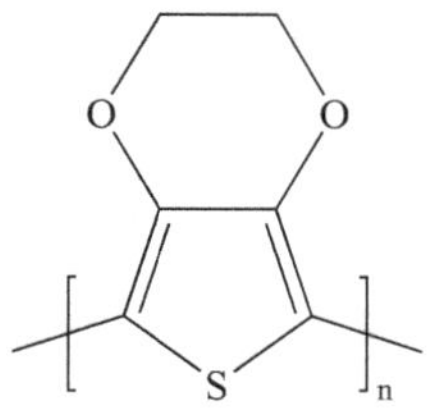

Figure 6.7 Structure of PEDOT

is the case of ammonia. So when NH_3 reacts with polypyridine, its electric resistance decreases rapidly. The original resistance value is recovered when the system is purged with nitrogen or dry air. To explain this phenomenon, the following recognition mechanism has been proposed [64,65]:

$$PPy^+ + NH_3 \rightarrow PPy^0 + NH_3^+ \quad \text{(Adsorption)}$$

$$PPy^0 + NH_3^+ \rightarrow PPy^0 + NH_3 \quad \text{(Desorption)}$$

This proposed mechanism for the recognition of ammonia by means of p-type semiconductors does not seem to be generally applicable, as there are several examples that contradict it [66]. This demonstrates that further research needs to be done on this mechanism.

2. *Gases that are adsorbed and change the protonation state of the polymer.* Gas or vapour molecules with a significant acid–base character, such as HCl, HBr and NH_3, are capable of changing the net charge located on the polymer. They do this by inducing exchange processes of protons or hydronium ions.

   Doped PANI is an example of a material widely used for the detection of these kinds of gas and vapour. PANI films and some composites obtained from it have been used for the detection of nitrogen oxide (NO_2) by employing optical detection techniques (UV–Vis and ellipsometry) and resistive measuring techniques with response times of only 10 s [67]. This same polymer has been used to detect ammonia by using the same UV–Vis technique and measuring the electrical resistance of the film. The studied concentration range which exhibits a linear response is large, from 180 to 18,000 ppm, and exhibits a sensitivity of 1 ppm and a response time of 2 min. The mechanism that has been proposed to justify the obtained results can be seen in Figure 6.8. The oxidised polymer undergoes a doping loss (dedoping) as a consequence of deprotonation [68].

   According to the paper presented by Kukla *et al.*, whose study was based on resistive measurements, the adsorption and desorption of ammonia are reversible events when the polymer has been prepared recently, but they detected a signal drift as the sensors aged [69]. In this study, they used heaters to elevate the temperature of the polymer up to 150 °C, in order to favour the gas desorption. This might have been one of the causes of such observed drift.

3. *Detection of absorbed gas molecules and vapours by absorption.* Many volatile organic molecules such as acetonitrile, benzene, toluene, alcohols and many other VOCs are not reactive at ambient temperature from the redox or acid–base

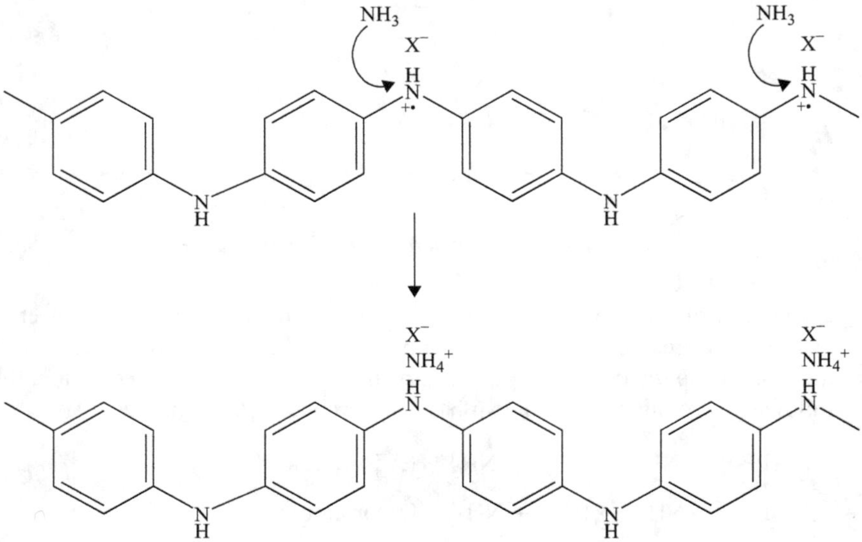

Figure 6.8 Dedoping process of polyaniline film

point of view. But there are a series of favourable physical interactions that assist the superficial adsorption of the polymer or its reticular absorption, producing a swelling phenomenon which substantially modifies the electrochemical behaviour of the sensor. Amrani *et al.* proved that it is possible to detect and quantify the vapour concentration of methanol, ethyl acetate and acetone in dry air by using impedance spectroscopy as the measuring technique, with a frequency scan from 0.1 to 100 Hz [70].

Some experimental results showed that some vapours, especially alcohols [71] and low molecular weight ketones, were capable of changing the crystallinity of polymers after being absorbed by them. The same researchers obtained a highly specific sensor for detecting methanol by using a Pd nanocomposite and PANI; they were also able to describe the mechanism of this selective recognition [72].

In addition to the resistance variations that take place when a redox-active gas is adsorbed, there is also other evidence of the alteration of the semiconductor polymer properties when gases of low redox activity are absorbed. Some gases such as carbon monoxide are not considered to be active redox gases at room temperature, but the conductivity of PANI in the presence of carbon monoxide decreased significantly in one study [73]. According to results published by Josowicz and her collaborators [74], those variations are justified based on a partial charge transfer, which depends on the electronegativity of the adsorbed molecule and on the work function value of the adsorbent polymer.

Nevertheless, Sirivat and his co-workers compared the UV–Vis spectra and X-ray diffraction patterns of the PANI films before and after exposure to carbon

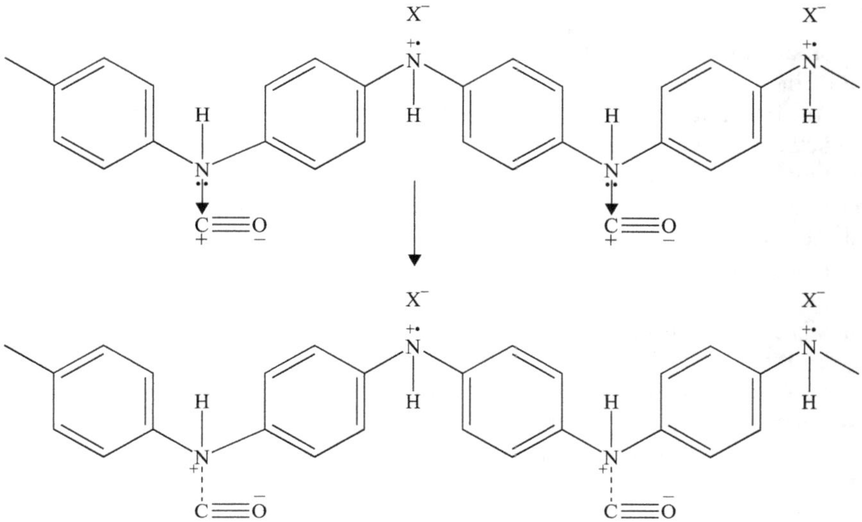

Figure 6.9 Proposed PANI–CO interaction mechanism by Sirivat et al.

monoxide without finding significant differences between both scenarios [75]. These authors proposed the recognition mechanism shown in Figure 6.9.

Some other explanations have been given to understand the conductivity changes in this polymer. For example, if the detection is performed in the presence of O_2, then the increase of CO in the mixture makes the O_2 adsorbed onto the polymer decrease, which causes the conductivity reduction.

In order to explain the physical interactions between the analyte and the conductive polymeric membrane, it has been proposed that vapour absorption can produce a series of swelling phenomena, which are responsible for the change in the conducting properties of the absorbent. The interaction can be a simple adsorption process, and if it occurs at the active centres of the exterior surface of the polymer, it can modify the charge distribution of its surface, thus modifying the surface conductivity of the membrane [76]. This process should be reversible in order to rapidly recover the original conductivity values of the polymer when the gas or vapour is desorbed by, for instance, the passing of an inert gas. If the physical process is a vapour or gas absorption process inside the structure of the polymeric membrane, then the process will be slower and it will be associated with a swelling phenomenon and therefore the polymer conductive properties will increase. But it will require more time to recover the original conductivity values when desorbing the absorbate or when applying the proper temperature values in order to favour desorption. High crystallinity polymers may possibly respond in a more natural way to superficial adsorption phenomena, while the amorphous polymeric phases will preferably respond to internal absorption.

6.4.4 Adsorbent organic conducting polymer composites

The production of composites from polymers and carbon nanotubes or graphene is one of the most promising applications of carbon nanotubes. With these composites, the intention is to take advantage of the ease of the processing procedures that conducting organic polymers offer in addition to the exceptional features of carbon nanotubes. In situ polymerisation is especially convenient and effective in the preparation of composites. The π–π interactions between the monomer molecules and the walls of the carbon nanotubes seem to facilitate their dispersion in the reaction media and favour the polymeric reaction in the proximity of the carbon nanotube surface. This strategy is commonly used in the preparation of carbon nanotubes and graphene composites with several types of conducting polymers such as PPy [77,78], PT [79] and PANI [80].

These composites have been successfully used in the preparation of gas sensors based on the combination of a polymer or resin and carbon in the form of nanotubes [81], carbon black [82], etc. because of their absorbent properties.

6.5 Sensor configurations

6.5.1 Chemiresistors

Chemiresistors are the most common type of sensor. They can be manufactured by means of simple and cheap processes [83]. These devices are resistors whose resistivity depends on the chemical environment to which the sensor is subjected [84,85]. They are formed by one or more couples of electrodes and a layer of polymer attached to them, as illustrated in Figure 6.10.

Measuring the electrical resistance changes of the chemiresistor only requires a simple resistivity meter. The resistance value of the polymeric membrane can be

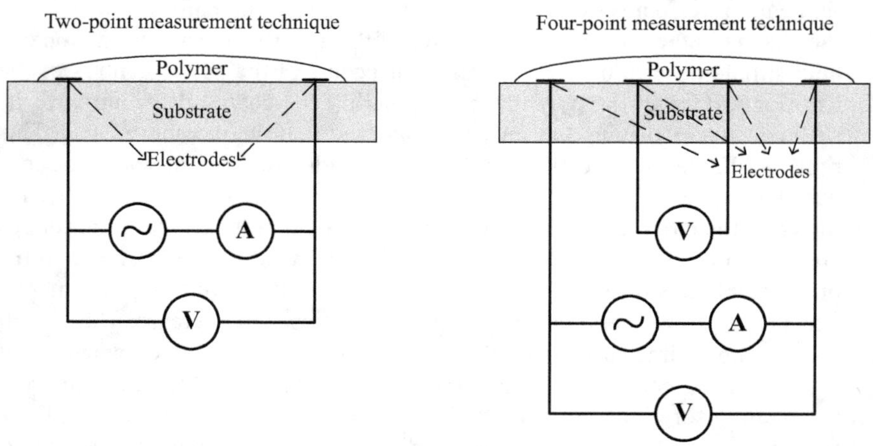

Figure 6.10 Two- and four-point measuring techniques

determined by using either DC or AC. If using the former, it is necessary to consider that redox processes can take place at the electrode–polymer interface and that these redox reactions can modify the characteristics of the polymer–electrode bond, which is a phenomenon that should be studied thoroughly.

Alternating current has also been used in determining the resistivity of organic conducting polymers [86–88]. In this situation, capacitance and even induction have to be considered in order to analyse the equivalent circuit from which the resistance value of the material is to be obtained. The advantage of working with AC and, even more conveniently, using an impedance spectroscopy scan at medium to high frequencies, is that no interfacial redox processes will take place caused by redox reactions of the vapour at the metallic contacts of the sensor with the polymer. In addition, three values are available to characterise the gaseous system: resistance, inductance and capacitance. The dissipation factor values obtained (the resistance divided by the reactance module) change with the frequency of the signal used. Furthermore, when the maximum dissipation is divided by the frequency, the resulting quotient changes depending on the gas in contact with the sensor, which is useful for the identification of gases or mixtures thereof.

Moreover, several studies on the importance of contact resistance on the measurement have been performed and the conclusion has been drawn that the most reliable working methodology is the four-point technique [89]. In order to go into further detail on the topic of contact resistance, the use of interdigitated electrodes has been studied because they seem to exhibit improvements in the precision of the measurements taken. This improvement is caused by the reduction of the metal–polymer contact resistance, which simplifies the data gathering because with interdigitated electrodes it is not essential to use the four-point measuring method. It is advantageous because of the technological simplification of the procedure.

The results of the studies with conducting nanofibres point in the same direction; they exhibit significantly worse characteristics than those shown by polymeric films, probably because the contact resistance value of these sensors is high [90].

Chemiresistive sensors are very efficient and robust devices. This is the reason they are one of the most commonly employed configurations in commercial equipment. Thin films, fibres and conducting polymeric materials can be used as chemiresistors.

The handicap of these sensors lies in the fact that their electrical resistance depends on many ambient factors and the resistance of the polymer is affected by a great amount of interferences, apart from problems such as the interfacial contact resistance, which should be further investigated.

6.5.2 Transistors and diodes

The devices known as organic field-effect transistors (OFETs) have been used in sensing practically since their discovery [91–93]. In generic terms, it can be said that an OFET consists of an active semiconducting layer (channel) in contact with two electrodes (source and drain) and a third electrode (gate) which is insulated from the active layer by means of a thin film of insulating material. When it is in active mode, a potential difference is applied between the source and the drain and the current flowing between them is measured. The voltage at the gate controls the amount of current flowing from the source to the drain. The advantage of this type of sensor is that either the voltage at the gate changes or the current flowing through the channel changes when the conducting polymer interacts with certain analytes.

Basically, two types of OFETs can be considered; first, those in which the active conducting polymeric layer (acting as channel) is in direct contact with the gas or vapour (Figure 6.11(a)) or in which the gate of the transistor is the conducting polymer (Figure 6.11(b)) and second, those in which the active conducting polymeric layer (channel) is insulated (insulated gate field-effect transistors) and the detection takes place through a chemosensitive gate in the presence of the gas (Figure 6.11(c)).

The modulation of the current between the source and the drain depends on the gate's voltage, which, in the semiconducting region close to the gate, favours or inhibits the charge transfer through the channel connecting the source and the drain. The operation of an organic transistor is modulated by the capacitance of the gate, whose potential difference is due to the accumulated charge between the silicon layer and the polymer layer [94].

When two different surfaces are put in electronic contact, an electrical field appears at their interface. This electrical field is proportional to the difference in the

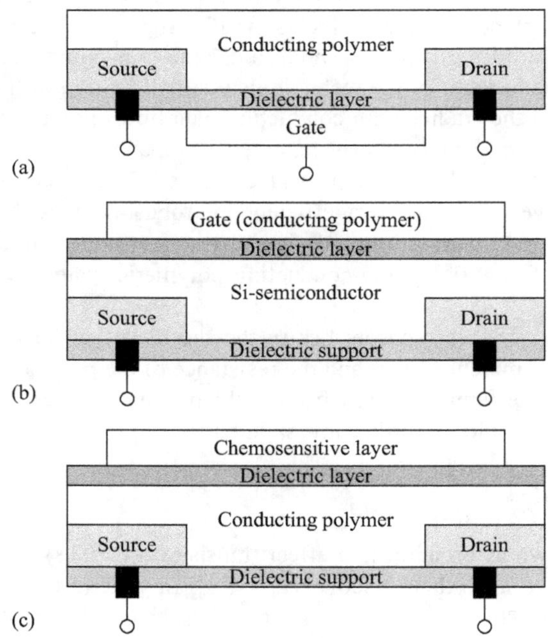

Figure 6.11 OFETs and IGFET basic schemes

work functions of both materials [95], and both the current between the source and the drain and the activation voltage are controlled by this field. The electron transfer between the analyte molecules and the sensitive polymer layer is capable of changing the value of the work function of the polymer. This, in turn, causes a response potential on both the gate potential and the current value between the source and the drain.

Diodes are much simpler devices than transistors are, as can be seen in Figure 6.12. They can also be used in the detection of gases.

Conducting polymers that are stable in air are generally doped p-type semiconducting polymers. When a conducting polymeric film is put into contact with an n-type semiconductor or with a suitable metallic surface (Schottky diode, which is characterised by its high switching speed and its low forward voltage drop), a contact potential difference is generated at the interfacial region. This potential difference is a characteristic of the materials the diode is made of. Due to the fact that the conducting characteristics of the polymers used in the manufacturing of these devices depend on the interactions with the analyte, the observed variations in the current response of the device could be correlated with the chemical activity of the vapour surrounding it.

As examples for the application of this type of device in gas and vapour sensing, some metal/semiconductor devices have been used. For instance, sensors made of gold and aluminium with PANI were used for the detection of methane gas [96]. Based on this same technology, a diode made with PANI and gold for the detection of nitrogen oxides was employed with good reproducibility and short response times [97]. Also, nanodiodes made from positively doped PANI with camphorsulfonic acid in contact with a negatively doped silicon semiconductor were used to detect ammonia vapour and to assess their reversibility and time response [98].

The primary advantage of using a chemiresistive device obtained from a conducting polymer lies essentially in its ease of manufacture. On the other hand, the disadvantage of transistor or diode devices based on conducting polymers is that their preparation is quite complex because specific tools are required. Moreover, the characterisation of transistors is more complex than the characterisation of resistors.

Unlike chemiresistors, transistors and diodes can be considered hybrid sensing devices because they provide more useful parameters for the characterisation of the

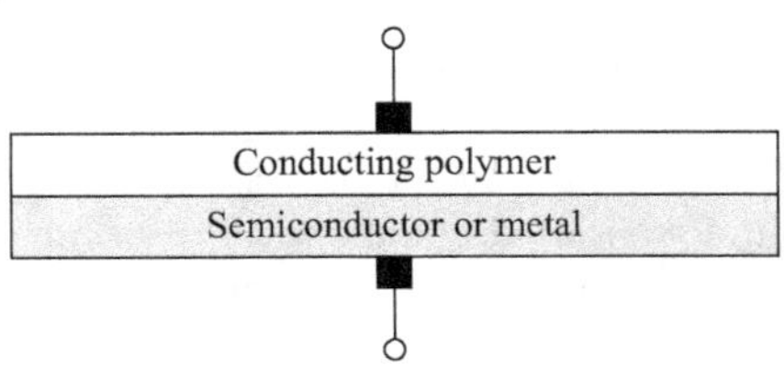

Figure 6.12 Typical configuration of a diode used for gas and vapour sensing

system and more detailed information on the response of the sensing polymer when it interacts with the gas mixture. Furthermore, the detection limits of OFET devices or diodes are better because of the amplified signal which is obtained when an appropriate potential difference is applied.

6.5.3 Optodes

Optodes are sensing devices that use the variations in the optical properties of the sensing material caused by their interaction with the analyte. The optical properties to be analysed can be absorbance, transmittance, reflectance, fluorescence, etc. The configuration of the optodes used in sensing applications for the monitoring of absorbance or fluorescence is the same as that normally used by other classical optical techniques. It consists of three parts: where light is conducted from the source to the sample or to a recognition phase in contact with the sample, the area where the radiation interacts with the sample and the signal sampling device and its conduction towards the detector, where it is transformed into a treatable electrical signal.

The intrinsic advantages of this technique compared to other transduction techniques are the following:

- It does not require a reference signal to perform the measurement (in electrodics, a common reference signal would be the one provided by a saturated calomel electrode).
- It is a non-destructive measuring technique.
- It widens the variety of applications when the signal is processed through optic fibre.
- It facilitates miniaturisation and allows the use of disposable devices.

This type of transduction has been widely used in semiconducting polymer sensing, due to the significant changes in colour that the polymeric films undergo when interacting with the analytes they are sensitive to.

In a study carried out by Gallazzi *et al.*, polyalkoxythiophene conductors were used as detectors with resistive and optical measurements of different organic vapours (*o*-xylene, *n*-butanol, benzene, methanol, *n*-heptane and 3-pentanone) in the concentration range between 50 and 66,500 ppm using argon as a carrier gas [99]. Each sensor was exposed to each vapour at least 40 times over 25 days with good sensitivity, reversibility and response time (from 15 to 90 s).

However, the PPy films used for the detection and quantification of organic vapours exhibited a degenerative evolution, as revealed in a paper by Vrkoslav *et al.* [100]. They attributed this to swelling processes of the polymer, produced by the retention of residual vapours.

Another application example would be an optode fabricated based on a Prussian blue composite with PANI, used in the optical determination of pH [101]. The obtained pH value was not affected by the ionic strength of the solution, the presence of alkali cations or typical oxidising or reducing agents.

6.6 Practical application examples

Research into gas recognition nowadays is mainly focused on the search for films made from selective or specific chemical or biochemical materials for the recognition of gases and is working towards designing arrays of devices with partially selective sensors which, by means of multivariate analysis or pattern-recognition techniques, are able to determine the composition of complex samples of gases or vapours [102]. As a summary of the papers published in this field, the following chapters provide some examples of applications developed in recent years.

6.6.1 Humidity sensors

Lately, many sensing prototypes have been described for monitoring ambient humidity and controlling air conditioning devices for both domestic and industrial use. To achieve this, many compounds have been used, such as electrolytic compounds, ceramic materials [103] or polymeric materials [104], thus correlating their electrical properties with the vapour content in the system.

However, many problematic issues still arise with the use of these devices: hysteresis dependence, stability of the sensors at high temperature in the continuous presence of humid air and the durability of polymers facing attack by organic vapours. These reasons make many of these sensors inappropriate for practical applications [33,105].

6.6.2 Ammonia sensors

Nohria *et al.* used surface acoustic wave sensors (SAWs) to detect ammonia in vapour phase [106]. PPy films, prepared by using the Langmuir–Blodgett technique, were deposited onto the surface of a SAW device as gas absorbent layers. The intention was to simultaneously analyse the phase and speed attenuation of the signal. The sensor responses revealed a very high sensitivity towards ammonia and good selectivity in relation to the interfering gases studied (CO, CH_4, H_2 and O_2) at ambient temperature. Their low cross-selectivity permitted the recognition of NH_3 in different mixtures of interfering gases.

Studies with hybrid sensors formed by PPy and SnO_2 have been conducted and these sensors presented elevated sensitivity (1 ppm) and shorter response times than PPy itself. This improvement in the response time performance has been attributed to a synergic effect between the conducting polymer and tin oxide [107]. PANI nanocomposites with TiO_2, MoO_3, In_2O_3, etc. have also been studied for the same application and had similar results. The nanocomposite PANI/Tio_2 synthesised at low temperature exhibited higher reversibility and durability [108].

Water vapours interfere significantly in the measured signal values. It is necessary, though, to compensate for this principal interfering factor in order to reliably evaluate the real ammonia content of the gas samples.

6.6.3 Nitrogen oxide sensors

Nitrogen oxides, together with carbon monoxide, sulphur oxides, ozone and aromatic hydrocarbons, are contaminant gases whose concentration levels must be controlled because they are toxic and imply a risk for human health. Gas sensors based on metal oxides (SnO_2, WO_3, In_2O_3, etc.) have been used extensively in the analysis of combustion gases which may contain these contaminant gases.

Conducting polymers (PPy, PT and PANI), conducting polymer composites with metal oxides and other mixed devices and nanocomposites [109] have been developed to detect nitrogen oxides.

6.6.4 Hydrogen chloride sensors

Hydrogen chloride is not only a source of dioxins in incineration plants or an additional element of acid rain; it has also been identified as an occupational hazard with a permissible exposure limit of just 5 ppm per day. Nakagawa developed a sensor for this compound based on a polymeric composite with tetraphenyl porphyrin, which exhibited great sensitivity [110,111].

PANI nanofibres have turned out to be very versatile in application because they can be used for the detection of HCl, NH_3 and NO_2 by only making small preparative modifications [112].

6.6.5 Detection of TNT and other nitroaromatic explosives

Swager and his collaborators used poly(phenylene ethynylene) to detect 2,4,6-trinitrotoluene (TNT) [113]. The recognition of this compound requires an electron transfer from the photoinduced excited state of the donor electron (the polymer) to the acceptor electron (TNT). Other conducting polymers have been used to detect the presence of TNT and other nitroaromatic explosives [114]. These conducting polymers have high fluorescence efficiency, so when they react with the acceptor electron a fluorescence quenching occurs. The limit of detection in these conditions is very low (just a few ppb).

6.7 Advantages of using sensors fabricated with organic polymers

To summarise, conducting polymers have been widely used in the fabrication of gas and vapour sensors. Although they still present some problems, the principal advantages of using conducting polymers in the fabrication of sensor devices and electronic noses can be summed up as follows:

1. *Ambient temperature workability.* As discussed in the previous sections of this chapter with respect to recognition principles, the interaction of a conducting polymer and the gaseous molecules of an analyte are not strong enough at ambient temperature to allow their proper detection. Therefore, sensors based on semiconducting polymers can provide perceptible signals at ambient temperature, while sensors based on metal oxides are not capable of it. This makes

it so the conducting polymers have low energetic requirements and, even in some applications (chemiresistors and optodes), a simple device configuration.

2. *Easy modulation of the receptor properties.* The structure of the majority of conducting polymers is based on aromatic rings, which can be easily implanted by using electrophilic reactive groups. These modifications can be used to modulate the polymer's selectivity towards different types of gas or vapour molecules, based on their size, shape, polarity or chemical reactivity.
3. *High sensitivity and short response time.* The detection limits for the fabricated sensors based on conducting polymers are quite low. For the detection of active redox analytes or with acid–base reactions, the limits of detection are lower than 1 ppm in many cases. For analytes with an inert behaviour, the limit is of a few tens of ppm or lower. Time responses in this type of sensors are usually of a few hundreds of seconds and, especially when working with ultrathin film sensors, their time responses are reduced to only a few seconds.
4. *Simple fabrication of devices.* The fabrication of sensing devices based on conducting polymeric materials is much simpler than that of other sensing materials, such as silicon-based technologies. Manufacturing polymeric films using techniques such as electrolysis can be considered a simple processing procedure.

Furthermore, conducting polymers possess the inherent mechanical properties of polymers and so the majority of the processing techniques for this family of materials can also be used on them.

In addition, due to the modification of the backbone of the polymer, its physicochemical characteristics can change, such as its solubility. Because of this, for example, conducting polymers can be processed by deposition techniques such as spin coating, layer-by-layer deposition or the Langmuir–Blodgett technique, etc. These techniques are easily applicable and provide good repeatability of the characteristics of the polymeric films formed, which makes their characteristics superior to those of inorganic sensors.

6.8 Disadvantages of polymeric sensors

Some of the main problems with polymeric sensors are the low long-term stability, the existence of several adsorption or absorption processes of low reversibility and in general their low specificity. In the following paragraphs, these issues will be discussed.

1. *Instability and irreversibility of polymeric membranes*
   Long-term instability is the main handicap of sensors based on conducting or semiconducting polymers. This is something that has been said in many papers; nevertheless, little research covers extensive, systematic studies on the lifetime of this type of sensor [115]. The performance of these devices diminishes abruptly when they are stored in air during relatively long periods. This phenomenon can be explained by the oxidation over time of the doping

agents added during the fabrication of the polymers and before processing. However, the most systematic studies conducted on the degeneration of the recognition characteristics have been performed on well-known polymers such as PANI or PT [116].

The second problem is the irreversibility of these sensors, generally caused by the slow kinetics related to electrochemical reactions and to the desorption of the absorbed substances. These issues can cause the signal measured from the sensors to gradually degenerate when a series of consecutive readings are made [117].

2. *Low selectivity*

The majority of sensors used in these sensing devices present the problem of low selectivity. With only a single sensor, the detection of a determined gas or vapour cannot be assured, nor can a correct quantification be established. Fortunately, multivariate statistical analysis can use the information provided by an array of different sensors of cross-selectivity, and thus solve what was initially considered a problem.

A priori, it is necessary to keep in mind that almost all conducting or semiconducting polymers are sensitive to the presence of water vapour. On the other hand, polymeric sensors respond to the presence of redox-active vapours or gases such as NH_3, SH_2, SO_2, SO_3, NO_2 and generally to organic vapours.

One of the drawbacks to this type of sensor is the swelling effect of the conducting or semiconducting polymer due to the absorption of vapours; it causes variations in the electrical resistivity of the material. If it is not taken into consideration it can be one of the causing effects of hysteresis cycles or of poor repeatability of the initial signal of the sensing system.

6.9 Conclusions

Considering the discussion in the previous paragraphs, the work that lies ahead is to deepen knowledge about electronic noses based on conducting and semiconducting polymers, putting emphasis on the following aspects:

First, study in depth the recognition mechanisms of the molecules adsorbed or absorbed by the polymer. Many of the mechanisms proposed in the literature available are hypotheses made by the authors of the published research papers, and which are not always confirmed experimentally.

Second, expand knowledge on what causes the drifts in sensing results; for instance, the non-recovery of the original signal when the sensors are subjected to successive sensing cycles or the appearance of hysteresis phenomena.

Third, elucidate the causes of irreversibility of the recognition mechanisms, in order to search for the adequate solutions of each particular problem. Study the electrochemical kinetics of sensing and the vapour and gas desorption from the polymers.

Fourth, take into account the dependence of the physical and chemical properties of the polymer on ambient conditions such as temperature and humidity in order to design systems which present as little dependency as possible on these factors.

Fifth, prolong sensor lifetime (the main challenge for researchers working with this type of device).

Sixth, conduct further studies on gases and vapours that are more complex mixtures than those studied so far. This can be achieved by studying cross-sensitivity and utilising the excellent hybrid properties (conducting, semiconducting and optical) of this type of sensing polymers.

List of Abbreviations

3D: three dimensions
AC: alternating current
DC: direct current
FET: field-effect transistor
IR: infrared, normally referred to absorption spectroscopy
MOS: metal-oxide semiconductor
OFET: organic field-effect transistor
PA: polyacetylene
PANI: polyaniline
PEDOT: poly(ethylenedioxythiophene)
PLS: partial least square
PPy: polypyrrole
PTs: polythiophene.
SAW: surface acoustic wave sensor.
UV–Vis: ultraviolet–visible, normally referred to absorption spectroscopy
VOC: volatile organic compounds

References

[1] Buck, L.B.: 'The molecular architecture of odor and pheromone sensing in mammals'. *Cell*, 2000, **100**, pp. 611–618.

[2] Walker, D.B., Walker, J.C., Cavnar, P.J., *et al.*: 'Naturalistic quantification of canine olfactory sensitivity'. *Applied Animal Behaviour Science*, 2006, **97**, (2–4), pp. 241–254.

[3] Craven, B.A., Paterson, E.G., Settles, G.S.: 'The fluid dynamics of canine olfaction: unique nasal airflow patterns as an explanation of macrosmia'. *Journal of the Royal Society Interface*, 2010, **7**, (47), pp. 933–943.

[4] Furton, K.G., Myers, L.J.: 'The scientific foundation and efficacy of the use of canines as chemical detectors for explosives'. *Talanta*, 2001, **54**, pp. 487–500.

[5] Furton, K.G., Hong, Y.-C., Hsu, Y.-L., Luo, T., Rose, S., Walton, J.: 'Identification of odor signature chemicals in cocaine using solid-phase microextraction-gas chromatography and detector-dog response to isolated

compounds spiked on U.S. paper currency'. *Journal of Chromatographic Science*, 2002, **40**, (3), pp. 147–155.

[6] Cornu, J.-N., Cancel-Tassin, G., Ondet, V., Girardet, C., Cussenot, O.: 'Olfactory detection of prostate cancer by dogs sniffing urine: a step forward in early diagnosis'. *European Urology*, 2011, **59**, (2), pp. 197–201.

[7] Persaud, K., Dodd, G.: 'Analysis of discrimination mechanisms in the mammalian olfactory system using a model nose'. *Nature*, 1982, **299**, (5881), pp. 352–355.

[8] Gardner, J.W., Bartlett, P.N.: 'A brief history of electronic noses'. *Sensors and Actuators B: Chemical*, 1994, **18**, (1–3), pp. 210–211.

[9] Holmberg, M., Eriksson, M., Krantz-Rülcker, C., *et al.*: '2nd workshop of the second network on artificial olfactory sensing (NOSE II)'. *Sensors and Actuators B: Chemical*, 2004, **101**, (1–2), pp. 213–223.

[10] Pravdová, V., Pravda, M., Guilbault, G.G.: 'Role of chemometrics for electrochemical sensors'. *Analytical Letters*, 2002, **35**, (15), pp. 2389–2419.

[11] Ciosek, P., Wróblewski, W.: 'The analysis of sensor array data with various pattern recognition techniques'. *Sensors and Actuators B: Chemical*, 2006, **114**, (1), pp. 85–93.

[12] Howard, I.P., Rogers, B.J. *Binocular Vision and Stereopsis*. New York, Oxford University Press; 1995.

[13] Bakken, G.S., Krochmal, A.R.: 'The imaging properties and sensitivity of the facial pits of pitvipers as determined by optical and heat-transfer analysis'. *Journal of Experimental Biology*, 2007, **210**, (16), pp. 2801–2810.

[14] Henson, O.W.: 'The activity and function of the middle-ear muscles in echo-locating bats'. *The Journal of Physiology*, 1965, **180**, (4), pp. 871–887.

[15] Winquist, F., Holmin, S., Krantz-Rülcker, C., Wide, P., Lundström, I.: 'A hybrid electronic tongue'. *Analytica Chimica Acta*, 2000, **406**, (2), pp. 147–157.

[16] Gutiérrez, M., Domingo, C., Vila-Planas, J., *et al.*: 'Hybrid electronic tongue for the characterization and quantification of grape variety in red wines'. *Sensors and Actuators B: Chemical*, 2011, **156**, (2), pp. 695–702.

[17] Di Natale, C., Paolesse, R., Macagnano, A., *et al.*: 'Application of a combined artificial olfaction and taste system to the quantification of relevant compounds in red wine'. *Sensors and Actuators B: Chemical*, 2000, **69**, (3), pp. 342–347.

[18] Apetrei, C., Apetrei, I.M., Villanueva, S., de Saja, J.A., Gutierrez-Rosales, F., Rodriguez-Mendez, M.L.: 'Combination of an e-nose, an e-tongue and an e-eye for the characterisation of olive oils with different degree of bitterness'. *Analytica Chimica Acta*, 2010, **663**, (1), pp. 91–97.

[19] Brattain, W.H., Bardeen, J.: 'Surface properties of germanium'. *Bell System Technical Journal*, 1953, **32**, (1), pp. 1–41.

[20] Yamazoe, N.: 'New approaches for improving semiconductor gas sensors'. *Sensors and Actuators B: Chemical*, 1991, **5**, (1–4), pp. 7–19.

[21] Sberveglieri, G.: 'Recent developments in semiconducting thin-film gas sensors'. *Sensors and Actuators B: Chemical*, 1995, **23**, (2–3), pp. 103–109.

[22] Meixner, H., Lampe, U.: 'Metal oxide sensors'. *Sensors and Actuators B: Chemical*, 1996, **33**, (1–3), pp. 198–202.

[23] Tobias, P., Mårtensson, P., Baranzahi, A., *et al.*: 'Response of metal–insulator–silicon carbide sensors to different components in exhaust gases'. *Sensors and Actuators B: Chemical*, 1998, **47**, (1–3), pp. 125–130.

[24] Lloyd-Spetz, A., Tobias, P., Ekedahl, L.-G., Martensson, P., Lundstrom, I.: 'Fast chemical sensors for emission control'. *Journal of The Royal Society Interface*, 1998, **7**, (4), pp. 34–38.

[25] Pearce, T.C., Schiffman, S.S., Nagle, H.T., Gardner, J.W. (eds.): *Handbook of Machine Olfaction: Electronic Nose Technology*. Wiley-VCH Verlag GmbH & Co. KGaA, Weinheim, 2004.

[26] Tsumura, A., Koezuka, H., Ando, T.: 'Macromolecular electronic device: field-effect transistor with a polythiophene thin film'. *Applied Physics Letters*, 1986, **49**, (18), p. 1210.

[27] Tsumura, A., Koezuka, H., Ando, T.: 'Polythiophene field-effect transistor: its characteristics and operation mechanism'. *Synthetic Metals*, 1988, **25**, (1), pp. 11–23.

[28] Miasik, J.J., Hooper, A., Tofield, B.C.: 'Conducting polymer gas sensors'. *Journal of the Chemical Society, Faraday Transactions 1: Physical Chemistry in Condensed Phases*, 1986, **82**, (4), pp. 1117–1126.

[29] Jiakun, W., Hirata, M.: 'Research into normal temperature gas-sensitive characteristics of polyaniline material'. *Sensors and Actuators B: Chemical*, 1993, **12**, (1), pp. 11–13.

[30] Bartlett, P.N., Gardner, J.W. (Eds.): *'Odour Sensors for an Electronic Nose - Sensors and Sensory Systems for an Electronic Nose'*. Springer Netherlands: NATO ASI Series, 1992, pp. 31–51.

[31] Fratoddi, I., Venditti, I., Cametti, C., Russo, M.V.: 'Chemiresistive polyaniline-based gas sensors: a mini review'. *Sensors and Actuators B: Chemical*, 2015, **220**, pp. 534–548.

[32] Barisci, J.N., Wallace, G.G., Andrews, M.K., Partridge, A.C., Harris, P.D.: 'Conducting polymer sensors for monitoring aromatic hydrocarbons using an electronic nose'. *Sensors and Actuators B: Chemical*, 2002, **84**, (2–3), pp. 252–257.

[33] Matsuguchi, M., Umeda, S., Sadaoka, Y., Sakai, Y.: 'Characterization of polymers for a capacitive-type humidity sensor based on water sorption behavior'. *Sensors and Actuators B: Chemical*, 1998, **49**, (3), pp. 179–185.

[34] Crone, B., Dodabalapur, A., Gelperin, A., *et al.*: 'Electronic sensing of vapors with organic transistors'. *Applied Physics Letters*, 2001, **78**, (15), p. 2229.

[35] Kim, K.-C., Cho, S.M., Choi, H.-G.: 'Detection of ethanol gas concentration by fuel cell sensors fabricated using a solid polymer electrolyte'. *Sensors and Actuators B: Chemical*, 2000, **67**, (1–2), pp. 194–198.

[36] Patra, D., Mishra, A.K.: 'Fluorescence quenching of benzo[k]fluoranthene in poly(vinyl alcohol) film: a possible optical sensor for nitro aromatic compounds'. *Sensors and Actuators B: Chemical*, 2001, **80**, (3), pp. 278–282.

[37] Brédas, J.L., Scott, J.C., Yakushi, K., Street, G.B.: 'Polarons and bipolarons in polypyrrole: evolution of the band structure and optical spectrum upon doping'. *Physical Review B*, 1984, **30**, (2), pp. 1023–1025.

[38] Guadarrama, A., Fernández, J., Íñiguez, M., Souto, J., de Saja, J.: 'Discrimination of wine aroma using an array of conducting polymer sensors in conjunction with solid-phase micro-extraction (SPME) technique'. *Sensors and Actuators B: Chemical*, 2001, **77**, (1–2), pp. 401–408.

[39] Scorsone, E., Pisanelli, A.M., Persaud, K.C.: 'Development of an electronic nose for fire detection'. *Sensors and Actuators B: Chemical*, 2006, **116**, (1–2), pp. 55–61.

[40] Chang, Z.-T., Cherry, B., Marosszeky, M.: 'Polarisation behaviour of steel bar samples in concrete in seawater. Part 1: experimental measurement of polarisation curves of steel in concrete'. *Corrosion Science*, 2008, **50**, (2), pp. 357–364.

[41] Randles, J.E.B.: 'The determination of kinetic parameters of redox reactions from current–potential curves'. *Canadian Journal of Chemistry*, 1959, **37**, (1), pp. 238–246.

[42] Heeger, A.J.: 'Semiconducting and metallic polymers: the fourth generation of polymeric materials (nobel lecture)'. *Angewandte Chemie International Edition*, 2001, **40**, (14), pp. 2591–2611.

[43] Shirakawa, H., Louis, E.J., MacDiarmid, A.G., Chiang, C.K., Heeger, A.J.: 'Synthesis of electrically conducting organic polymers: halogen derivatives of polyacetylene, (CH)x'. *Journal of the Chemical Society, Chemical Communications*, 1977, (**16**), pp. 578–580.

[44] Bartlett, P.N., Ling-Chung, S.K.: 'Conducting polymer gas sensors part II: response of polypyrrole to methanol vapour'. *Sensors and Actuators*, 1989, **19**, (2), pp. 141–150.

[45] Koezuka, H., Tsumura, A., Ando, T.: 'Field-effect transistor with polythiophene thin film'. *Synthetic Metals*, 1987, **18**, (1–3), pp. 699–704.

[46] Nishio, K., Fujimoto, M., Yoshinaga, N., Ando, O., Ono, H., Murayama, T.: 'Electrochemical characteristics of polyaniline synthesized by various methods'. *Journal of Power Sources*, 1995, **56**, (2), pp. 189–192.

[47] Genies, E.M., Tsintavis, C.: 'Redox mechanism and electrochemical behaviour of polyaniline deposits'. *Journal of Electroanalytical Chemistry and Interfacial Electrochemistry*, 1985, **195**, (1), pp. 109–128.

[48] Huang, W.-S., Humphrey, B.D., MacDiarmid, A.G.: 'Polyaniline, a novel conducting polymer. Morphology and chemistry of its oxidation and reduction in aqueous electrolytes'. *Journal of the Chemical Society, Faraday Transactions 1: Physical Chemistry in Condensed Phases*, 1986, **82**, (8), pp. 2385–2400.

[49] Kang, E.T., Neoh, K.G., Tan, K.L.: 'Polyaniline: a polymer with many interesting intrinsic redox states'. *Progress in Polymer Science*, 1998, **23**, (2), pp. 277–324.

[50] Virji, S., Huang, J., Kaner, R.B., Weiller, B.H.: 'Polyaniline nanofiber gas sensors: examination of response mechanisms'. *Nano Letters*, 2004, **4**, (3), pp. 491–496.

[51] Vernitskaya, T. V, Efimov, O.N.: 'Polypyrrole: a conducting polymer; its synthesis, properties and applications'. *Russian Chemical Reviews*, 1997, **66**, (5), p. 443.

[52] Kanazawa, K.K., Diaz, A.F., Geiss, R.H., *et al.*: '"Organic metals": polypyrrole, a stable synthetic "metallic" polymer'. *Journal of the Chemical Society, Chemical Communications*, 1979, (19), pp. 854–855.

[53] Waghuley, S.A., Yenorkar, S.M., Yawale, S.S., Yawale, S.P.: 'Application of chemically synthesized conducting polymer–polypyrrole as a carbon dioxide gas sensor'. *Sensors and Actuators B: Chemical*, 2008, **128**, (2), pp. 366–373.

[54] Cho, J.-H., Yu, J.-B., Kim, J.-S., Sohn, S.-O., Lee, D.-D., Huh, J.-S.: 'Sensing behaviors of polypyrrole sensor under humidity condition'. *Sensors and Actuators B: Chemical*, 2005, **108**, (1–2), pp. 389–392.

[55] Pirsa, S., Alizadeh, N.: 'Design and fabrication of gas sensor based on nanostructure conductive polypyrrole for determination of volatile organic solvents'. *Sensors and Actuators B: Chemical*, 2010, **147**, (2), pp. 461–466.

[56] Mazur, M.: 'Preparation of poly(3-methylthiophene) thin films by in situ polymerization'. *Thin Solid Films*, 2005, **472**, (1–2), pp. 1–4.

[57] Si, P., Mortensen, J., Komolov, A., Denborg, J., Møller, P.J.: 'Polymer coated quartz crystal microbalance sensors for detection of volatile organic compounds in gas mixtures'. *Analytica Chimica Acta*, 2007, **597**, (2), pp. 223–230.

[58] Li, B., Santhanam, S., Schultz, L., *et al.*: 'Inkjet printed chemical sensor array based on polythiophene conductive polymers'. *Sensors and Actuators B: Chemical*, 2007, **123**, (2), pp. 651–660.

[59] Gonçalves, V.C., Balogh, D.T.: 'Optical VOCs detection using poly(3-alkylthiophenes) with different side-chain lengths'. *Sensors and Actuators B: Chemical*, 2009, **142**, (1), pp. 55–60.

[60] Ma, X., Li, G., Xu, H., Wang, M., Chen, H.: 'Preparation of polythiophene composite film by in situ polymerization at room temperature and its gas response studies'. *Thin Solid Films*, 2006, **515**, pp. 2700–2704.

[61] Liao, F., Yin, S., Toney, M.F., Subramanian, V.: 'Physical discrimination of amine vapor mixtures using polythiophene gas sensor arrays'. *Sensors and Actuators B: Chemical*, 2010, **150**, (1), pp. 254–263.

[62] Chiang, C.K., Fincher, C.R., Park, Y.W., *et al.*: 'Electrical conductivity in doped polyacetylene'. *Physical Review Letters*, 1977, **39**, (17), pp. 1098–1101.

[63] McCall, R.P., Ginder, J.M., Leng, J.M., *et al.*: 'Spectroscopy and defect states in polyaniline'. *Physical Review B*, 1990, **41**, (8), pp. 5202–5213.

[64] Bhat, N.V., Gadre, a. P., Bambole, V. a.: 'Structural, mechanical, and electrical properties of electropolymerized polypyrrole composite films'. *Journal of Applied Polymer Science*, 2001, **80**, (13), pp. 2511–2517.

[65] Yoon, H., Chang, M., Jang, J.: 'Sensing behaviors of polypyrrole nanotubes prepared in reverse microemulsions: effects of transducer size and transduction mechanism'. *Journal of Physical Chemistry B*, 2006, **110**, (29), pp. 14074–14077.

[66] Saxena, V., Choudhury, S., Gadkari, S.C., Gupta, S.K., Yakhmi, J.V.: 'Room temperature operated ammonia gas sensor using polycarbazole Langmuir–Blodgett film'. *Sensors and Actuators B: Chemical*, 2005, **107**, (1), pp. 277–282.

[67] Xie, D., Jiang, Y., Pan, W., Li, D., Wu, Z., Li, Y.: 'Fabrication and characterization of polyaniline-based gas sensor by ultra-thin film technology'. *Sensors and Actuators B: Chemical*, 2002, **81**, (2–3), pp. 158–164.

[68] Jin, Z., Su, Y., Duan, Y.: 'Development of a polyaniline-based optical ammonia sensor'. *Sensors and Actuators B: Chemical*, 2001, **72**, (1), pp. 75–79.

[69] Kukla, A.L., Shirshov, Y.M., Piletsky, S.A.: 'Ammonia sensors based on sensitive polyaniline films'. *Sensors and Actuators B: Chemical*, 1996, **37**, (3), pp. 135–140.

[70] Hassan Amrani, M.E., Payne, P. a., Persaud, K.C.: 'Multi-frequency measurements of organic conducting polymers for sensing of gases and vapours'. *Sensors and Actuators B: Chemical*, 1996, **33**, (1–3), pp. 137–141.

[71] Athawale, A.A., Kulkarni, M. V.: 'Polyaniline and its substituted derivatives as sensor for aliphatic alcohols'. *Sensors and Actuators B: Chemical*, 2000, **67**, (1), pp. 173–177.

[72] Athawale, A.A., Bhagwat, S.V., Katre, P.P.: 'Nanocomposite of Pd-polyaniline as a selective methanol sensor'. *Sensors and Actuators B: Chemical*, 2006, **114**, (1), pp. 263–267.

[73] Dixit, V., Misra, S.C.K., Sharma, B.S.: 'Carbon monoxide sensitivity of vacuum deposited polyaniline semiconducting thin films'. *Sensors and Actuators B: Chemical*, 2005, **104**, (1), pp. 90–93.

[74] Blackwood, D., Josowicz, M.: 'Work function and spectroscopic studies of interactions between conducting polymers and organic vapors'. *Journal of Physical Chemistry A*, 1991, **95**, (14), pp. 493–502.

[75] Watcharaphalakorn, S., Ruangchuay, L., Chotpattananont, D., Sirivat, A., Schwank, J.: 'Polyaniline/polyimide blends as gas sensors and electrical conductivity response to CO–N_2 mixtures'. *Polymer International*, 2005, **54**, (8), pp. 1126–1133.

[76] Tan, C.K., Blackwood, D.J.: 'Interactions between polyaniline and methanol vapour'. *Sensors and Actuators B: Chemical*, 2000, **71**, (June), pp. 184–191.

[77] Fan, J.H., Wan, M.X., Zhu, D.B., Chang, B.H., Pan, Z.W., Xe, S.S.: 'Synthesis, characterizations, and physical properties of carbon nanotubes coated by conducting polypyrrole'. *Journal of Applied Polymer Science*, 1999, **74**, (11), pp. 2605–2610.

[78] Sahoo, N.G., Jung, Y.C., So, H.H., Cho, J.W.: 'Polypyrrole coated carbon nanotubes: Synthesis, characterization, and enhanced electrical properties'. *Synthetic Metals*, 2007, **157**, (8–9), pp. 374–379.

[79] Guo, H., Zhu, H., Lin, H., Zhang, J., Yu, L.: 'Synthesis and characterization of multi-walled carbon nanotube/polythiophene composites'. *Journal of Dispersion Science and Technology*, 2008, **29**, (5), pp. 706–710.

[80] Al-Mashat, L., Shin, K., Kalantar-Zadeh, K., *et al.*: 'Graphene/polyaniline nanocomposite for hydrogen sensing'. *Journal of Physical Chemistry C* 2010, **114**, (39), pp. 16168–16173.

[81] Zhang, T., Mubeen, S., Myung, N.V, Deshusses, M.A.: 'Recent progress in carbon nanotube-based gas sensors'. *Nanotechnology*, 2008, **19**, (33), p. 332001.

[82] Lonergan, M.C., Severin, E.J., Doleman, B.J., Beaber, S. a, Grubbs, R.H., Lewis, N.S.: 'Array-based vapor sensing using chemically sensitive, carbon black–polymer resistors'. *Chemistry of Materials*, 1996, **8**, (14), pp. 2298–2312.

[83] Janata, J., Josowicz, M.: 'Conducting polymers in electronic chemical sensors'. *Nature Materials*, 2003, **2**, (1), pp. 19–24.

[84] Ogura, K., Saino, T., Nakayama, M., Shiigi, H.: 'The humidity dependence of the electrical conductivity of a soluble polyaniline–poly (vinyl alcohol) composite film'. *Journal of Materials Chemistry*, 1997, **7**, (12), pp. 2363–2366.

[85] Dobay, R., Harsányi, G., Visy, C.: 'Conducting polymer based electrochemical sensors on thick film substrate'. *Electroanalysis*, 1999, **11**, (10–11), pp. 804–808.

[86] Musio, F., Amrani, M.E., Persaud, K.C.: 'High-frequency a.c. investigation of conducting polymer gas sensors'. *Sensors and Actuators B: Chemical*, 1995, **23**, (2–3), pp. 223–226.

[87] Musio, F., Ferrara, M.C.: 'Low frequency a.c. response of polypyrrole'. *Sensors and Actuators B: Chemical*, 1997, **41**, (1–3), pp. 97–103.

[88] Kriván, E., Visy, C., Dobay, R., Harsányi, G., Berkesi, O.: 'Irregular response of the polypyrrole films to H_2S'. *Electroanalysis*, 2000, **12**, (15), pp. 1195–1200.

[89] Hao, Q., Kulikov, V., Mirsky, V.M.: 'Investigation of contact and bulk resistance of conducting polymers by simultaneous two- and four-point technique'. *Sensors and Actuators B: Chemical*, 2003, **94**, (3), pp. 352–357.

[90] Li, G., Martinez, C., Janata, J., Smith, J.A., Josowicz, M., Semancik, S.: 'Effect of morphology on the response of polyaniline-based conductometric gas sensors: nanofibers vs. thin films'. *Electrochemical and Solid-State Letters*, 2004, **7**, (10), pp. H44–H47.

[91] Mabeck, J.T., Malliaras, G.G.: 'Chemical and biological sensors based on organic thin-film transistors'. *Analytical and Bioanalytical Chemistry*, 2006, **384**, (2), pp. 343–353.

[92] Covington, J. a., Gardner, J.W., Briand, D., De Rooij, N.F.: 'A polymer gate FET sensor array for detecting organic vapours'. *Sensors and Actuators B: Chemical*, 2001, **77**, (2001), pp. 155–162.

[93] Dimitrakopoulos, C.D., Mascaro, D.J.: 'Organic thin-film transistors: a review of recent advances'. *IBM Journal of Research and Development*, 2001, **45**, (1), pp. 11–27.

[94] Janata, J.: 'Electrochemical microsensors'. *Proceedings of the IEEE*, 2003, **91**, (6), pp. 864–869.

[95] Janata, J., Josowicz, M.: 'Chemical modulation of work function as a transduction mechanism for chemical sensors'. *Accounts of Chemical Research*, 1998, **31**, (5), pp. 241–248.

[96] Campos, M., Bulhões, L.O.S., Lindino, C. a.: 'Gas-sensitive characteristics of metal/semiconductor polymer Schottky device'. *Sensors and Actuators A: Physical*, 2000, **87**, (1–2), pp. 67–71.

[97] Nguyen, V.C., Potje-Kamloth, K.: 'Electrical and chemical sensing properties of doped polypyrrole/gold Schottky barrier diodes'. *Thin Solid Films*, 1999, **338**, (1–2), pp. 142–148.

[98] Pinto, N.J., Gonzalez, R., Johnson, A.T., MacDiarmid, A.G.: 'Electrospun hybrid organic/inorganic semiconductor Schottky nanodiode'. *Applied Physics Letters*, 2006, **89**, (3), pp. 4–7.

[99] Gallazzi, M.C., Tassoni, L., Bertarelli, C., Pioggia, G.: 'Poly(alkoxy-bithiophenes) sensors for organic vapours'. *Sensors and Actuators B: Chemical*, 2003, **88**, pp. 178–189.

[100] Vrkoslav, V., Jelínek, I., Broncová, G., Král, V., Dian, J.: 'Polypyrrole-functionalized porous silicon for gas sensing applications'. *Materials Science and Engineering: C*, 2006, **26**, (5–7), pp. 1072–1076.

[101] Koncki, R., Wolfbeis, O.S.: 'Composite films of Prussian blue and N-substituted polypyrroles – fabrication and application to optical determination of pH'. *Analytical Chemistry*, 1998, **70**, (13), pp. 2544–2550.

[102] Hagleitner, C., Hierlemann, a., Lange, D., *et al.*: 'Smart single-chip gas sensor microsystem'. *Nature*, 2001, **414**, pp. 293–296.

[103] Traversa, E.: 'Ceramic sensors for humidity detection: the state-of-the-art and future developments'. *Sensors and Actuators B: Chemical*, 1995, **23**, (2–3), pp. 135–156.

[104] Sakai, Y.: 'Humidity sensors using chemically modified polymeric materials'. *Sensors and Actuators B: Chemical*, 1993, **13**, (1–3), pp. 82–85.

[105] Nohria, R., Khillan, R.K., Su, Y., Dikshit, R., Lvov, Y., Varahramyan, K.: 'Humidity sensor based on ultrathin polyaniline film deposited using layer-by-layer nano-assembly'. *Sensors and Actuators B: Chemical*, 2006, **114**, (1), pp. 218–222.

[106] Penza, M., Milella, E., Anisimkin, V.I.: 'Monitoring of NH_3 gas by LB polypyrrole-based SAW sensor'. *Sensors and Actuators B: Chemical*, 1998, **47**, pp. 218–224.

[107] Jun, Z., Shurong, W., Mijuan, X., *et al.*: 'Polypyrrole-coated sno2 hollow spheres and their application for ammonia sensor'. *Journal of Physical Chemistry C*, 2009, **113**, (5), pp. 1662–1665.

[108] Tai, H., Jiang, Y., Xie, G., Yu, J., Chen, X., Ying, Z.: 'Influence of polymerization temperature on NH3 response of PANI/TiO_2 thin film gas sensor'. *Sensors and Actuators B: Chemical*, 2008, **129**, (1), pp. 319–326.

[109] Ram, M.K., Yavuz, O., Aldissi, M.: 'NO_2 gas sensing based on ordered ultrathin films of conducting polymer and its nanocomposite'. *Synthetic Metals*, 2005, **151**, (1), pp. 77–84.

[110] Nakagawa, K., Sadaoka, Y., Supriyatno, H., Kubo, A., Tsutsumi, C., Tabuchi, K.: 'Optochemical HCl gas detection using alkoxy substituted tetraphenylporphyrin-polymer composite films effects of alkoxy chain length on sensing characteristics'. *Sensors and Actuators B: Chemical*, 2001, **B76**, (1–3), pp. 42–46.

[111] Supriyatno, H., Nakagawa, K., Sadaoka, Y.: 'Optochemical HCl gas detection using mono-substituted tetraphenylporphin-polymer composite films'. *Sensors Actuators B Chem.*, 2001, **76**, (1–3), pp. 36–41.

[112] Li, D., Huang, J., Kaner, R.B.: 'Polyaniline nanofibers: a unique polymer nanostructure for versatile applications'. *Accounts of Chemical Research*, 2009, **42**, (1), pp. 135–145.

[113] Yang, J.S., Swager, T.M.: 'Fluorescent porous polymer films as TNT chemosensors: electronic and structural effects'. *Journal of the American Chemical Society*, 1998, **120**, (46), pp. 11864–11873.

[114] Toal, S.J., Trogler, W.C.: 'Polymer sensors for nitroaromatic explosives detection'. *Journal of Materials Chemistry*, 2006, **16**, (28), p. 2871.

[115] Kondratowicz, B., Narayanaswamy, R., Persaud, K.C.: 'An investigation into the use of electrochromic polymers in optical fibre gas sensors'. *Sensors and Actuators B: Chemical*, 2001, **74**, (1–3), pp. 138–144.

[116] Bai, H., Shi, G.: 'Gas sensors based on conducting polymers'. *Sensors*, 2007, **7**, (3), pp. 267–307.

[117] Hao, Q., Wang, X., Lu, L., Yang, X., Mirsky, V.M.: 'Electropolymerized multilayer conducting polymers with response to gaseous hydrogen chloride'. *Macromolecular Rapid Communications*, 2005, **26**, (13), pp. 1099–1103.

Chapter 7

Ion sensors based on organic transistors

Luis Gil Sanchez[1], *Laura Contat Rodrigo*[1] *and Eduardo Garcia-Breijo*[1]

Abstract

This chapter presents the ion sensors which convert the activity of an ion dissolved in a solution into an electrical potential. This type of sensors is well-known and they are widely used in analytical chemistry and biochemistry. The use of organic polymers as active part of the ion sensors has led to a new way of research in these sensors. The chapter is divided into two parts: pH sensors and general ion sensors. The two parts are further divided based on the type of organic semiconductor used in the device fabrication: P3MT, PANI, PEDOT, etc. As a result of the organic electronic applications into the field of organic sensors, a wide variety of devices have been developed, for instance, ISOFET, OFET, EGOFET. For this reason, some applications of these devices have been explained for each organic semiconductor.

7.1 Introduction

The discovery of conducting conjugated polymers in the 1970s paved the way for a new type of Electronics based on organic semiconductors instead of silicon, called Organic Electronics. Among the various organic electronic devices that have been developed, organic transistors stood out quickly. The transistor, from an electronic point of view, is on the basis of present Electronics industry. Transistor manufacturing has always been complex and has traditionally required expensive resources. The development of organic transistors has opened up a range of new possibilities owing to their relatively easy manufacture and the wide variety of organic semiconductors that the chemical industry can offer. One of the most promising applications of organic transistors is their use as sensors.

[1]Group of Electronic Development and Printed Sensors +(ged+ps), Instituto Interuniversitario de Reconocimiento Molecular y Desarrollo Tecnológico (IDM), Universitat Politècnica de València, Camí de Vera, s/n. 46022 Valencia, Spain

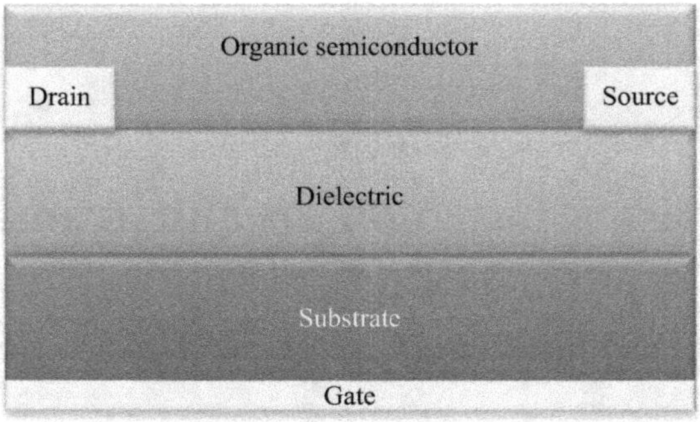

Figure 7.1 Organic thin-film transistor structure

In general, silicon-based transistors can be divided into bipolar transistors and field effect transistors. The first field-effect transistor (FET) was patented in 1930, and later, the first metal-oxide-semiconductor FET (MOSFET) was designed in 1960. In 1987, Koezuka reported the first organic FET (OFET) based on a thiophene polymer. Nowadays, most of the OFETs are designed like thin-film transistors (TFTs) which allow for the miniaturization of their design. These transistors are known as organic TFTs (OTFTs).

As its counterpart silicon-based TFT, the OTFT is a three-terminal electrical device with a similar architecture. The gate is connected to the substrate, which is usually thermally oxidized silicon. The active layer (semiconductor polymer or organic semiconductor) is deposited on the top of the Isolator layer that covers the gate. Source and drain are connected in opposite sides of the active layer, forming the channel (Figure 7.1).

When a source–drain voltage (V_{DS}) is applied to the transistor, an electric current (I_{DS}) flows between these electrodes by charge carrier transport. This current is modulated by the electric field applied to the organic semiconductor through the gate electrode (V_{GS}). In organic semiconductors, the charge carrier can be electrons (n-type), holes (no electrons) (p-type) or both (ambipolar).

Depending on the mechanism used to achieve the current modulation, the OTFTs can be classified into two main types [1]: OFETs and organic electrochemical transistors (OECTs).

1. *Organic field-effect transistors.* As TFTs, OFETs are composed of three electrodes (source, drain and gate), an organic semiconductor and a dielectric (between the gate and the active layer). Four basic OFET structures can be distinguished, depending on the position of the gate and source–drain electrodes: (i) bottom-contact/bottom-gate, (ii) bottom-contact/top-gate, (iii) top-contact/bottom-gate and (iv) top-contact/top-gate (Figure 7.2). In most of the OFETs used in sensor applications, the active semiconductor layer is exposed to the analyte that causes current modulation by charge doping or trapping. Therefore, OFET-based sensors are usually designed with bottom-contact/bottom-gate or top-contact/bottom-gate configurations.

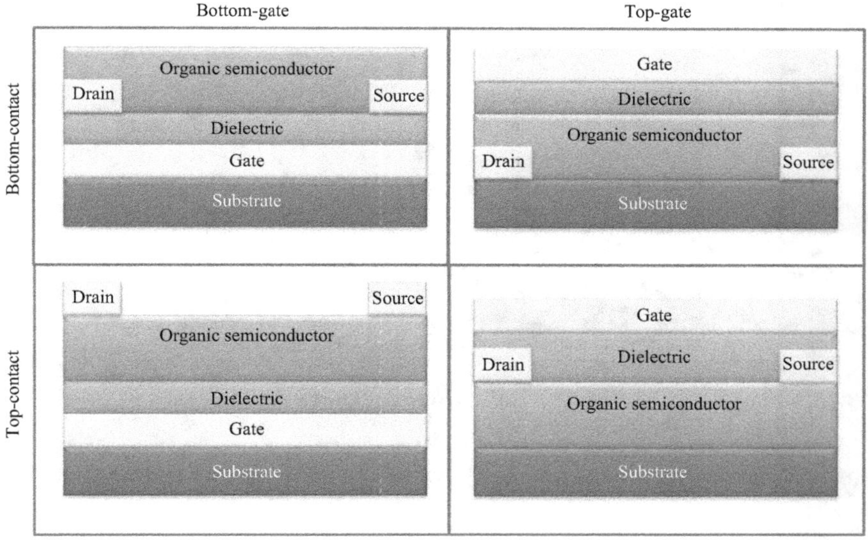

Figure 7.2 OFET standard architectures

2. *Organic electrochemical transistors.* In these transistors, an electrolyte medium instead of a dielectric is used between the gate and the organic semiconductor (Figure 7.3). OECTs can also be regarded as OFETs that operate by inducing a reduction or oxidation reaction, which leads to changes in the source–drain current, when a gate voltage is applied. The OECTs can work as three-terminal electrochemical cells in which the source, drain and gate electrodes play the roles of the working electrode, counter-electrode and reference electrode, respectively.

   Based on these two main types of OTFTs, other more specific types of OTFTs have arisen: electrolyte-gated OFETs (EGOFETs), ion-sensitive OFETs (ISOFETs), hygroscopic-insulator FETs (HIFETs) and organic charge-modulated FETs (OCMFETs).
3. *Electrolyte-gated OFETs (EGOFETs).* Like in conventional OECTs, in these transistors, the gate is separated from the organic semiconductor by an electrolyte (solid or liquid) (Figure 7.4). Due to the generation of an electric double layer at the electrolyte–organic semiconductor interface, EGOFETs exhibit a very high capacitance that allows for low-voltage operation.
4. *Hygroscopic-insulator FETs (HIFETs).* This device has the ability of taking advantage of ambient moisture by using a hygroscopic gate dielectric that allows ions to move freely through it. These ions can interact electrochemically and/or electrostatically with the semiconductor at the dielectric–semiconductor interface. This, in turn, results in current modulation through the semiconductor, by varying V_{GS}.

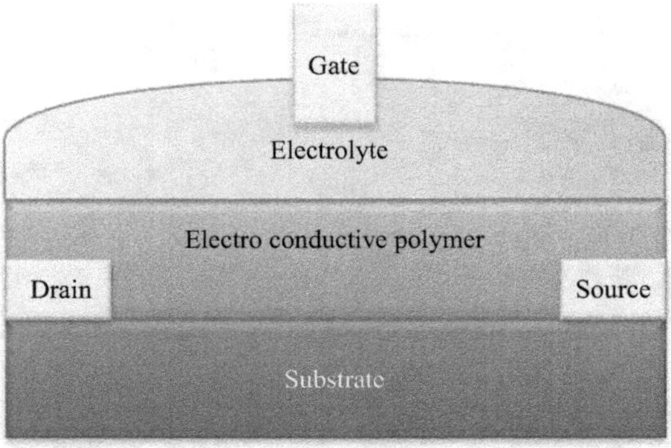

Figure 7.3 OECT architecture

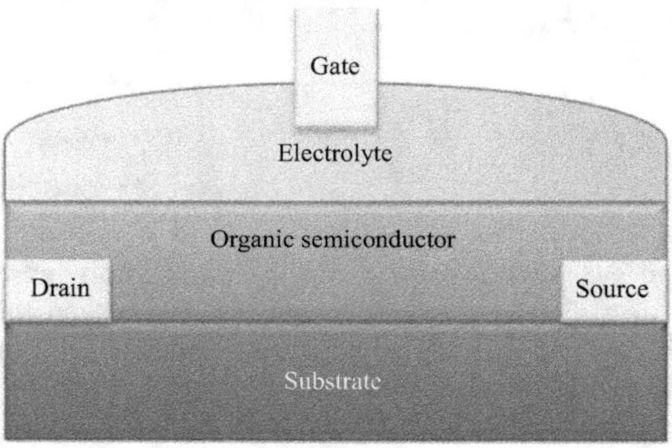

Figure 7.4 EGOFET architecture

5. *Ion-sensitive OFETs (ISOFETs).* The structure of these devices is very similar to that of EGOFETs, but ISOFETs have also a dielectric layer that isolates the electrolyte from the semiconductor (Figure 7.5). This structure allows overcoming the degradation of the semiconductor.
6. *Organic charge-modulated FETs (OCMFETs).* The architecture of OCMFETs is based on the ISOFET configuration. The difference lays on the use of two gate electrodes. One of them, called control capacitor or reference gate, bias the device. The other one, called floating gate, is connected to the sensing area, upon which the charge accumulates depending on the amount of the analyte (Figure 7.6).

All of these organic transistors can be used as ion sensors. This requires that a component of the device has sensitivity and selectivity towards a certain analyte. In recent years, OFETs have been investigated as ion sensors owing to the dependence of the conductivity of organic semiconductors on ion-doping. On the other hand,

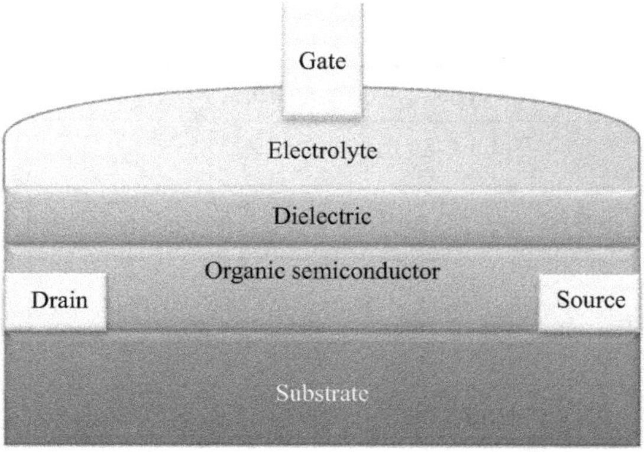

Figure 7.5 ISOFET architecture

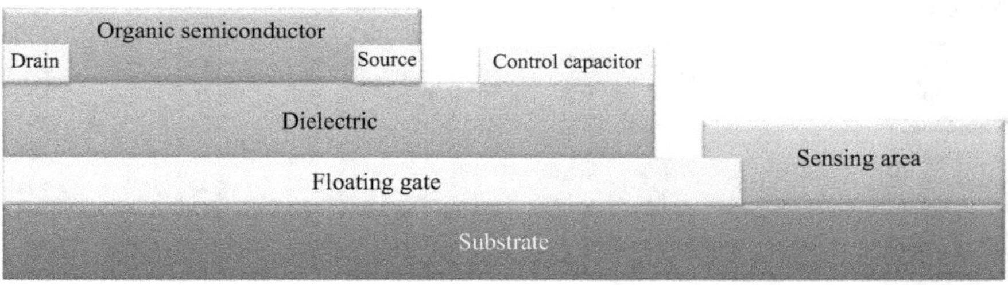

Figure 7.6 OCMFET architecture

OCETs have also attracted considerable attention for ion sensor applications since the 1980s. These sensors operate making use of the conductivity changes undergone by the organic semiconductor, either due to its oxidation/reduction or to changes in its configuration that occur in the presence of the analyte.

Sometimes chemical functionalization of one component of these organic electronic devices is additionally required for their use as ion sensors. Chemical tailoring in order to achieve specific properties can be applied not only to the semiconductor, but also to the metallic and insulating layers.

The following sections will be devoted to review the state-of-the-art of the main ion sensor applications of OTFTs. This review begins by describing pH sensors based on OTFTs. Next, OTFTs that have been successfully used for the detection of biologically and chemically relevant ions in aqueous medium are discussed.

7.2 pH sensors

Knowledge of the concentration of hydrogen ions is one of the most common and critical measurements in many applications. Many pH sensors have been developed along the Science history. As it can be expected, pH sensors were also one of the

Table 7.1 Summary of OTFT-based pH sensors

Semiconductor layer	Type	Range (pH)	Ref.	Year
P3MT	OECT	1–9	[2]	1985
P3MT	OECT	1–6.5	[3]	1986
PANI	OECT	1–6	[4]	1985
PANI	ISOFET	4–7	[5]	2015
PPy	OECT	3–11	[6]	1992
P3HT	OFET	2–10	[7]	2000
P3HT	OFET	2–10	[8]	2002
P3HT	OFET	4–10	[9,10]	2003
P3HT	OFET	6.5–7.5	[11]	2008
P3HT	ISOFET	6–8	[12]	2010
P3HT	ISOFET	3.5–5.5	[13]	2010
P3HT/PMMA	ISOFET	3.4–5.6	[14]	2012
P3HT	OFET	3.4–6.9	[15]	2013
P3HT	EGOFET	2.7–12	[16]	2014
Pentacene	OFET	4–10	[17]	2005
Pentacene	OFET	4–10	[18]	2006
Pentacene	OCMFET	2.5–5	[19]	2009
Pentacene	OCMFET	2.5–6	[20]	2009
Pentacene	ISOFET	2–12	[21]	2011
Pentacene	OFET	2–12	[22]	2014
PTAA	ISOFET	2–10	[23]	2010
DDFTTF	OFET	3–11	[24]	2008
DDFTTF	OFET	3–11	[25]	2009
αT6	EGOFET	2–7	[26]	2011
PEDOT:Tos/PSS	OEFT	1–3	[27]	2015

first types of sensors that were developed with organic transistors. Next, pH sensors based on OTFTs grouped by the type of the semiconductor layer are discussed in detail in chronological order (Table 7.1).

7.2.1 Poly(3-methylthiophene)

In 1985, only 8 years after the discovery of conducting polymers, the first OECT was used as a pH sensor. The OECT was based on poly(3-methylthiophene) (P3MT), a polymer with good stability in aqueous solutions, fast response and high transconductance. Over the past 20 years, P3MT and its derivatives have continued to be widely used in organic electronic devices.

Thackeray *et al.* [2] reported the optical and electrical properties of anodically grown P3MT (Figure 7.7). P3MT can be grown by the oxidation of 3-methylthiophene and deposited onto Au or Pt electrode surfaces. The devices displayed a high stability in aqueous electrolytes, in the pH range of 1–9.

Later, the same authors [3] fabricated a device based on platinized P3MT that was sensitive to H_2 and O_2 in aqueous electrolyte at room temperature and that was

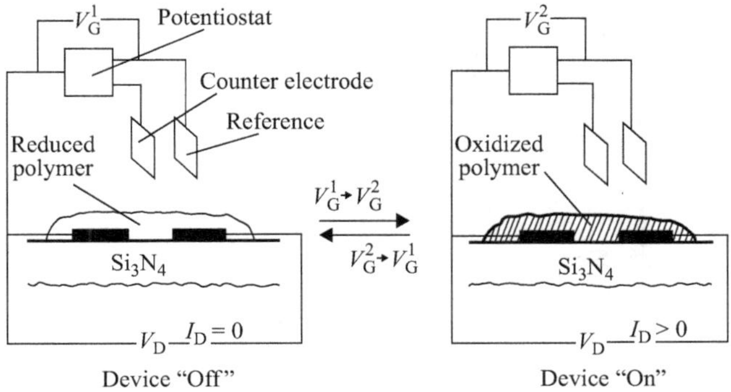

Figure 7.7 Poly(3-methylthiophene)-based microelectrochemical transistor. Reprinted with permission from [2]. Copyright (1985) American Chemical Society

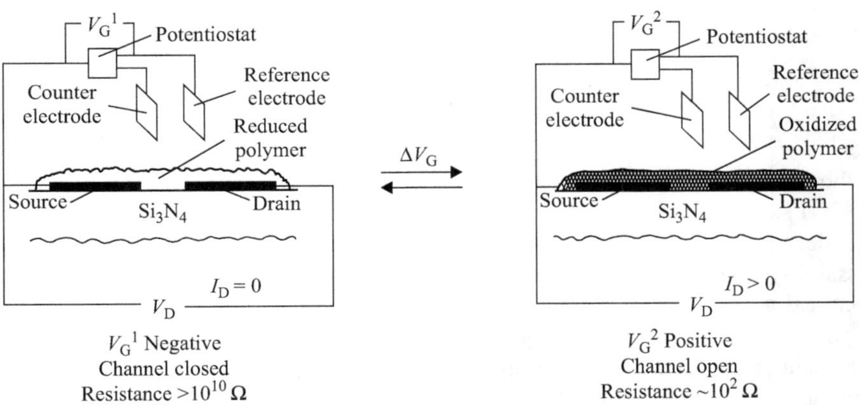

Figure 7.8 Representation of [poly(3-methylthiophene)/Pt]-based transistor. Reprinted with permission from [3]. Copyright (1986) American Chemical Society

able to detect the pH from 0 to 12 under aerobic conditions (Figure 7.8). Figure 7.9 shows the drain current (I_D) of a two-terminal P3MT/Pt device vs. pH for an O_2-saturated aqueous electrolyte. The oxidizing power of the O_2/H_2O couple is reduced by 59 mV/pH upon moving to more basic conditions. Thus, the drain current is reduced dramatically by shifting from pH 1 to pH 12. Above pH 12, the device shows no response to O_2-saturated solutions. It was concluded that P3MT/Pt is chemically sensitive in a way that allows for the demonstration of microelectrochemical devices responsive to pH at fixed O_2 pressure.

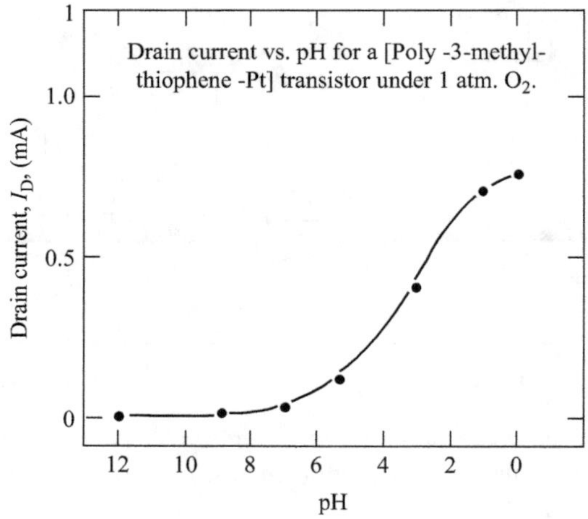

Figure 7.9 *I_D vs. pH for a [P3MT/Pt]-based device exposed to 1 atm of O_2/0.1M $NaClO_4$/H_2O at $V_D = 0.200$ V. Reprinted with permission from [3]. Copyright (1986) American Chemical Society*

7.2.2 Polyaniline

Although known since 1862, it was not until the early 1980s when polyaniline (PANI) started being used in organic electronic devices, due to its easy synthesis, excellent stability in electrolytes, high electrical conductivity, good solution processability and biocompatibility. Since then, polyaniline has been widely used in chemical and biological sensors.

In 1985, Paul *et al.* [4] reported a PANI-based OECT which was sensitive to pH in the range of 1–6. Polyaniline can be deposited onto an Au microelectrode array in a manner that allows for the demonstration of diode-like and transistor-like electronic devices (Figure 7.10). The large change in electronic conductivity of polyaniline with change in electrochemical potential is the basis of this electronic device. Results show that the response of a transistor-like device to pH changes can be achieved based on the pH dependence of the conductivity. From pH 1 to pH 6, the drain current (I_D) declines by a factor of $\approx 10^2$. However, return to pH 1 does not restore the original value of I_D. Apparently, chemically irreversible changes occur at high pH that preclude the use of PANI-based devices for pH sensing over a wide pH range.

Kaisti *et al.* [5] investigated the characteristics of a PANI-based field-effect transistor (OFET), including its pH sensitivity. They used a singular platform based on a commercial MOSFET which was functionalized with polyaniline by drop casting over the gate pad. pH sensitivity was shown to be 9.1 mV/pH in the range of 4–7.

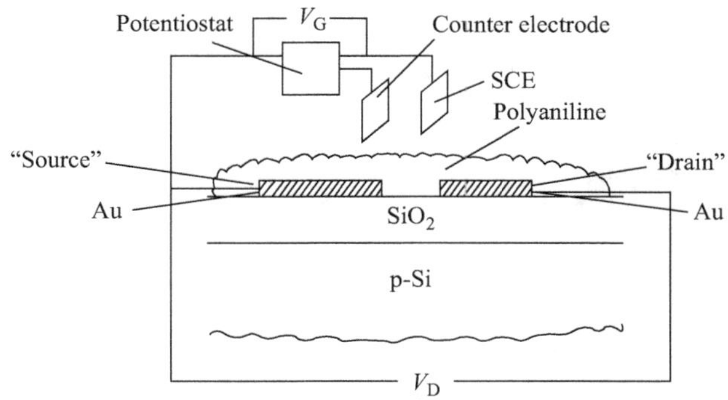

Figure 7.10 Configuration of polyaniline-based microelectrode device. Reprinted with permission from [4]. Copyright (1985) American Chemical Society

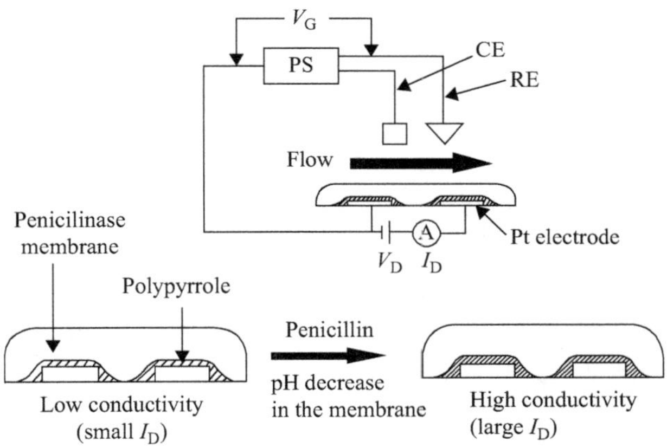

Figure 7.11 Principle of the penicillin sensor and the configuration of the electrochemical apparatus for in situ conductivity measurement. PS, potentiostat; CE, Pt counter electrode; RE, reference electrode (SCE). Reprinted with permission from [6]. Copyright (1992) American Chemical Society

7.2.3 Polypyrrole

Polypyrrole (PPy) is another conducting polymer that was discovered in 1968. However, the first highly conductive, stable and manageable PPy film under controlled electrochemical conditions was not reported until the 1980s.

Nishizawa *et al.* [6] found that the conductivity of PPy films is sensitive to pH when they were developing a penicillin sensor (Figure 7.11). This finding suggests that a PPy-coated electrode can serve as a pH-sensitive transducer.

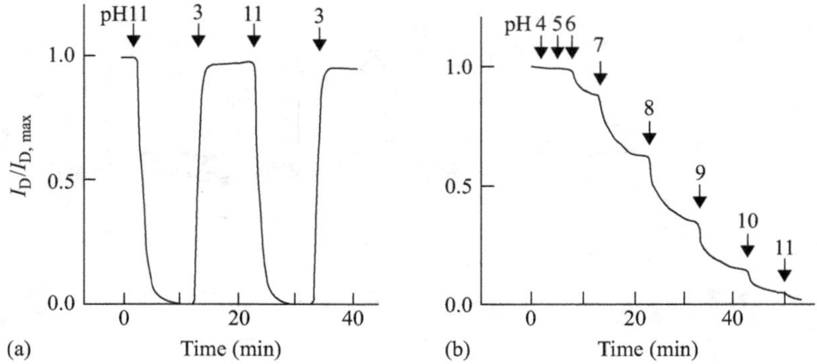

Figure 7.12 I_D responses of the polypyrrole-deposited microarray electrode to pH changes: (a) I_D responses to pHs changed repeatedly between pHs 11 and 3; (b) I_D responses to pHs changed stepwise from pHs 4 to 11. Reprinted with permission from [6]. Copyright (1992) American Chemical Society

The microarray electrode coated with PPy works as a pH-sensitive switching device, because the electrical conductivity of this polymer reversibly changes with the pH of the solution. Figure 7.12(a) depicts the reversible switching responses upon pH changes at $V_G = 0$ mV vs. reference electrode (SCE). This change in the I_D value demonstrates that the conductivity of polypyrrole at pH 11 is lower by two orders of magnitude than that at pH 3. The response times were within 3 min for the pH change from 3 to 11 (or 11 to 3) while a longer response time was observed for the pH changes in the intermediate region (Figure 7.12(b)).

7.2.4 Poly(3-hexylthiophene-2,5-diyl)

Thiophene-based conjugated polymers are other semiconductor polymers that have found their place in the development of OFETs. P3HT is the most widely studied thiophene-based polymer, due to its high carrier mobility. P3HT-based OFETs have showed mobilities higher than 0.1 cm^2/Vs and high on/off ratios ($>10^5$). P3HT is an excellent material for a large range of sensing applications based on organic transistors.

Bartic *et al.* reported the first OFET-based pH sensor at the beginning of this century [7]. They developed a P3HT-based OFET with proton sensitivity. The drain current in the P3HT channel is influenced by a chemistry-dependent potential, which is generated at the solution–dielectric interface. In this case, silicon nitride was used as dielectric, due to its proton sensitivity. The characterization of this device was performed with an Ag/AgCl reference electrode in different pH buffer solutions (pH 2–10). The sensitivity obtained was about 30 nA/pH.

The same authors described later an OFET-based pH sensor in which the semiconducting layer was not in direct contact with the aqueous solutions [8].

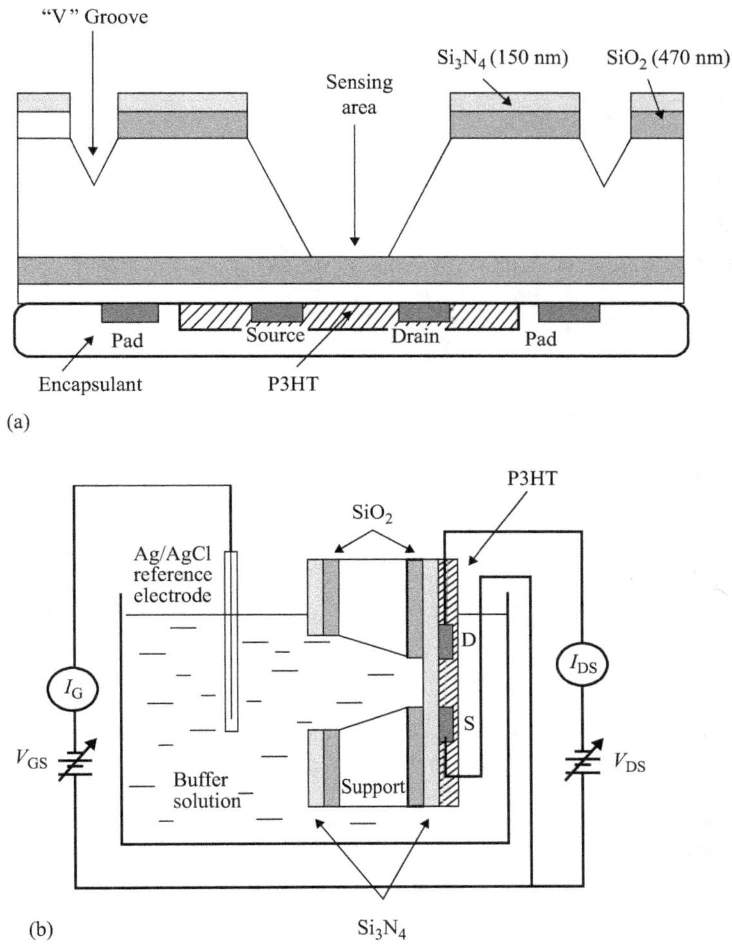

Figure 7.13 (a) Structure of the P3HT-based OFET with silicon nitride gate dielectric sensitive to protons. (b) Schematic representation of the measurement setup. Reprinted from [8]. Copyright (2002) with permission from Elsevier

The OFET was fabricated on a silicon wafer. Two layers of silicon oxide and silicon nitride were deposited on both sides of the silicon wafer (Figure 7.13(a)). Silicon nitride is used as the gate dielectric material sensitive to protons. At the same time, it acts as a stopping layer for the anisotropic KOH etching that is used to remove silicon from the sensitive area. In order to investigate the pH response, the OFET was immersed in different pH buffers (from 2 to 10) together with an Ag/AgCl reference electrode (Figure 7.13(b)). The OFET exhibited a linear pH dependence with a slope of 62 mV/pH (Figure 7.14).

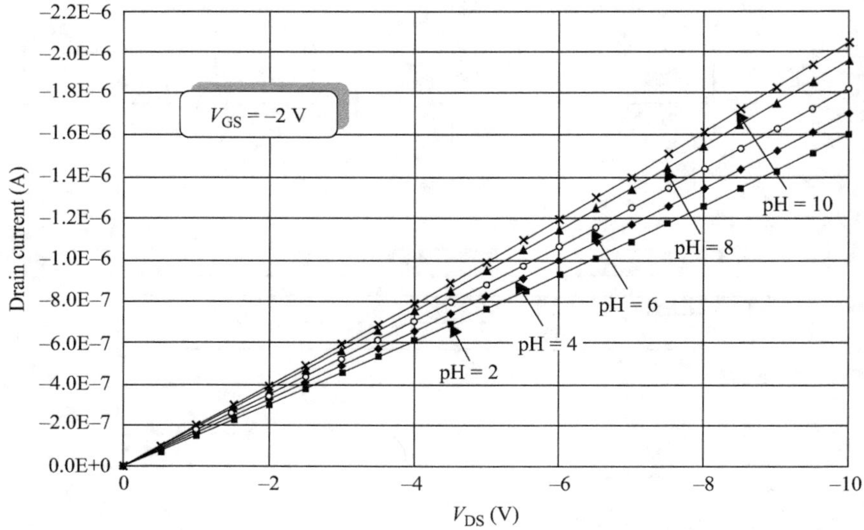

Figure 7.14 Output characteristics recorded in different pH buffers, using $V_{GS} = -2$ V. Reprinted from [8]. Copyright (2002) with permission from Elsevier

One of the problems associated with the first organic transistors was that the operating voltages required to achieve good performances were too high (~100 V). This was due to the mobility dependence on the charge accumulation in the channel. Since this charge is proportional to both the dielectric constant and the gate voltage, it was suggested that the use of high dielectric constant materials will allow for the necessary charge accumulation at lower voltages. In order to find a solution to this problem, Bartic *et al.* [9] fabricated OTFTs with tantalum pentoxide (Ta_2O_5) thin films using two different configurations (Figure 7.15). The "inverted" staggered configuration, in which the gate electrode is deposited first, is commonly used for organic transistors. The staggered configuration proves the perfect compatibility of the dielectric deposition method with plastic materials. It was found that the operation voltages are reduced to less than −3 V for both configurations.

These devices were later investigated as pH sensors [10], achieving a reproducible pH-dependent response when small voltages (<1 V) were applied (Figure 7.16). The pH response is also reflected in the output characteristics recorded at constant V_{GS} (Figure 7.17(a)). It was observed that a pH change of one unit leads to a drain current change of about 0.3 mA. The pH response is also revealed by the transfer characteristics recorded at constant V_{DS} (Figure 7.17(b)). A change in the drain current of about 0.3 mA /pH was observed.

Based on this OTFT structure, Ji *et al.* [11] used tantalum pentoxide (Ta_2O_5) deposited on the top of a P3HT layer to design an OFET-based pH sensor. The pH

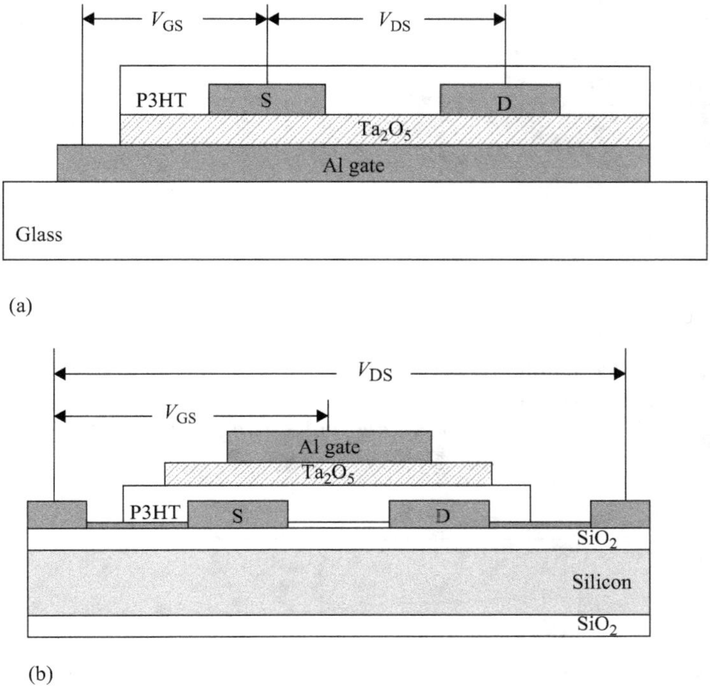

Figure 7.15 *(a) Inverted staggered configuration and (b) staggered configuration of P3HT-based OTFTs with tantalum pentoxide (Ta_2O_5) as gate dielectric. Reprinted from [9]. Copyright (2002) with permission from Elsevier*

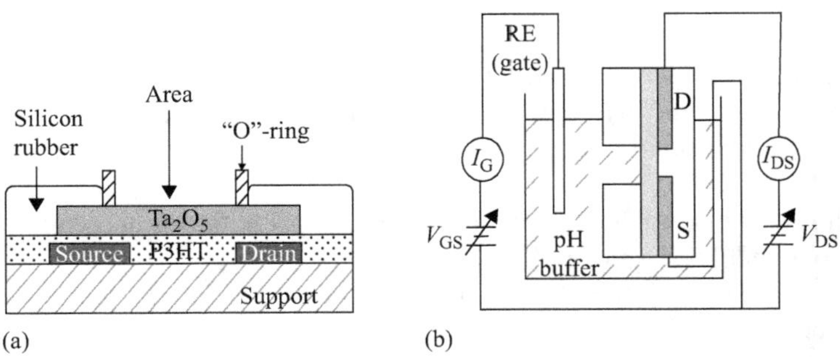

Figure 7.16 *(a) Schematic representation of the structure of the OTFT-based pH sensor with tantalum pentoxide (Ta_2O_5) and (b) scheme of the measurement setup. Reprinted with permission from [10]. Copyright [2003], AIP Publishing LLC*

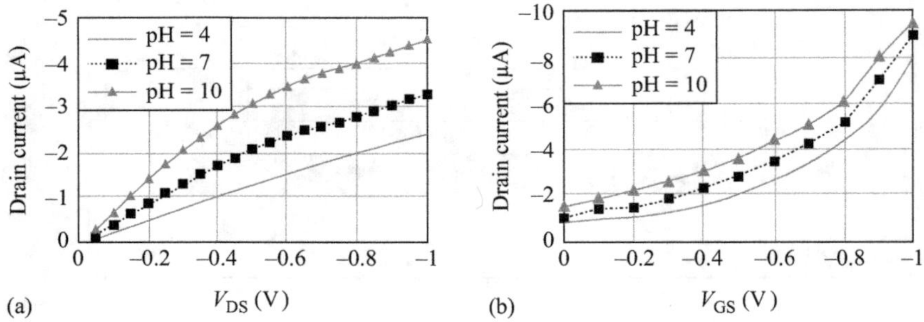

Figure 7.17 (a) Output characteristics of the Ta_2O_5/P3HT transducer recorded in buffer solutions with pH 4, 7, and 10 at $V_{GS} = -0.5$ V constant and (b) transfer characteristics recorded in the same pH buffers at $V_{DS} = -0.8$ V constant. Reprinted with permission from [10]. Copyright [2003], AIP Publishing LLC

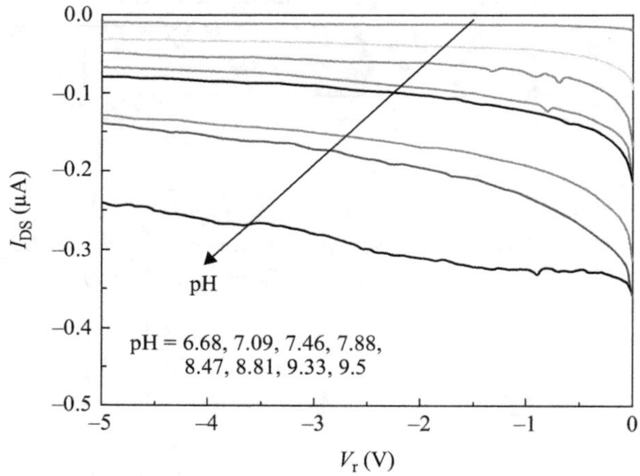

Figure 7.18 Drain current readout for pH sensor based on P3HT/Ta_2O_5 OFET. Reprinted with permission from [11]. Copyright [2003], AIP Publishing LLC

sensor showed a discernable variation of 0.071 μA/pH in the drain current at values between 6.5 and 7.5 (Figure 7.18).

Scarpa *et al.* [12] reported an ISOFET based on P3HT, for operation in aqueous electrolytes at low voltages. It was shown that this device displays a pH-dependent response (Figure 7.19). In the pH range from 6 to 8 (see the marked area within the figure), the dependence of the transistor current on the pH was found to be almost linear. In this pH range, the changes in the pH were greater than in the range beyond pH 8 and below pH 6, while adding the same amount of acid/base. This was attributed to the presence of HEPES (4-(2-hydroxyethyl)piperazine-1-ethanesulfonic acid) in the buffer solution.

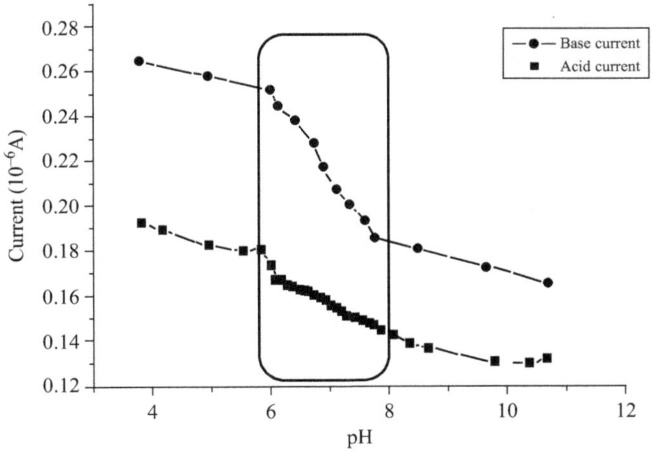

Figure 7.19 Transistor current behavior under various pH conditions. The transistor current was measured in DC mode with the drain-source voltage at 1 V and open gate. The enclosed area in the figure depicts the linear working range of the transistors from pH 6 to pH 8 [12]. Reprinted with permission from [12]. Copyright [2010], AIP Publishing LLC

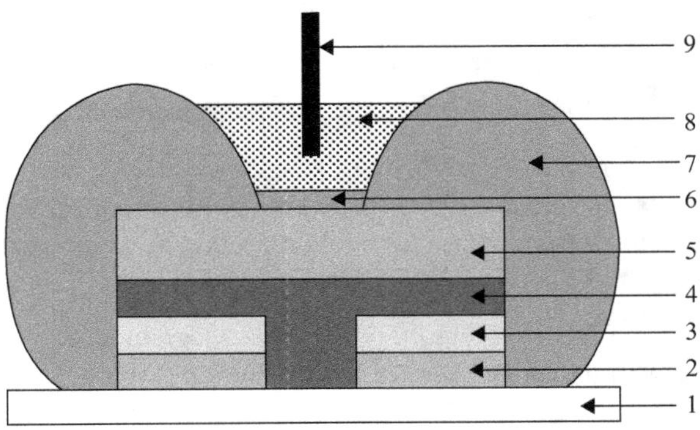

Figure 7.20 A typical structure for ISOFET: 1 – glass substrate, 2 – chromium, 3 – gold, 4 – P3HT, 5 – PMMA, 6 – LB film, 7 – epoxy resin, 8 – electrolyte solution, 9 – reference (gate) electrode. Reprinted from [13]. Copyright (2002) with permission from Elsevier

In order to increase the pH sensor sensitivity, Ritjareonwattu *et al.* [13] fabricated ISOFETs in which both the semiconducting layer and the gate insulator were organic materials. The devices used P3HT as the semiconductor and polymethylmethacrylate (PMMA) as the gate insulator. It was shown that the pH sensitivity of these ISOFETs is enhanced by coating the gate insulator with a few layers of a fatty acid

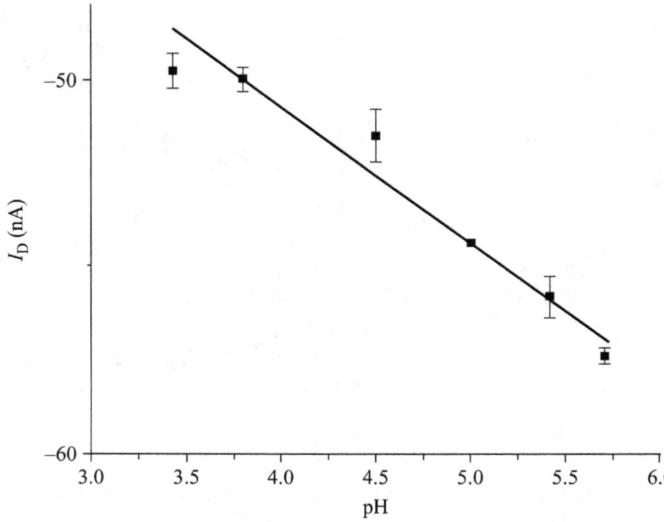

Figure 7.21 *pH response of the ISOFET with the LB layer ($V_g = -0.2$ V, $V_d = -0.2$ V). Reprinted from [13]. Copyright (2002) with permission from Elsevier*

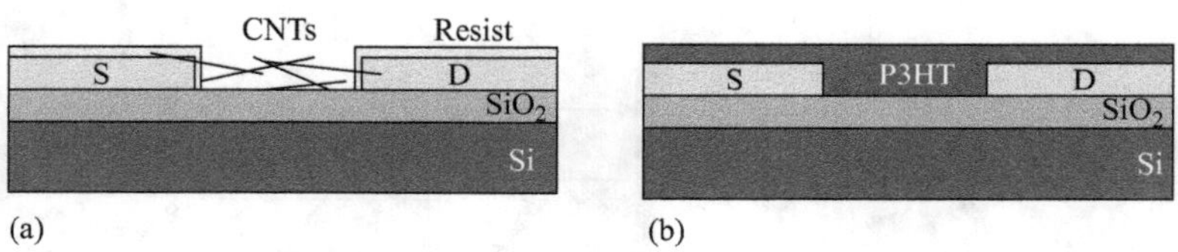

Figure 7.22 *Schematic layout of a CNT transistor (a) and a P3HT-FET (b). Reprinted from [15]. Copyright (2002) with permission from Elsevier*

Langmuir–Blodgett (LB) film (Figure 7.20). The magnitude of the drain current clearly decreased with increasing acidity of the solution (Figure 7.21).

Later, Ritjareonwattu *et al.* investigated the pH responses of ISOFETs coated either with fatty acid/valinomycin mixtures or pure valinomycin LB membranes [14]. Results showed that these devices display improved pH sensitivity compared to that of the ISOFETs without LB membrane.

Recently, Münzer *et al.* [15] compared the suitability of electrolyte-gated carbon nanotubes (CNT)-FETs and P3HT-FETs as pH sensors. Both semiconductors can be deposited on flexible substrates and used as cost-effective one-way disposable sensors (Figure 7.22). Especially, P3HT-FETs stand out due to the very fast fabrication process.

Even though the electrical stability of P3HT-devices is behind that of CNT transistors, both FETs perform well as pH sensors. The corresponding transfer

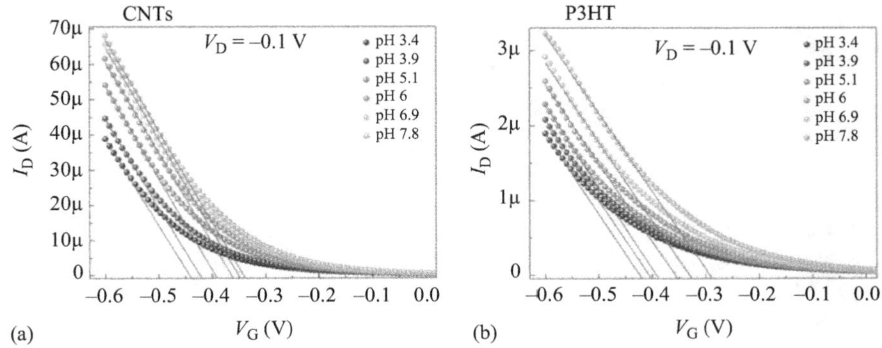

Figure 7.23 Transfer curves measured in buffers with different pH for CNTs (a) and P3HT (b). Reprinted from [15]. Copyright (2002) with permission from Elsevier

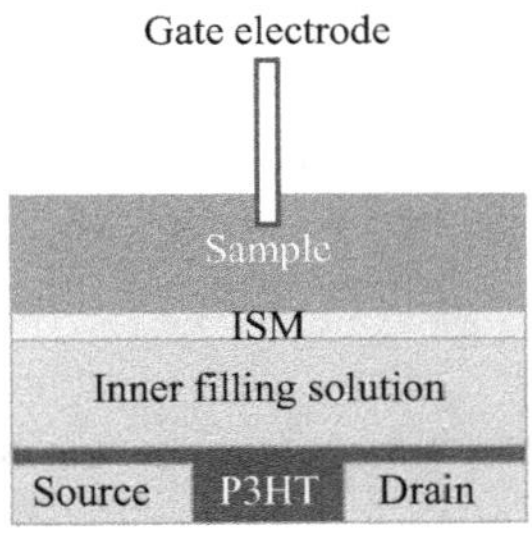

Figure 7.24 Structure of the EGOFET-based pH sensor with ion-selective membrane (ISM). Reprinted with permission from [16]. Copyright [2003], AIP Publishing LLC

curves are shown in Figure 7.23. The straight line fits in the linear regime of the curves indicate that sensing occurs via shifts in the threshold voltage. A threshold shift of almost 28 mV/pH was observed for P3HT-FETs. A shift of *ca.* 27 mV/pH was estimated for CNT-FETs, assuming a linear dependence for this device from pH 3.4 to pH 6.9.

Finally, Kofler *et al.* [16] reported a pH sensor that relies on the combination of an integrated EGOFET used as a transducer, and an ionophore-doped polymeric ion-selective membrane serving as a sensing element (Figure 7.24). The pH sensor exhibited a linear near-Nernstian response from pH 6 to pH 12 and a sub-Nernstian response from pH 2.7 to pH 6 (Figure 7.25(a)). A broad detection range (pH 2.7–12) was possible through a dynamic measurement protocol comprising a readjustment of the gate voltage, which ensures a stable device operation at a constant working point (Figure 7.25(b)). The effectiveness of this dynamic approach is confirmed by stability investigations. On the basis of this pH sensor

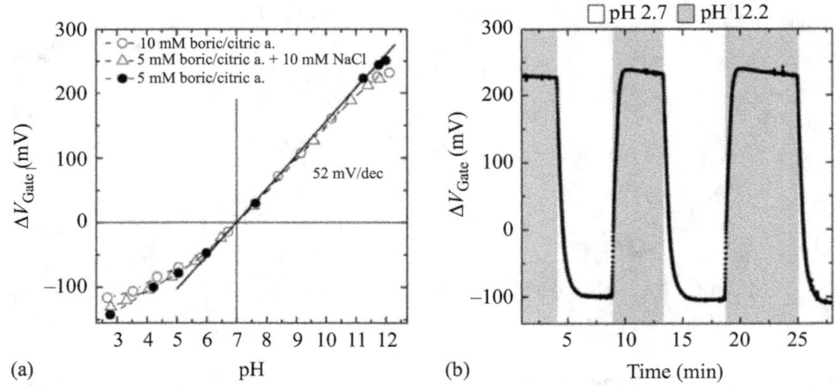

Figure 7.25 (a) pH value calibration curves for the pH sensor depicted in Figure 7.24: gate voltage change as a function of the pH value; (b) response curve upon a pH-value change from pH 2.7 to pH 12. Reprinted with permission from [16]. Copyright [2003], AIP Publishing LLC

concept, the importance of an appropriate gating electrolyte is highlighted, giving insights into the working mechanism of EGOFETs.

7.2.5 Pentacene

Pentacene is by far the most popular organic semiconductor for the fabrication of OFETs with high mobilities. In spite of its popularity, the use of pentacene is hindered by its high sensitivity to light and oxygen. That is why other pentacene derivatives have been developed to improve its functionality.

Loi *et al.* [17] reported an ion-sensitive field-effect transistor assembled on flexible plastic films. A thin Mylar^{TM} foil acts both as a substrate and as a gate dielectric. The device was shown to be sensitive to the pH of the electrolyte solution. A decrease of the current with basic solutions and an increase with acid solutions were observed. These results clearly indicate that ions accumulate at the interface between the insulating layer and the solution. When negative charge density is increased (as for basic solutions), holes accumulate in the channel with higher density than in the case of neutral or acid solutions.

Later, Manunza *et al.* proposed an alternative device consisting of a pentacene-based OFET that combines both switching and sensing functions [18]. A Mylar^{TM} sheet in a plastic frame was used as an insulator. Bottom-contact Au source and drain electrodes were patterned on one side of the dielectric. pH sensitivity was observed as proven by the I_D decrease in basic solutions and its increase in acid solutions.

A slightly different approach was described by Caboni *et al.* [19]. The device was based on an OCMFET configuration, which exhibited a good pH response between pH 2.5 and pH 5, with good reproducibility.

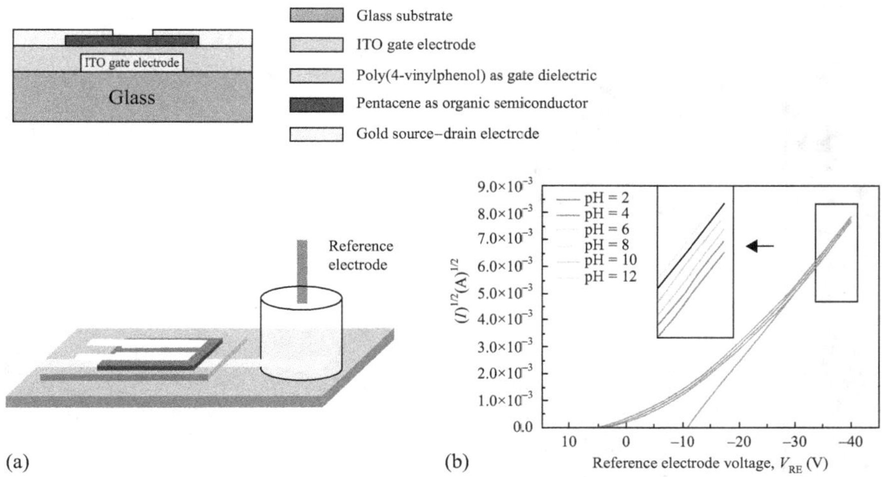

Figure 7.26 *(a) Schematic of the cross-section and top view of the ITO-extended gate OFET device and (b) ISD1/2 vs. V_{RE} with different pH solutions. Reprinted from [21] Copyright (2002) with permission from Elsevier*

In 2009, Caboni *et al.* developed a pH sensor based on a dual-gate OFET, using pentacene as the semiconducting layer [20]. The pH sensitivity was achieved by functionalization of the floating gate with thioamine groups which protonize proportionally to the concentration of H_3O^+ ions in the solution. A good pH response was observed between pH 2.5 and pH 6, with a maximum current recorded at pH 6.

Nguyen *et al.* [21] developed a transistor that uses indium tin oxide (ITO) as an extended-gate electrode, as well as an active layer for H^+ sensing. The threshold voltage of the fabricated ion-selective OFET was varied by the changes in the electrochemical potential at the ITO electrode surface, upon its exposure to buffer solutions with different pH. The sensor showed excellent linearity and a high sensitivity of 57–59 mV/pH in the pH range of 2–12 (Figure 7.26).

Yun *et al.* [22] fabricated an OFET-based pH sensor using a blend of TIPS-pentacene and PTAA, and a bilayer gate dielectric composed of CYTOP and Al_2O_3 layer (Figure 7.27(a)). The amplitude of ΔI_{DS} for basic solutions was found to be consistently smaller than the one found for acid solutions with equivalent ionic concentrations, that is, for pH 12 and 2, pH 4 and 10 or pH 6 and 8 (Figure 7.27(b)).

7.2.6 Polytriarylamine

Polytriarylamine (PTAA) is an amorphous p-type semiconductor polymer that is used to fabricate field-effect transistors. Spijkman *et al.* [23] used PTAA as the active layer in a dual-gate OFET-based pH sensor. The semiconductor layer

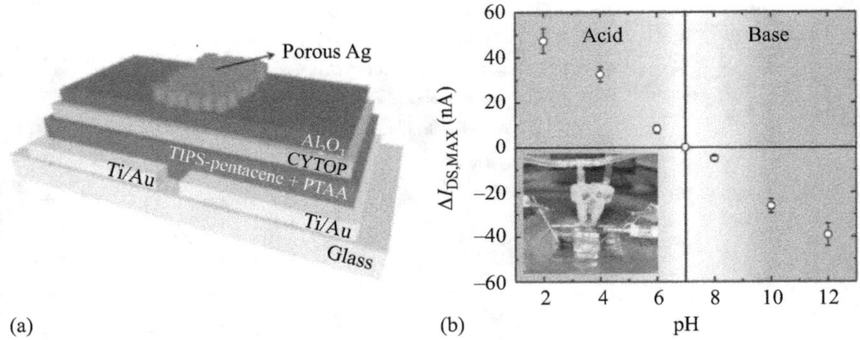

Figure 7.27 *(a) Schematic structure of the top-gate OFET sensors (b) $\Delta I_{DS,MAX}$ for all the pH values analyzed. Inset shows the microfluidic flow cell laminated on the top-gate OFET sensor. The error bars represent the standard deviation calculated over a set of five devices. Reprinted with permission from [22]. Copyright (2014) American Chemical Society*

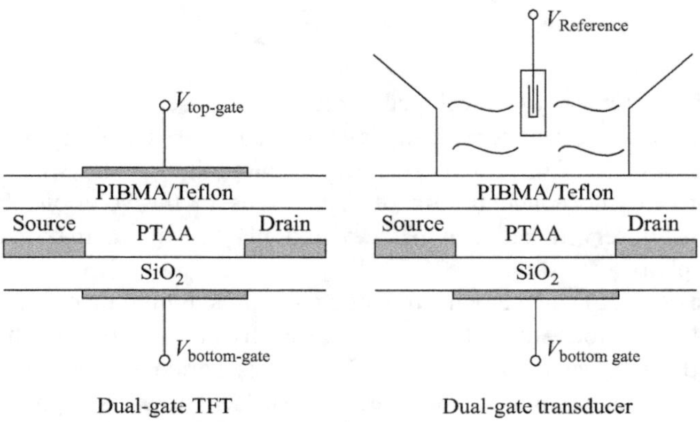

Figure 7.28 *Schematic representation of a dual-gate field-effect transistor and the corresponding transducer. Reprinted with permission from [23] Copyright (2010) John Wiley and Sons*

was protected by a hydrophobic barrier layer consisting of a stack of poly-isobutylmethacrylate and the Teflon derivative AF-1600. The corresponding dual-gate transducer is indeed a classical ISFET with an additional gate dielectric to enhance the sensitivity (Figure 7.28). The channel current shifted to a lower value when the pH was varied from 2 to 10.

It was found that the sensitivity of the dual-gate OFETs was strongly influenced by the capacitance ratio between the top and bottom gates. It was also observed that the threshold voltage shift of the dual-gate transducers depends

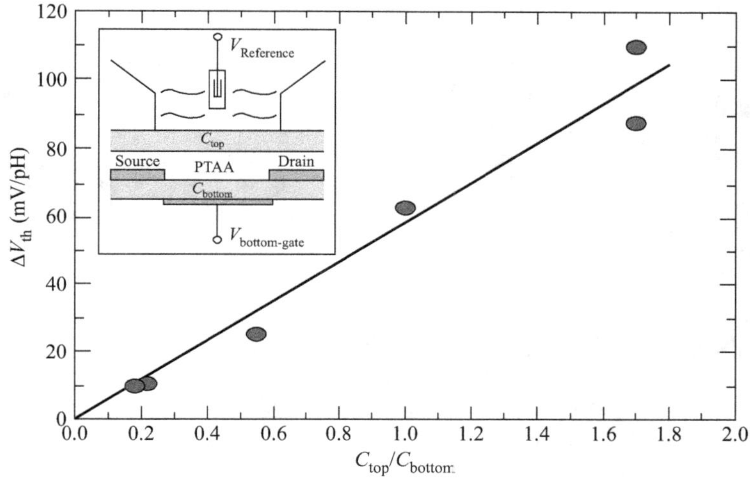

Figure 7.29 *Sensitivity of dual-gate transducers. Shift in the threshold voltage in mV per pH unit as a function of the capacitive coupling, C_{top}/C_{bottom}. The inset shows the experimental geometry. The solid line shows that the sensitivity increases linearly with capacitive coupling as $\Delta V_{th} = (C_{top}/C_{bottom}) \times 58$ mV per pH unit. Reprinted with permission from [23] Copyright (2010) John Wiley and Sons*

linearly on the pH: $\Delta V_{th} = (C_{top}/C_{bottom}) \times 58$ mV per pH unit (Figure 7.29), which is in good agreement with the value expected for an ideal Nernstian redox reaction (59 mV). An enhanced sensitivity was achieved by using a capacitive coupling larger than unity.

7.2.7 5,5′-Bis-(7-dodecyl-9H-fluoren-2-yl)-2,2′-bithiophene

The composite 5,5′-bis-(7-dodecyl-9*H*-fluoren-2-yl)-2,2′-bithiophene (DDFTTF) is a p-channel semiconductor for OFETs that exhibits high stability in aqueous environments.

Roberts *et al.* [24] reported low-operating voltage OTFTs with a cross-linked polymer gate dielectric (Figure 7.30). These devices displayed stable operation under aqueous conditions over $>10^4$ electrical cycles, using DDFTTF as a semiconductor.

The influence of pH on the drain current was examined in order to understand how water influences the electrical behavior of the organic semiconductor. In particular, an increase of the drain current was observed when the devices operate in water. It was concluded that this current change is directly related to the H_3O^+ concentration, since I_{DS} increases with decreasing pH (increasing $[H_3O^+]$) (Figure 7.31).

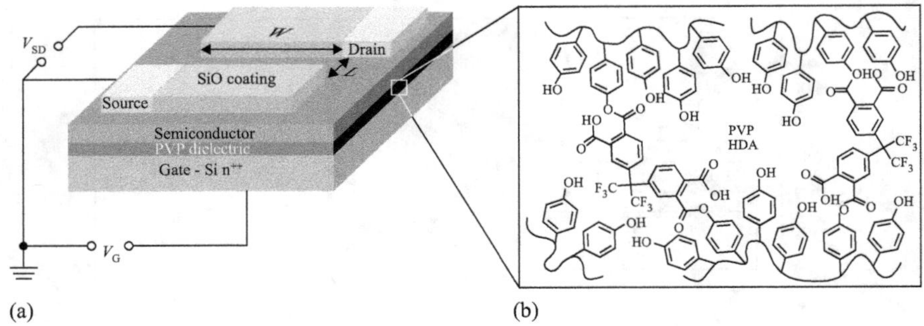

Figure 7.30 Cross-linked polymer gate insulator and its corresponding electrical properties. (a) Structure of the top-contact OTFT sensor. (b) Chemical structure of cross-linked PVP (poly(4-vinylphenol)) with HDA (diphthalic anhydride). [24] Copyright (2008) National Academy of Sciences, USA

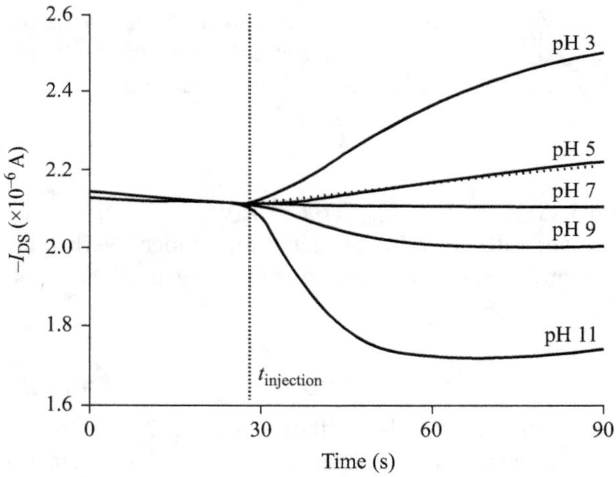

Figure 7.31 Drain current (I_{DS}) response to pH of a DDFTTF-based OTFT. I_{DS} response curves for the pH 5 solution were measured before (solid) and after (dashed) the exposure to other pH solutions [24]. Copyright (2008) National Academy of Sciences, USA

Later, Roberts *et al.* [25] continued investigating the influence of the pH on the performance of these OTFTs (Figure 7.32(a)). They observed a change in the drain current of less than 3% when exposing the polymer electrode to solutions of pH 3 (based on HCl) and pH 11 (based on NaOH) (Figure 7.32(b)).

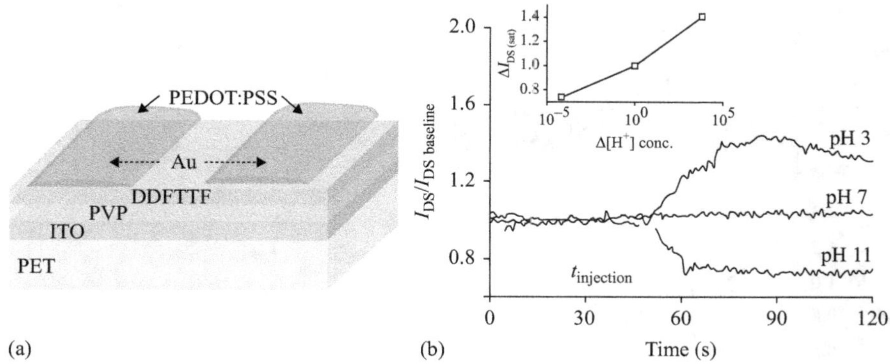

Figure 7.32 *(a) Structure of the DDFTTF-based OTFT used as pH sensor. (b) ΔI_{DS} with solution pH on ITO/PET substrates with PEDOT:PSS electrodes, (inset) ΔI_{DS} as a function of the change in $[H^+]$ concentration. Reprinted from [25]. Copyright (2002) with permission from Elsevier*

7.2.8 α-Sexithiophene

α-Sexithiophene (αT6) is a thiophene oligomer containing 12 double bonds that has been extensively studied.

Buth *et al.* [26] reported on electrolyte-gated OTFTs based on α6T, in which the organic semiconductor is in direct contact with the electrolyte. These transistors were shown to be sensitive to pH changes as a result of a pH-dependent surface charge, which modulates the threshold voltage. The resulting pH sensitivity was around 9 mV/pH vs. Ag/AgCl.

7.2.9 Poly(3,4-ethylenedioxythiophene)

Poly(3,4-ethylenedioxythiophene) (PEDOT), invented in 1988 by Bayer AG, is probably one of the best available conducting polymers in terms of conductivity, processability and stability.

Mitraka *et al.* [27] fabricated a new electrochemically active printable composite combining the properties of a conducting polymer (PEDOT:TOS) and a liquid metal alloy (GaInSn) (Figure 7.33). Because of the difference in the work function, when the oxide of the liquid metal dissolves in a specific acid pH range, it activates a spontaneous electron transfer from GaInSn to PEDOT. The pH of the medium triggers/activates the change in the electrochemical potential and produces a voltage up to 1.2 V in a conducting polymer electrochemical cell. The sensitivity of the pH threshold is further amplified by four orders of magnitude by coupling the PEDOT–GaInSn electrode to the gate of a fully printed PEDOT-based electrochemical transistor.

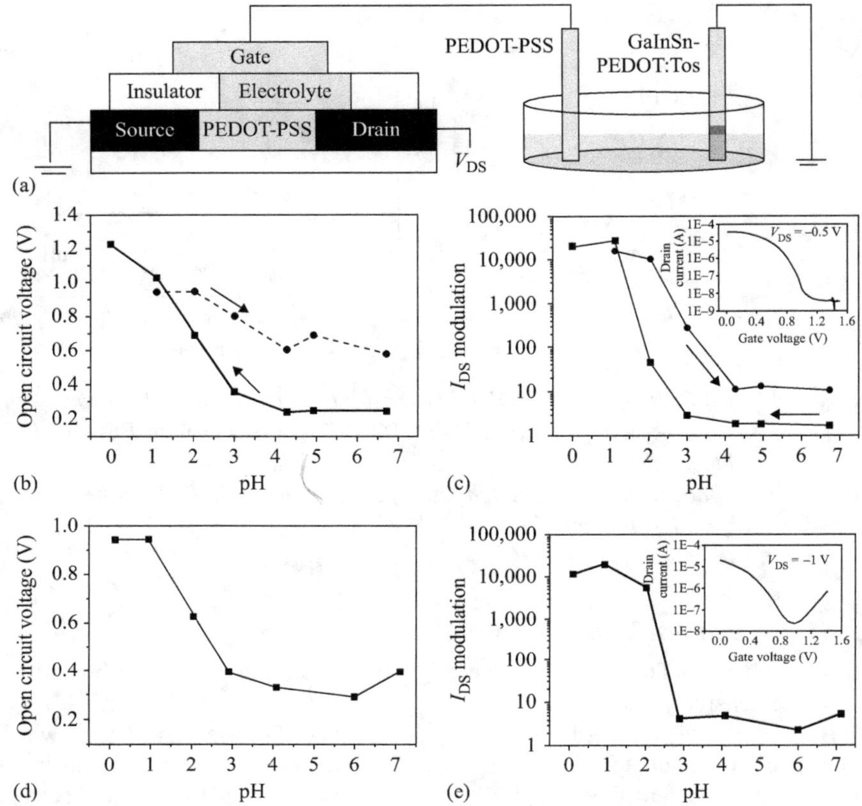

Figure 7.33 (a) Schematic description of the pH indicator composed of the electrochemical transistor (left) and the open circuit potential (V_{oc}) arrangement (right). The V_{oc} produced is applied to the transistor gate. (b) V_{oc} measured when the same set of electrodes was immersed sequentially from pH 7 to pH 0 and back. (c) Modulation of the OECT drain current (I_{DS}) when the V_{oc} of (b) is applied to the gate. The same transistor was used throughout the experiment. The OECT transfer curve is also included. (d) V_{oc} measured when a new device (device: one transistor and a set of electrodes) was used for each pH value. (e) I_{DS} modulation when the V_{oc} of (d) is applied to the gate of the OECT. The transfer curve presented here is a representative of one of the transistors used. Note that the transistor geometry used in (e) is different than in (c). [27] – Published by The Royal Society of Chemistry

7.3 Ion sensors in aqueous solution

In many applications, it is of interest to detect analytes in an aqueous medium. For example, ion sensing in aqueous solution is required in medical diagnosis, food monitoring, environmental monitoring and control of industrial processes, among others. Many organic semiconductors have proven to have good stability in aqueous media allowing for the development of OTFTs with stable operation in aqueous solution with relatively low working voltages. Therefore, the use of organic transistors for ion sensing has been widely developed these last few years. These include ion sensors based on both OECTs and OFETs. Next, ion sensors based on these electronic devices grouped by the organic semiconductor material are described in detail in chronological order (Table 7.2).

7.3.1 Poly(3-methylthiophene)

Thackeray *et al.* [2] first reported the use of OECTs as ion sensors in 1985. They fabricated microelectrochemical transistors based on P3MT microelectrodes for pH and ion sensing. They showed that these P3MT-based transistors can be turned "on" and "off" by an oxidant ($IrCl_6^{2-}$) or a reductant ($Fe(CN)_6^{4-}$), respectively, due to the conductivity change of P3MT induced by its reversible oxidation. Transistors based on this polymer could be used to detect the chemical oxidant $IrCl_6^{2-}$ up to $<10^{-15}$ mol.

Table 7.2 Summary of OTFTs-based ion sensors

Organic Semiconductor	Analyte	Transistor	Ref.	Year
P3MT	$IrCl_6^{2-}$, $Fe(CN)_6^{4-}$	OECT	[2]	1985
Polyaniline	$Fe(CN)_6^{3-}$, $Ru(NH_3)_6^{2+}$	OECT	[4]	1985
Polycarbazole	Cu^{2+}	ISMET	[28]	2000
PEDOT:PSS	Ca^{2+}	IS-OECT	[29]	2008
PEDOT:PSS	K^+, Ca^{2+}, Ag^+	IS-OECT	[30]	2009
PEDOT:PSS	H^+, K^+, Na^+, Ca^{2+}, Al^{3+}	OECT	[31]	2010
PEDOT:PSS	PAA, PAH	OECT	[32]	2012
PEDOT:PSS	K^+	IS-OECT	[33]	2014
P3HT	K^+	IS-OFET	[11]	2008
P3HT	Na^+, K^+, Ca^{2+}	OFET	[12]	2010
P3HT	K^+	IS-OFET	[13]	2012
P3HT	Na^+	IS-OFET	[34]	2013
P3HT	K^+, Ca^{2+}	IS-OFET	[35]	2014
PBI	Hg^{2+}	OFET	[36]	2013
Pentacene/P	Cu^{2+}, CN^-	Bilayer OFET	[37]	2014
pBTTT-C16	F^-	Extended-gate OFET	[38]	2015

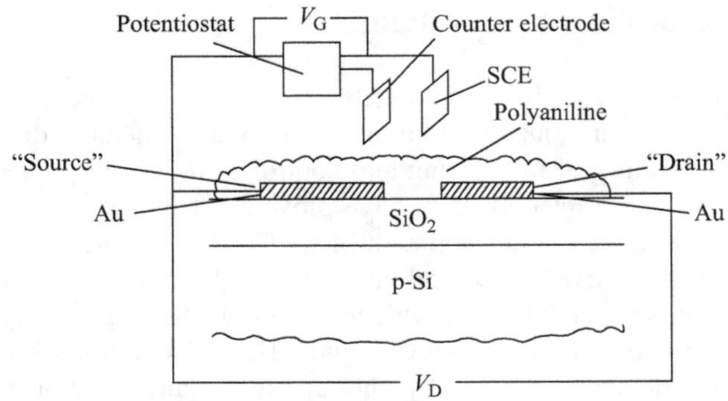

Figure 7.34 Configuration of polyaniline-based microelectrode device. Reprinted with permission from [4]. Copyright (1985) American Chemical Society

7.3.2 Polyaniline

Polyaniline (PANI) has been successfully used in OTFTs-based sensors since the 1980s. Paul *et al.* [4] reported in this decade the first PANI-based OECT as pH and ion sensor. They fabricated PANI-based microelectronic transistors that were successfully used for sensing redox reagents (Figure 7.34), owing to the large changes in the conductivity of polyaniline with change in its electrochemical potential. They reported changes in the conductivities of polyaniline upon immersion into $NaHSO_4$ solutions containing the oxidant $Fe(CN)_6^{3-}$ and the reductant $Ru(NH_3)_6^{2+}$. By the addition of these reagents, the PANI-based transistors could be switched to "on" and "off" states.

7.3.3 Polycarbazole

Polycarbazole (PCz) is an aromatic conjugated polymer with good conductivity that has been used later in organic electronic devices for sensing applications. In 2000, Saxena *et al.* [28] developed an ion-selective microelectrochemical transistor (ISMET) for the measurement of Cu^{2+} ions in aqueous medium. The OECT consisted of two platinum electrodes bridged with an anodically grown film of polycarbazole. These electrodes were used as the source and drain of the device, and an Ag/AgCl wire was used as the gate which was in contact with the PCz film through the electrolyte. It was observed that the conductivity of polycarbazole changed upon addition of Cu^{2+} ions. The device turned "on" by adding Cu^{2+} ions and exhibited a linear response (I_D vs. $\log[Cu^{2+}]$) in the concentration range of 2.5×10^{-6} M to 10^{-4} M. It was also shown that the addition of other metal ions (such as Cu^{+}, Ni^{2+}, Co^{2+}, Fe^{2+}, Fe^{3+}, Zn^{2+} and Pb^{2+}) did not give any response. Changes in the conductivity of polycarbazole were explained by changes in its conformation caused by the presence of the copper ions.

7.3.4 Poly(3,4-ethylenedioxythiophene): polystyrene sulfonate

Poly(3,4-ethylenedioxythiophene) doped with polysterene sulfonate (PEDOT:PSS) is a degenerately doped p-type organic semiconductor that is commercially available, and can be readily prepared by using conventional solution processing techniques. In addition, it exhibits high conductivity, excellent thermal stability and good stability in a wide pH range. Due to these properties, PEDOT:PSS has become one of the most widely used conducting polymers in OECTs. Upon the application of a positive gate voltage in PEDOT:PSS-based OECTs, cations from the electrolyte medium migrate into the conducting polymer. This, in turn, de-dopes PEDOT:PSS and leads to a decrease of the channel current. De-doping occurs by reduction of the highly conducting form of $PEDOT^+$ to the less conducting form of $PEDOT^0$, according to the following electrochemical reaction:

$$n(\mathrm{PEDOT^+{:}PSS^-}) + M^{n+} + ne^- \leftrightarrows \mathrm{PEDOT^0} + M^{n+}{:}n\mathrm{PSS^-}$$

where M^{n+} is a cation in the electrolyte medium, n is the number of charge of the cation and e^- is an electron from the source electrode. When the gate voltage is removed, cations migrate back into the electrolyte medium, and the original conductivity of the organic semiconductor is restored.

Breggen *et al.* [29] reported in 2008 the use of an ion-selective OECT (IS-OECT) for ion sensing. The device consisted of a PEDOT:PSS-based OECT with a Ca^{2+}-selective membrane coating the channel (Figure 7.35(a)). Both the channel and the gate electrode were immersed in an aqueous solution of $CaCl_2$ and KCl (0.1 M) background electrolyte. The response curves of this device for different Ca^{2+} concentrations showed the dependence of the drain current (I_{DS}) on the concentration at a constant gate voltage (Figure 7.35(b)).

Later, Mousavi *et al.* [30] fabricated K^+, Ca^{2+} and Ag^+ ion-selective OECTs by means of the incorporation of ionophore-based polymeric membranes to PEDOT:PSS-based transistors. Both K^+- and Ca^{2+}-selective OECTs were shown to be sensitive and selective to the target ions (K^+ and Ca^{2+}) in the concentration range of 10^{-1}–10^{-4} M in the presence of an interfering ion. Ag^+-selective OECTs exhibited a different response which was attributed to the oxidation of the PEDOT: PSS polymer by the silver ions. Oxidation occurred spontaneously in Ag^+ solutions also in the absence of the silver ionophore. These OECTs showed selectivity to Ag^+ ions in the 10^{-1}–10^{-5} M concentration range in the presence of an interfering ion, with and without applying any ionophore membrane on the channel.

The ion-sensitive properties of OECTs based on PEDOT:PSS have been studied by Lin *et al.* [31] in aqueous solutions of H^+, K^+, Na^+, Ca^{2+} and Al^{3+} with different concentrations. They found that the transfer curves ($I_{DS} - V_G$) of these OECTs shift to lower gate voltage horizontally with the increase of the concentration of cations in the electrolyte (Figure 7.36(a)). Furthermore, all these transfer curves could be merged into a universal curve, by shifting them horizontally to the right. They also showed that the relative change of the gate voltage $|\Delta V_G|$ exhibits a logarithmic dependence on the concentration of these cations (Figure 7.36(b)).

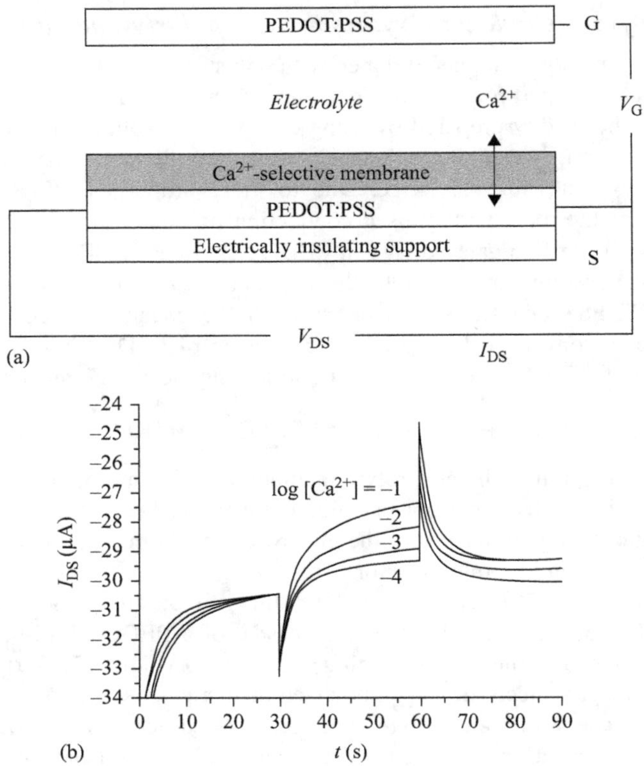

Figure 7.35 (a) Description of the Ca^{2+}-selective organic electrochemical transistor. (b) Response curves for different Ca^{2+} concentrations (at a constant background electrolyte of 0.1 M KCl) ($V_G = 0.15V$). Reprinted with permission from [29]. Copyright (2008) Springer

These authors also demonstrated in this work that the gate electrode of the PEDOT:PSS-based OECTs plays an important role on their ion-sensitive properties. Devices with the Ag/AgCl gate electrode followed Nernstian relationships between the shift of the gate voltage and the concentration of cations. In contrast, the ion sensitivity of OECTs with metal gate electrodes (Pt and Au) was found to be higher than that given by the Nernst equation, which was attributed to the different properties of the interface between the metal gate electrode and the electrolyte.

Toccoli *et al.* [32] successfully used PEDOT:PSS-based OECTs for the detection of polyelectrolyte polymeric shells coating gold nanoparticles (NP) for drug delivering. Measurements on the sensitivity and detection limit were performed comparing the response of these OECTs in a $CaCl_2$ saline solution containing either two polymer salts (poly(acrylic acid) sodium salt (PAA) and poly (allylamine hydrochloride) chlorine salt (PAH)), or gold NPs coated with one of these two polymers. They found a different response of the OECTs to PAA and

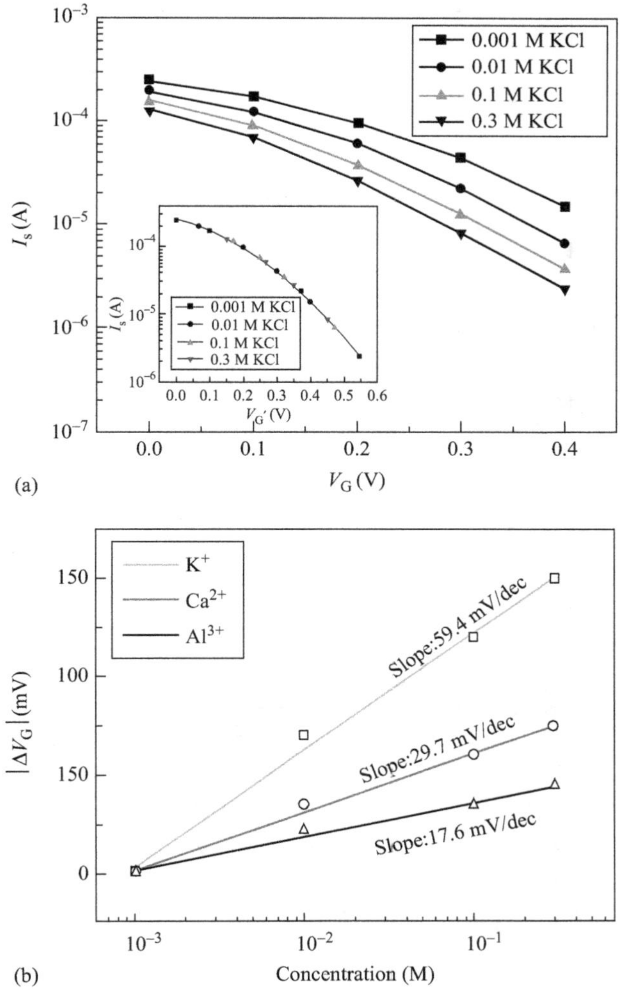

Figure 7.36 *(a) Transfer characteristics (I_S—V_G, V_{DS} = −0.1 V) of an OECT measured in KCl solutions with different concentrations. Inset shows transfer characteristics of the OECT where the curves are scaled in the x-axis to yield a universal curve. (b) Dependence of |ΔV_G| on the concentration of cations in the aqueous solutions of KCl, $Ca(NO_3)_2$ and $Al_2(SO_4)_3$. Reprinted with permission from [34]. Copyright (2010) American Chemical Society*

PAH saline solutions, due to their different interactions with the ions present in the electrolyte. The negatively charged polymer (PAA) decreased the OECTs modulation, whereas modulation did not change in the presence of the positively charged polymer (PAH). The correlation of the response of the OECTs to the

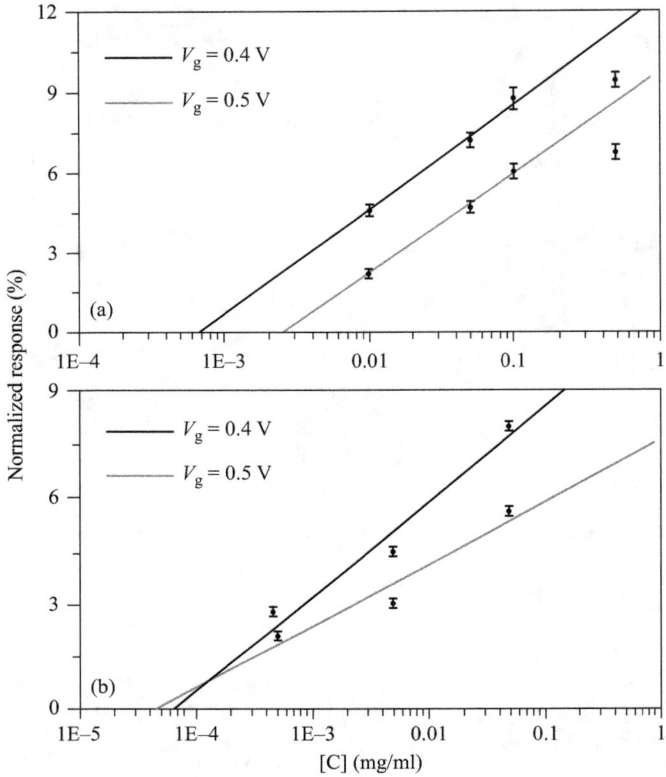

Figure 7.37 Normalized response of the PEDOT:PSS OECT at different concentrations using gate potentials of 0.4 and 0.5 V: (a) PAA and (b) gold NPs covered with a PAA shell. Reprinted from [32]. Copyright (2012) with permission from Elsevier

concentration of polymer and that of the gold NPs with the polymeric shell was found to be linear (Figure 7.37). Although the sensitivities obtained for the gold NPs and the polymer alone were similar, the detection limit was lower for the NPs, which was attributed to the large number of polymeric chains attached to them.

More recently, Sessolo *et al.* [33] have demonstrated the use of an all-solid-state ion-selective OECT (IS-OECT) for sensing of potassium ions with low detection limit. The IS-OECT was prepared by placing a K^+-selective membrane on the top of a hydrogel electrolyte in a PEDOT:PSS-based OECT (Figure 7.38). They obtained potassium selective OECTs with a sensitivity of 47 μA dec^{-1} in the concentration range of 10^{-4}–10^{-1} M and a detection limit that approaches 10^{-5}–10^{-6} M. The sensitivity of these OECTs to K^+ ions was also confirmed in the presence of Na^+ interfering ions (Figure 7.39(a)). It was also found that, for certain working regimes,

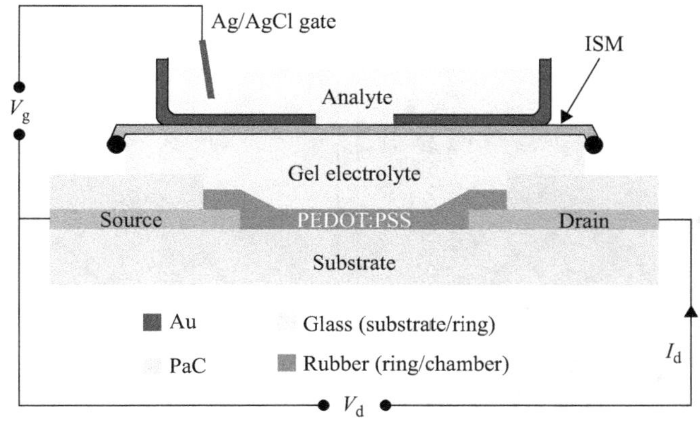

Figure 7.38 Cross-section schematic and wiring of the all-solid-state IS-OECT. Reprinted with permission from [33]. Copyright (2014) John Wiley and Sons

the current modulation of these devices is increased to the maximum transconductance (Figure 7.39(b)).

7.3.5 Poly 3-hexylthiophene

P3HT is an excellent candidate for its use in organic electronic devices, due to its high carrier mobility and good solution processability. P3HT-based OFETs were first successfully used as pH sensors and, later, as ion sensors. In 2008, Ji *et al.* [11] designed an ion-sensitive OFET (IS-OFET) for potassium sensing with high specificity. The IS-OFET was fabricated on a polyethylene naphtalene substrate with bottom contact architecture of thin film transistor with gold source–drain electrodes (Figure 7.40(a)). P3HT was used as the semiconductor and tantalum pentoxide (Ta_2O_5) as the gate dielectric. A valinomycin ionophore monoloyer, with high selectivity for K^+ ions, was formed over the gate electrode. The drain current measurements showed that this device is sensitive to K^+ ions in the range of low concentrations (33–50 mM) (Figure 7.40(b)) and has a high signal-to-noise ratio that approaches 6.5–8.5.

Scarpa *et al.* [12] fabricated P3HT-based OFETs that can operate at low voltages as ion sensors in liquid electrolytes. The design of the OFETs was based on a bottom-gate, bottom-contact configuration with boron-doped silicon wafer (p^+-Si) as substrate and gate. A thick silicon dioxide (SiO_2) film was formed on the top of the p^+-Si layer. P3HT was used as a semiconductor polymer and titanium/gold was used for the contact electrodes. The ion sensitivity of these OFETs was measured in Na^+, K^+ and Ca^{2+} solutions with concentration varying from 10^{-2} to 10^{-5}M. Results showed that the transistors are sensitive to these cations. It was also found that the change in the drain–source current was the highest for the calcium ions.

Ritjareonwattu *et al.* [13] reported the use of IS-OFETs based on P3HT for the detection of K^+ ions with top-gate bottom-contact architecture. The IS-OFETs were

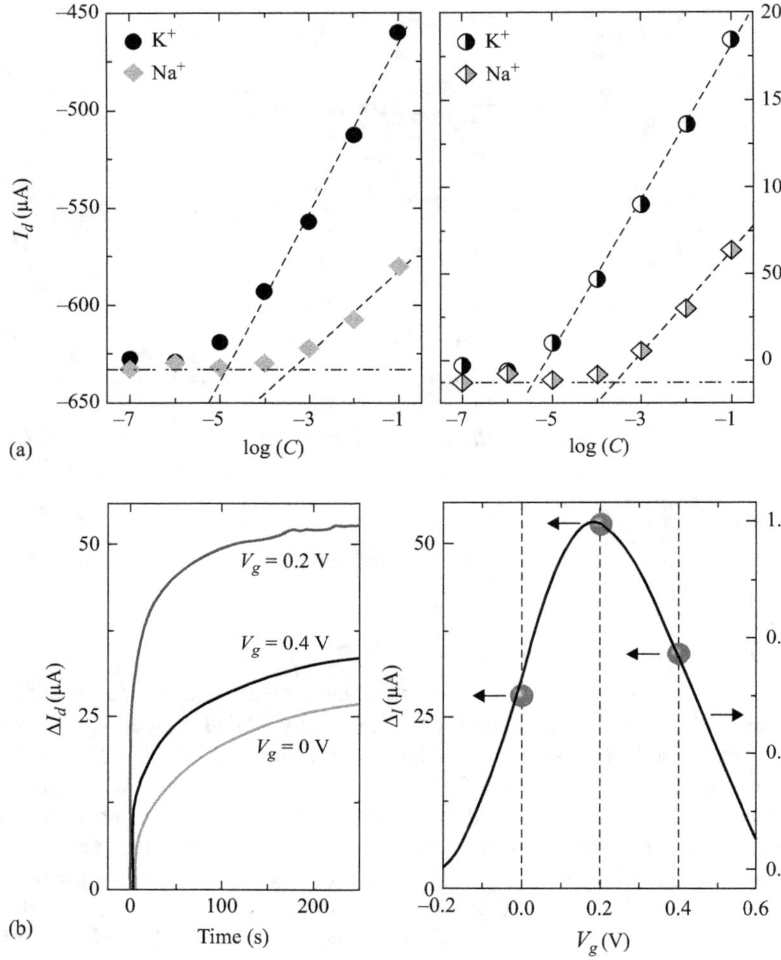

Figure 7.39 (a) Measured current response (left) and calculated membrane voltage (right) of the IS-OECTs in potassium and sodium solutions. (b) Drain current modulation to a sudden change in potassium concentration from 10^{-4} to 10^{-3} M, for increasing applied gate voltages (left). Current modulation at different gate voltages (right axis) superposed to the transconductance curve (left axis) of the device measured at $[K^+] = 10^{-3}$ M (right). Reprinted with permission from [33]. Copyright (2014) John Wiley and Sons

fabricated using chromium and gold for the source and drain electrodes, PMMA for the gate dielectric and a standard Ag/AgCl electrode for the gate. A LB membrane was deposited on the top of the PMMA layer. This consisted either on a pure fatty acid (arachidic acid), pure ionophore valinomycin (K^+-selective) or a mixture of the

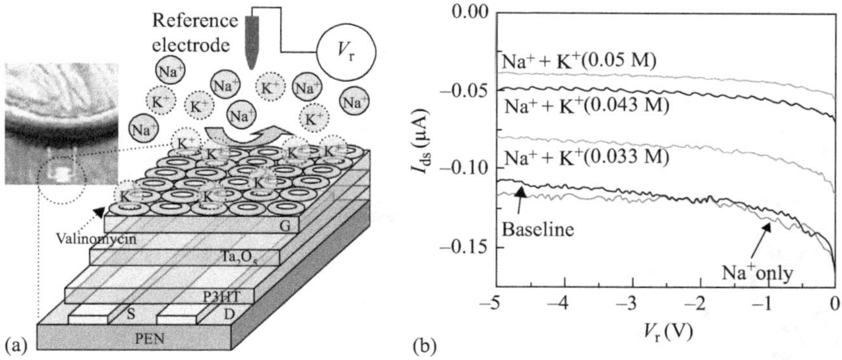

Figure 7.40 (a) Scaled up schematics of the ISFET for K^+ sensing where valinomycin monolayer is mounted on the gate (G) followed by field effect transducer element consisting of Ta_2O_5, P3HT, source (S), and drain (D) on PEN flexible substrate. (b) Drain current readout for potassium ion sensor. Reprinted with permission from [11]. Copyright (2008) AIP Publishing LLC

fatty acid and the ionophore. The IS-OFETs with pure fatty acid membrane only displayed a small sensitivity to potassium ions. Although the sensitivity to K^+ ions was enhanced when valinomycin was added to the fatty acid LB membrane, no selectivity was observed over Na^+ ions. It was suggested that the addition of the ionophore causes disruption of the fatty acid matrix in the membrane, leading to unselective interactions of the K^+ and Na^+ ions with the carboxylic acid groups. Selectivity to potassium ions was only achieved when a pure valinomycin LB membrane was used, as a consequence of the formation of valinomycin-K^+ complexes.

Later, Schmoltner *et al.* [34] described the modification of an EGOFET with an ion-selective polymeric membrane for the reversible and selective detection of sodium ions. The device was designed with bottom-contact architecture, using P3HT as organic semiconductor and gold source–drain electrodes (Figure 7.41). P3HT was in direct contact with an electrolyte (NaCl solution). A polyvinylchloride (PVC) Na^+-selective membrane (ISM) separated the electrolyte from the analyte, where an Ag/AgCl electrode was immersed. The IS-OFET was mounted in a flow cell. The device displayed a Nernstian response in the Na^+ concentration range of 10^{-6} to 10^{-1} M. It was also demonstrated that the flow cell configuration allows for a reversible response. Selectivity of this sensor over interfering K^+ ions was also successfully tested.

Recently, Melzer *et al.* [35] also used EGOFETs with different polymeric ion-selective membranes to detect K^+ and Ca^{2+} ions. The P3HT-based IS-OFETs were prepared using p-doped silicon wafers as the substrate covered with a thick silicon dioxide (SiO_2) layer. Gold source and drain electrodes were patterned on the top of

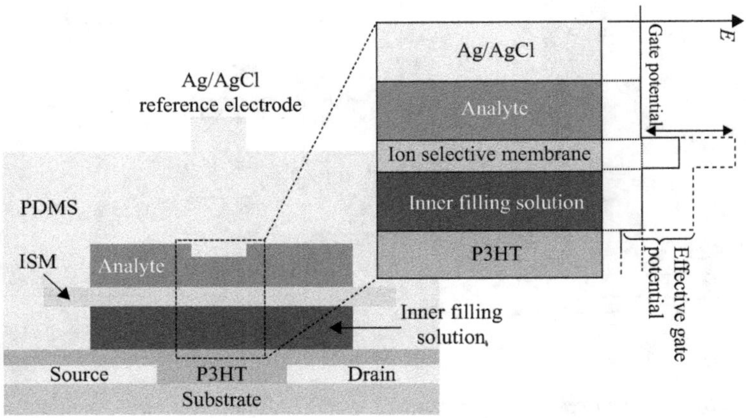

Figure 7.41 Cross-section of an ion-sensitive EGOFET: the bulk potentials in case of equal ion concentration in the analyte and the inner filling solution (solid line) as well as in case of lower concentration in the analyte (dashed line) are illustrated. Reprinted with permission from [34]. Copyright (2013) John Wiley and Sons

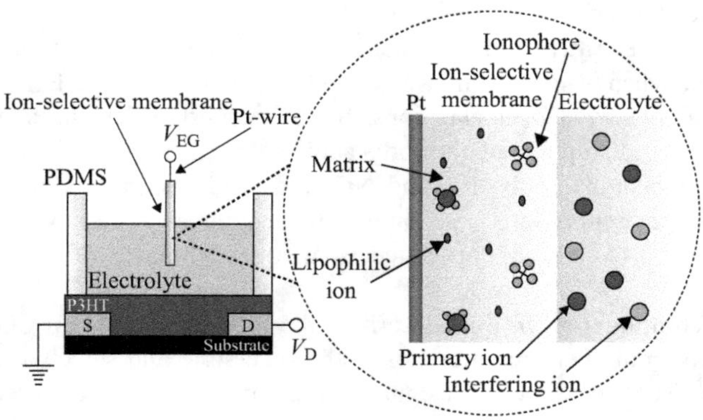

Figure 7.42 Schematic representation of an IS-OFET. The conventional electrolyte-gate (Pt-wire) is functionalized with an ion-selective polymeric membrane. Reprinted from [35]. Copyright (2014) with permission from Elsevier

the SiO_2 layer. Functionalization of the gate electrodes was performed by coating Pt-wires with the ion-selective membrane (Figure 7.42). K^+-selective membranes consisted of PVC-based membranes with valinomycin as ionophore, while Ca^{2+}-selective membrane were PVC-membranes with ETH-5234 *N,N*-dicyclohexyl-*N′*, *N′*-dioctadecyl-3-oxa-pentanediamide as ionophore.

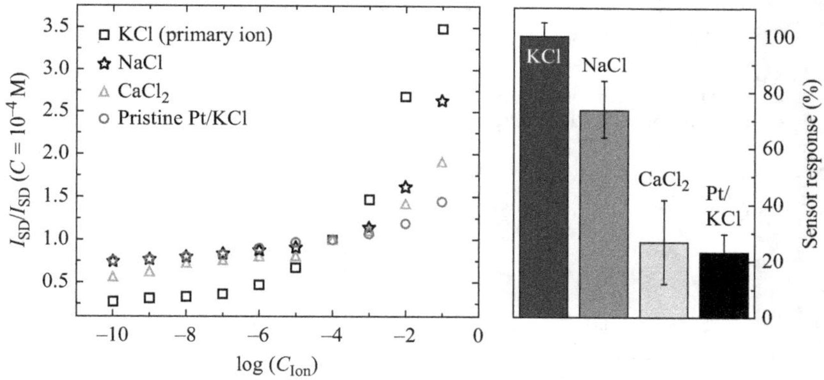

Figure 7.43 Drain current response of a K^+-selective electrolyte-gated OFET towards different concentrations of primary (K^+) and interfering ions (Na^+ and Ca^{2+}). The right panel shows the mean value and the standard deviation of the sensor response of three different devices to prove the selective response of the IS-OFETs. Reprinted from [35]. Copyright (2014) with permission from Elsevier

It was found that the sensitivity of the IS-OFETs towards primary ions (K^+ and Ca^{2+}) or interfering ions in solution was higher or similar to that of the OFETs with unfunctionalized gate electrode. Figure 7.43 shows as example the response of a K^+-selective IS-OFET. These sensors could selectively detect ions down to concentrations in the μM range, even in solutions with an ionic background of 150 mM NaCl (Figure 7.44).

7.3.6 Other semiconducting polymers

Ramesh *et al.* [36] reported an OTFT that was successfully used for the selective detection of mercury ions (Hg^{2+}). The OFET was fabricated with a top-contact structure from Si wafers covered with silicon dioxide (SiO_2) as a gate dielectric, using gold source–drain electrodes (Figure 7.45). The performance of the OFET was improved by depositing PTCDI-C8 (*N,N′*-Dioctyl-3,4,9,10-perylenedicarboximide) organic semiconductor between the dielectric (CYTOP) and the active layer (perylene bisimide, PBI). As a result, the electron mobility (μ) and the on/off ratio (I_{on}/I_{off}) of the OFET were increased by two orders of magnitude.

The sensitivity of this transistor towards Hg^{2+} ions was attributed to the formation of coordination complexes between mercury divalent ions and thymine ligands of the PBI, which leads to a significant change in the threshold voltage. This device could selectively detect Hg^{2+} ions down to concentrations in the μM range, even in solutions with other interfering metal ions.

Ramesh *et al.* [37] also developed bilayer (semiconductor/ligand) OFETs for the sensing of Cu^{2+} metal ions and CN^- anions. The OFETs were designed with top-contact architecture using gold source–drain electrodes (Figure 7.46). The

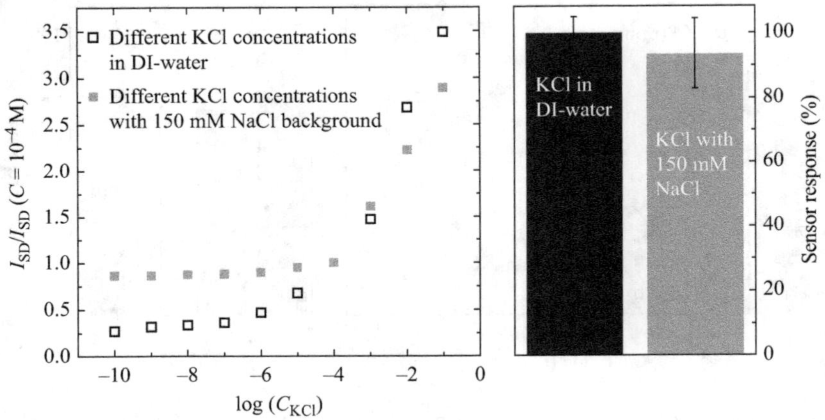

Figure 7.44 Response of a K^+-selective OFET towards different primary ion (K^+) concentrations either in DI-water or with an ionic background of 150 mM NaCl. The right panel shows the mean value and the standard deviation of the sensor response with or without ionic background for three different K^+-selective OFETs. Reprinted from [35]. Copyright (2014) with permission from Elsevier

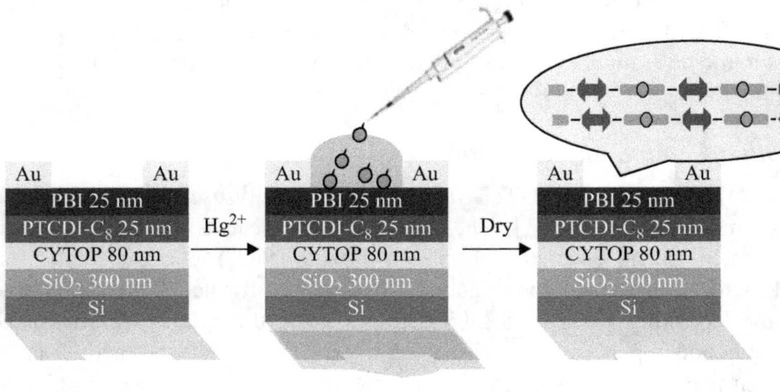

Figure 7.45 Side view of the structure of the PTCDI-C8/PBI OTFT and the operation of the OTFT sensor in the presence of $Hg(OAc)_2$ solution. Reprinted from [36]. Copyright (2013) with permission from Elsevier

OFETs were constructed on Si wafers covered with SiO_2. A layer of cross-linked poly(4-vinylpyridine) (C-PVP) was formed on the top of the Si/SiO_2 substrate. Pentacene was used as semiconductor and a Schiff base pyridine derivative (P) as a ligand. These pentacene/P bilayer OFETs could detect Cu^{2+} ions down to a

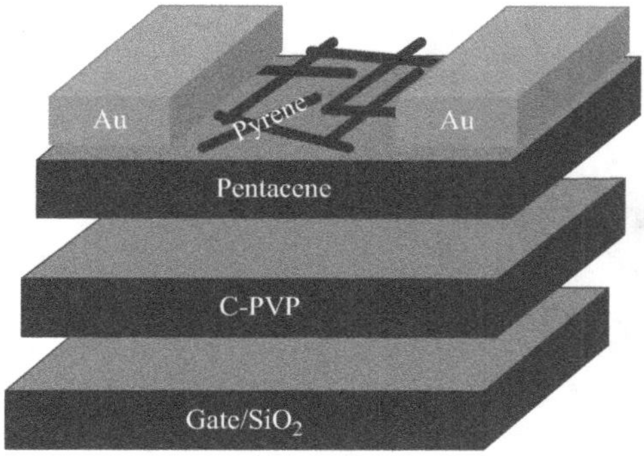

Figure 7.46 Structure of the device incorporating pentacene and the pyrene derivative. Reprinted from [37]. Copyright (2014) with permission from Elsevier

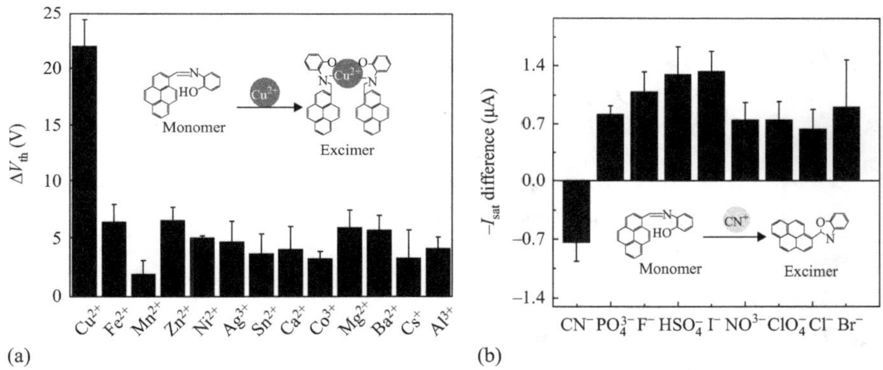

Figure 7.47 Differences in threshold voltage (ΔV_{th}) in the presence of (a) various single metal ions and (b) different anions. Reprinted from [37]. Copyright (2014) with permission from Elsevier

concentration of 50 μm. The excellent sensitivity of these devices to Cu^{2+} ions over other metal cations was attributed to the self-assembly of P–P* coordination species upon the addition of Cu^{2+} ions (Figure 7.47(a)).

These OFETs were also sensitive to CN^- anions down to a concentration of 150 μm. The sensitivity to CN^- anions over other anions was associated with the formation of benzoxazole rings upon the addition of CN^- anions (Figure 7.47(b)). Selectivity towards Cu^{2+} and CN^- ions in mixed ion solutions was also successfully tested.

Recently, Minami *et al.* [38] have designed an extended-gate type OFET for anion sensing. The extended-gate electrode consisted of a gold layer that is functionalized with a phenyl boronic acid (PBA) derivative as an anion receptor (Figure 7.48(a)). This extended gate was connected to the OFET through a copper

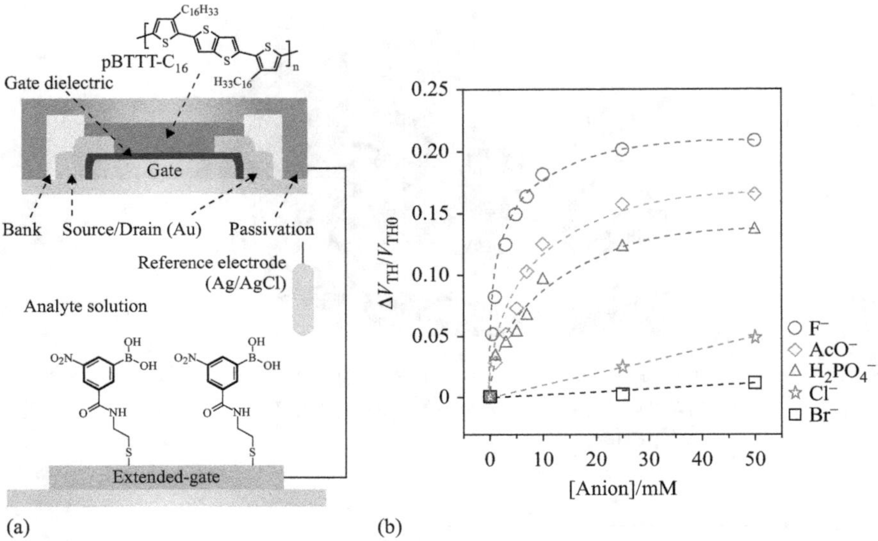

Figure 7.48 (a) Schematic structure of the anion sensor based on the extended-gate type OFET. The extended-gate electrode is functionalized with PBA. (b) Changes in the threshold voltage of the OFET device by anions at various concentrations in an MES buffer solution (100 mM) with NaCl (100 mM) at pH 5.5 at r.t. [anion] = 0–50 mM. Reprinted with permission from [38]. Copyright (2015) Royal Society of Chemistry

wire. The conducting polymer poly(2,5-bis(3-hexadecylthiophene-2-yl)thieno[3,2-*b*] thiophene) (pBTTT-C16) was employed as the active layer. The device was completed with gold source–drain electrodes and an Ag/AgCl gate electrode. This OFET was sensitive to F^- ions down to a concentration of 0.7mM. It was observed that the increase of F^- concentration in an MES (2-(N-morpholino)-ethanesulfonic acid) buffer solution (pH 5.5) leads to a positive shift of the transfer curves. Selectivity to F^- ions over Cl^-, Br^-, AcO^- and $H_2PO_4^-$ $H_2PO_4^-$ interfering anions was also investigated (Figure 7.48(b)). The device only showed significant responses to basic anions (F^-, AcO^- and $H_2PO_4^-$) due to the Lewis acidity of boron in PBA.

References

[1] D. Elkington, N. Cooling, W. Belcher, P. C. Dastoor and X. Zhou, "Organic Thin-Film Transistor (OTFT)-Based Sensors," Electronics, no. 3, pp. 234–254, 2014.

[2] J. W. Thackeray, H. S. White and M. S. Wrighton, "Poly(3-methylthiophene)-Coated Electrodes: Optical and Electrical Properties as a Function of Redox

Potential and Amplification of Electrical and Chemical Signals Using Poly(3-methylthlophene)-Based Microelectrochemical Transistors," Journal of Physical Chemistry, no. 89, pp. 5133–5140, 1985.

[3] J. W. Thackeray and M. S. Wrighton, "Chemically Responsive Microelectrochemical Devices Based on Platinized Poly(3-methylthiophene): Variation in Conductivity with Variation in Hydrogen, Oxygen or pH in Aqueous Solution," Journal of Physical Chemistry, no. 90, pp. 6674–6679, 1986.

[4] E. W. Paul, A. J. Ricco and M. S. Wrighton, "Resistance of Polyaniline Films as a Function of Electrochemical Potential and the Fabrication of Polyaniline-Based Microelectronic Devices," Journal of Physical Chemistry, no. 89, pp. 1441–1447, 1985.

[5] M. Kaisti, A. Knuutila, Z. Boeva, C. Kvarnström and K. Levon, "Low-Cost Chemical Sensing Platform with Organic Polymer Functionalization," IEEE Electron Device Letters, vol. 36, no. 8, pp. 844–846, 2015.

[6] M. Nishizawa, T. Matsue and I. Uchida, "Penicillin Sensor Based on a Microarray Electrode Coated with pH-Responsive Polypyrrole," Analytical Chemistry, vol. 64, no. 21, pp. 2042–2044, 1992.

[7] C. Bartic, A. Campitelli, K. Baert, J. Suls and S. Borghs, "Organic-based transducer for low-cost charge detection in aqueous media," in Electron Devices Meeting, 2000. IEDM '00. Technical Digest International, San Francisco, CA, USA, 2000.

[8] C. Bartic, B. Palam, A. Campitelli and G. Borghs, "Monitoring pH with Organic-Based Field Effect Transistors," Sensors and Actuators B, no. 83, pp. 115–122, 2002.

[9] C. Bartic, H. Jansen, A. Campitelli and S. Borghs, "Ta_2O_5 as Gate Dielectric Material for Low-Voltage Organic Thin-Film Transistors," Organic Electronics, no. 3, pp. 65–72, 2002.

[10] C. Bartic, A. Campitelli and S. Borghs, "Field-Effect Detection of Chemical Species with Hybrid Organic/Inorganic Transistors," Applied Physics Letters, no. 82, pp. 475–477, 2003.

[11] T. Ji, P. Rai, S. Jung and V. K. Varadan, "In Vitro Evaluation of Flexible pH and Potassium Ion-Sensitive Organic Field," Applied Physics Letters, no. 92, pp. 233304-1, 233304-3, 2008.

[12] G. Scarpa, A.-L. Idzko, A. Yadav and S. Thalhammer, "Organic ISFET Based on Poly (3-hexylthiophene)," Sensors, no. 10, pp. 2262–2273, 2010.

[13] S. Ritjareonwattu, Y. Yun, C. Pearson and M. C. Petty, "Enhanced Sensitivity of an Organic Field-Effect Transistor pH Sensor Using a Fatty Acid Langmuir–Blodgett Film," Organic Electronics, no. 11, pp. 1792–1795, 2010.

[14] S. Ritjareonwattu, Y. Yun, C. Pearson and M. C. Petty, "An Ion Sensitive Organic Field-Effect Transistor Incorporating the Ionophore Valinomycin," IEEE Sensors Journal, no. 5, pp. 1181–1186, 2012.

[15] A. Münzer, K. Melzer, M. Heimgreiter and G. Scarpa, "Random CNT Network and Regioregular Poly(3-hexylthiophen) FETs for pH Sensing

Applications: A Comparison," Biochimica et Biophysica Acta, no. 1830, pp. 4353–4358, 2013.

[16] J. Kofler, K. Schmoltner, A. Klug and E. J. W. List-Kratochvil, "Hydrogen Ion-Selective Electrolyte-Gated Organic Field-Effect Transistor for pH Sensing," Applied Physics Letters, no. 104, pp. 193305-1–193305-5, 2014.

[17] A. Loi, I. Manunza and A. Bonfiglio, "Flexible, Organic, Ion-Sensitive Field-Effect Transistor," Applied Physics Letters, no. 86, pp. 103512, 2005.

[18] I. Manunza, E. Orgiu, A. Caboni, M. Barbaro and A. Bonfiglio, "Producing smart sensing films by means of organic field effect transistors," in 28th IEEE EMBS Annual International Conference, New York City, USA, 2006.

[19] A. Caboni, E. Orgiu, M. Barbaro and A. Bonfiglio, "Flexible Organic Thin-Film Transistors for pH Monitoring," IEEE Sensors Journal, vol. 9, no. 12, 2009.

[20] A. Caboni, E. Orgiu, E. Scavetta, M. Barbaro and A. Bonfiglio, "Organic-Based Sensor for Chemical Detection in Aqueous Solution," Applied Physics Letters, no. 95, pp. 123304, 2009.

[21] T. Nguyen, Y. Seol and N. Lee, "Organic Field-Effect Transistor with Extended Indium Tin Oxide Gate Structure for Selective pH Sensing," Organic Electronics, no. 12, pp. 1815–1821, 2011.

[22] M. Yun, A. Sharma, C. Fuentes-Hernandez, *et al.*, "Stable Organic Field-Effect Transistors for Continuous and Nondestructive Sensing of Chemical and Biologically Relevant Molecules in Aqueous Environment," ACS Applied Material Interfaces, no. 6, pp. 1616–1622, 2014.

[23] M.-J. Spijkman, J. J. Brondijk, T. C. T. Geuns, *et al.*, "Dual-Gate Organic Field-Effect Transistors as Potentiometric Sensors in Aqueous Solution," Advanced Functional Materials, no. 20, pp. 898–905, 2010.

[24] M. E. Roberts, S. C. B. Mannsfeld, N. Queraltó, C. Reese, J. Locklin and W. Knoll, "Water-Stable Organic Transistors and Their Application in Chemical and Biological Sensors," PNAS, vol. 105, no. 34, pp. 12134–12139, 2008.

[25] M. E. Roberts, S. C. Mannsfeld, R. M. Stoltenberg and Z. Bao, "Flexible, Plastic Transistor-Based Chemical Sensors," Organic Electronics, no. 10, pp. 377–383, 2009.

[26] F. Buth, D. Kumar, M. Stutzmann and J. Garrido, "Electrolyte-Gated Organic Field-Effect Transistors for Sensing Applications," Applied Physics Letters, vol. 15, no. 98, pp. 153302, 2011.

[27] E. Mitraka, L. Kergoat, Z. U. Khan, S. Fabiano, O. Douhe′ret and P. Lecle're, "Solution Processed Liquid Metal-Conducting Polymer Hybrid Thin Films as Electrochemical pH-Threshold Indicators," Journal of Materials Chemistry C, no. 3, pp. 7604–7611, 2015.

[28] V. Saxena, V. Shirodkar and R. Prakash, "Copper(II) Ion-Selective Microelectrochemical Transistor," Journal of Solid State Electrochemistry, no. 4, pp. 234–236, 2000.

[29] M. Berggren, R. Forchheimer, J. Bobacka, *et al.*, "PEDOT:PSS-based electrochemical transistors for ion-to-electron transduction and sensor signal amplification," in Organic Semiconductors in Sensor Applications, edited by D. A. Bernards, R. Owens and G. Malliaras, Berlin, Springer, 2008, pp. 63–280.

[30] Z. Mousavi, A. Ekholm, J. Bobacka and A. Ivaska, "Ion-Selective Organic Electrochemical Junction Transistors Based on Poly(3,4-ethylenedioxythiophene) Doped with Poly(styrene sulfonate)," Electroanalysis, vol. 21, no. 3–5, pp. 472–479, 2009.

[31] P. Lin, F. Yan and H. Chan, "Ion-Sensitive Properties of Organic Electrochemical Transistors," ACS Applied Materials and Interfaces, no. 2, pp. 1637–1641, 2010.

[32] T. Toccoli, E. Borga, H. Blond, *et al.*, "Polyelectrolytes-Coated Gold Nanoparticles Detection by PEDOT:PSS Electrochemical Transistors," Organic Electronics, no. 13, pp. 1716–1721, 2012.

[33] M. Sessolo, J. Rivnay, E. Bandiello, G. G. Malliaras and H. J. Bolink, "Ion-Selective Organic Electrochemical Transistors," Advanced. Materials, no. 26, pp. 4803–4807, 2014.

[34] K. Schmoltner, J. Kofler, A. Klug and E. List-Kratochvil, "Electrolyte-Gated Organic Field-Effect Transistor for Selective Reversible Ion Detection," Advanced Materials, no. 25, pp. 6895–6899, 2013.

[35] K. Melzer, A. Mara Münzer, E. Jaworska, K. Maksymiuk, A. Michalska and G. Scarpa, "Polymeric Ion-Selective Membrane Functionalized Gate-Electrodes: Ion-Selective Response of Electrolyte-Gated Poly(3-hexylthiophene) Field-Effect Transistors," Organic Electronics, no. 15, pp. 595–601, 2014.

[36] M. Ramesh, H.-C. Lin and C.-W. Chu, "Organic Thin Film Transistors as Selective Sensing Platforms for Hg^{2+} Ions and the Amino Acid Cysteine," Biosensors and Bioelectronics, no. 42, pp. 76–79, 2013.

[37] M. Ramesh, Y.-R. You, M. Shellaiah, M.-C. Wu, H.-C. Lin and C.-W. Chu, "Star-Shaped Self-assembly of an Organic Thin Film Transistor Sensor in the Presence of Cu^{2+} and CN^{-} ions," Organic Electronics, no. 42, pp. 76–79, 2013.

[38] T. Minami, T. Minamiki and S. Tokito, "An Anion Sensor Based on an Organic Field Effect," Chemical Communications, no. 51, pp. 9491–9494, 2015.

Index

www.ingramcontent.com/pod-product-compliance
Lightning Source LLC
LaVergne TN
LVHW011916200926
848484LV00002B/637